Cable Television Technology and Operations

Other McGraw-Hill Reference Books of Interest

For more information about other McGraw-Hill materials, call 1-800-2-MCGRAW in the United States. In other countries, call your nearest McGraw-Hill office.

Cable Television Technology and Operations

HDTV and NTSC Systems

Eugene R. Bartlett

Tele-Media Corporation
Pleasant Gap, Pennsylvania

McGraw-Hill Publishing Company
New York   St. Louis   San Francisco   Auckland   Bogotá
Caracas   Hamburg   Lisbon   London   Madrid   Mexico
Milan   Montreal   New Delhi   Paris
San Juan   São Paulo   Singapore
Sydney   Tokyo   Toronto

Library of Congress Cataloging-in-Publication Data

Bartlett, Eugene R.
  Cable television technology and operations: HDTV & NTSC systems/
Eugene R. Bartlett.
    p. cm.
  ISBN 0-07-003957-7
  1. Cable television   2. High definition television.   3. Color
television.   I. Title.
TK6675.B37   1990
621.388'57—dc20   89-13365
                                                            CIP

1 2 3 4 5 6 7 8 9 0   DOCDOC   9 5 4 3 2 1 0

ISBN 0-07-003957-7

*The sponsoring editor for this book was Daniel A. Gonneau, the de-
signer was Naomi Auerbach, and the production supervisor was
Dianne L. Walber. Project supervision by The Total Book. It was set in
Century Schoolbook by the McGraw-Hill Publishing Company Professional & Reference Division composition unit.*

Printed and bound by R. R. Donnelley & Sons Company.

*For more information about other McGraw-Hill materials,
call 1-800-2-MCGRAW in the United States. In other
countries, call your nearest McGraw-Hill office.*

Contents

Preface

This book is a culmination of in-house training lecture notes and solutions to practical problems faced by actual technical cable television people. The first chapter introduces the need for a method to improve television reception and the cable television solution. Chapter 2 introduces many elementary electric circuit theorems and their application to cable television problems and systems. The material contained in Chapter 3 treats in detail the coaxial cable amplifier cascade, as well as signal transportation techniques such as super-trunking, microwave, and fiber-optic systems. Chapters 4, 5, 6, and 7 address the design and maintenance of the main sections of a cable television, namely the head-end source of television signals, cable system design and layout methods, system testing and measurements, and standards and records to be maintained.

Persons with reasonable proficiency in arithmetic and secondary school algebra should have no problem understanding this work. In areas such as elementary trigonometry and logarithms, tutorial sections are provided in condensed form. Illustrative problems all pertain to cable television subjects. New or seldom used techniques are introduced when providing some unique solutions to cable system problems. This whole work is aimed at practical solutions to problems and the necessary maintenance and testing procedures needed to design, build, and maintain a cable television system to today's specifications which are needed to carry National Television System Committee (NTSC) television and high-definition television (HDTV) signals.

The writing attempts to explain problems and solutions in plain language or today's vernacular. Entry-level technicians will find this work a useful handbook to guide their progress to the system technician and/or chief technician level. Other technical persons should find this book useful as a desk reference ready and able to provide answers to practical problems.

Although at the time of this writing many HDTV signal standards have been proposed, acceptance of such a standard does not appear to be forthcoming in the near future. However, cable systems should prepare themselves by maintaining present-day cable systems to proper bandwidth and lower noise and distortion specifications than previously maintained to be able to perform adequately in the future as a viable delivery method for HDTV signals. Information and procedures contained in this book will help cable television personnel achieve this goal.

Eugene R. Bartlett

Introduction to Cable Television Systems

1.1 The Transmit and Receive Path

As its name implies, a cable television system is simply a means of delivering television signals to a home television set via a cable instead of the usual rooftop antenna. When the television broadcast industry developed to the point that rural or suburban areas desired service, then more sophisticated rooftop antennas, antenna rotation schemes, and antenna amplifiers were employed by residents as their homes became more distant from the television transmitting antenna. Of course, as the distance from the transmitting antenna and the receiving antennas increased, the signal level decreased to unusable values. In mountainous country, people living in the valleys could not get sufficient signal level and often received the same signal twice, once directly and once reflected off some object. This is known as *multipath reception*. Some of the problems with television reception are shown in Fig. 1.1.

Notice that the home at *a* receives television signals from both paths 1 and 2. Path 2 is much longer than path 1; therefore, the signal from path 2 reflected off the water tank *W* arrives at home *a* much later than the signal from path 1. Two signals will be seen on the television receiver at home *a* displaced on the screen. This form of picture impairment, called *ghosting,* is caused by multipath reception.

At home *b* the television signal is received along path 3. However, path 3 is interrupted by the ground, thus causing the signal level to be extremely weak. The television receiver at home *b* will see a picture that is noisy, causing a form of picture impairment called *snow*. Noise as seen on a black-and-white (monochrome) television receiver resembles objects viewed through a snowstorm. Early color television sets

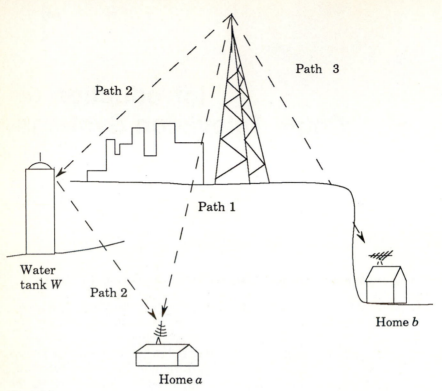

Figure 1.1

presented noise as a form of confetti falling on the screen. This neces-sitated installing circuits in the television receiver that on a loss of signal would cause the color television set to revert to black and white.

One other cause of poor television reception is called *co-channel*, in which the television receiver receives two television stations on the same channel. In the United States the Federal Communications Commission has meticulously placed television stations on the same channel far apart. Also, the frequency of the television stations is displaced ± 10 kilohertz (kHz). Therefore, the separation in frequency may be as much as 20 kHz (one $+10$ kHz, the other -10 kHz). This allows the primary television station to be received by the television receiver even though there is some picture impairment caused by the other station. Distant television stations on the same channel can be received because of peculiarities of the receiving path. Fig. 1.2 shows how co-channel can occur.

When the ocean is much warmer than the air, it can produce a moisture-laden cloud. As a result of impurities in the air, the moisture-

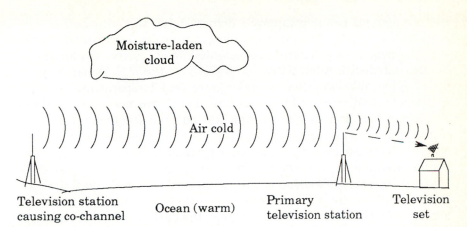

Figure 1.2

laden cloud is conductive and hovers over the ocean supporting it. This causes the transmitted signal from the distant television station to have its signal concentrated and funneled through the space between the ocean and the moisture-laden cloud to be received by the television receiver. Picture impairment caused by co-channel is similar in appearance to an object viewed through venetian blinds.

1.2   The Cable Television Solution

Early attempts to solve the difficult receiving problems in distant communities involved several schemes to obtain high-quality television signals at some good location. This receiving location was and to this day is known as the *head end*. At this location a steel tower of appropriate height with receiving antennas of high gain and high directivity which were aimed at the television transmitting antennas usually provided the high-quality signal desired. Sometimes wooden poles placed on a mountaintop with antennas mounted and aimed was the method of choice. The higher the receiving antenna, the higher the signal level until the receiving antenna is at the same point of elevation as the transmitting antenna.

The highly directive antennas in most cases are able to differentiate between the desired station and the offending one on the same television channel, thus eliminating co-channel. Multipath problems also diminished with use of highly directive receiving antennas.

A simple cable system connecting the received television signals from the antenna site to subscribers was fashioned by war surplus coaxial cable and broadband amplifiers. Early amplifiers were constructed by using vacuum tubes and were powered individually from the commercial power company's domestic 120-volt alternating current (ac) source. A potentially dangerous situation resulted. Of course, this happened long before the establishment of the Occupational Safety and Health Administration (OSHA) of the U.S. Department of Labor. The cable-amplifier cascades were very temperature-sensitive and constant adjustment by the technical staff was needed since these amplifiers had limited gain (amplification) and frequency response. Cable attenuation increased sharply as the television carrier frequency increased, causing these early systems to be able to carry television channels 2 through 6. These systems were known as five-channel systems. Television stations received on ultrahigh frequency (UHF) or on channels 7 to 13 were converted to channels in the band 2 to 6 at the head-end site. At that time, the early 1950s, five channels were a lot and subscribers were more tolerant of system outages and technical problems than they are today.

When surplus cable became scarce and more systems were built, the manufacturers responded with improved cable and amplifiers. The solid aluminum sheath cable filled with polyethylene foam and a copper-clad aluminum center conductor soon became the standard of the industry. During the 1960s and 1970s this type of cable was available mainly in two sizes: 0.412-inch (in) outside diameter (OD) and 0.500-in OD. The 0.412-inch cable was used for what was known as feeder and the 0.500-inch for trunking purposes. Of course, system design philosophy improved as well. Instead of connecting subscribers to a single cable-amplifier system, a signal transportation scheme in the form of a trunk-feeder system was developed. Here the trunk system fed the television signals from the head end to the extremities of the system, which varied in distances and numerous system splits. The feeder cables with subscribers connected to them were bridged out of the trunk system at the trunk amplifier locations, thus providing subscriber system isolation from the trunk cable system. With the development of the transistor, the cable amplifiers soon improved in performance and in low power consumption. This made practical the carrying of the power on the cable plant itself. A low voltage of 30 Volts ac (V_{ac}) provided by a step-down transformer (pole-mounted) fed power through a power inserter to the trunk cable. This power block isolated the signal from the cable to the power supply and allowed the power to be inserted on the cable. The amplifiers along the cascade extracted power from the cable to power themselves. Each power supply supplied power to only a segment of plant, limiting the line voltage drop problems. Power blocks were used to separate the segments. Power was also bridged out from the trunk system to power the line extender (feeder) amplifiers along the way.

Cable connectors also improved, and many design types were available during the 1960s and 1970s. The main problem at that time was to keep the connections tight and weatherproof, to prevent moisture from deteriorating the cable as well as limiting the reflections of signal and signal loss and creating power interuption problems. By then cable systems were constructed by improved techniques and improved products and provided better pictures more reliably. In addition, the very high frequency (VHF) channels, channels 7 to 13, could be used giving a total of 12 channels.

During the 1960s the feeder network provided subscriber signals through what is known as a *pressure tap*. Essentially a hole was cored in the outer aluminum sheath and the pressure tap center pin was centered in the hole to connect to the center conductor (stinger). Self-tapping threads and a clamp provided the pressure to clamp the tap to the aluminum tubing. The stinger pin was connected to the subscriber connector center socket. A small network of resistors and capacitors provided the 75-ohm (Ω) match to the cable and some isolation from the subscriber drop wire. As the system developed further, the directional coupler and signal splitters improved, causing the appearance of the subscriber multiport tap. In this case the tap housing was spliced into the feeder cable by using proper cable-to-tap housing connectors. These taps, in the beginning, usually had two or four subscriber ports. Today eight-port taps are very common, particularly in densely populated areas. Amplifiers also improved in performance and cost. Since amplifier distortion products from channels 2 to 6 falling in the midband were eliminated by modern amplifiers, the midband spectrum could be used to carry more programming. Many cable systems where the midband area was reasonably clean carried the frequency-modulated (FM) radio signals in the 88- to 108-MHz space. Still from 108 MHz on up to channel 7 (175 MHz) 9 more television signals could be carried, thus yielding a 21-channel system. The main problem at this time was that the subscriber television sets could not tune to these channels since all they had was channels 2 to 13 and UHF channels 14 to 83. To solve this problem, a block (up) converter was used. This converter selected this midband and converted the 9 channels to, for example, UHF channels 27 to 35. The block converter connections are shown in Fig. 1.3.

At this time (the 1970s), UHF television stations increased in number. Cable systems could only carry up through channel 13 (216 MHz) in frequency, so cable television system operators elected to convert at the head end (as they did earlier) some of these UHF stations to the nine available midband channels.

Early in the 1980s movie programming became available through a satellite-delivered channel. The originating taped or filmed movie was converted to a National Television System Committee (NTSC) video

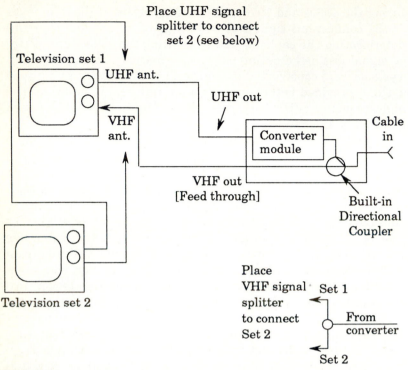

Place UHF signal
splitter to connect
set 2 (see below)

Television set 1

UHF ant.

UHF out

VHF
ant.

Converter
module

Cable
in

VHF out
[Feed through]

Built-in
Directional
Coupler

Television set 2

Place
VHF signal   Set 1
splitter
to connect   From
Set 2   converter

Set 2

Figure 1.3

signal that modulated a satellite up-link carrier channel to be trans-
lated by the satellite receiver-transmitter system (transponder) to a
ground receiving antenna station at a local cable television system.
The receiving system at this time employed large antennas (10 meters
[m]) because the microwave (4000-MHz) antenna amplifiers were
noisy and had limited gain. Through the 1980s one has seen improved
low noise receiving antenna preamplifiers, or low noise amplifiers
(LNAs), which were small and lower in cost. Receiving parabolic antennas appeared in many systems. The arrival of the low noise block
converters (LNBC) in the late 1980s improved signal quality and lowered cost. The LNBC was essentially a low noise preamplifier
mounted at the antenna that also contained a block-down converter.
The 4000-MHz (4-GHz) signal containing as many as 24 channels of
programming was converted to 24 channels in the band of, for example, 950 to 1450 MHz. Therefore, the cable from the antenna to the
receivers had much lower loss at 950 to 1450 MHz than at 3.7 to 4.2
GHz. Now the receivers needed only one conversion, thus lowering
their cost by some 60 percent. Since each channel of programming
needed its own receiver, the cost saving became substantial.

Because most satellite-delivered channels were premium or pay channels, some means of separating the signals at the subscriber's tap port was needed to block the signal to those subscribers unwilling to pay for the service. A network of resistors, capacitors, and inductors in the form of a sharp trapping filter was made and housed in a metal weather-tight enclosure and was connected in line with the subscriber's drop wire. This effectively removed the unwanted channel from the subscriber's home. It was known as a *negative-type signal trap*. Defeating this mechanism only required removal of the trap from the line. However, most people trying to steal the service had to mount the utility pole and remove the trap. Most people will not do anything so visible or hazardous.

More and more satellite programmers appeared, and soon more 24-channel satellite systems were employed, causing many cable television system operators to plan for adding channel capacity. Of course, better quality cable and amplifiers appeared and new systems were constructed to 30 channels (55 to 270 MHz), 35 channels (55 to 300 MHz), 40 channels (55 to 330 MHz), 52 channels (55 to 400 MHz), 62 channels (55 to 450 MHz), to 78 channels (55 to 550 MHz). Many systems did not fill all their channel capacity, and program choices and packages were numerous.

The channel security trap method soon became unwieldy because traps had to be stacked (placed end to end) to trap out the non-pay-service subscriber. To solve this problem, signal scrambling techniques were developed. One approach was to place an interfering signal over the pay-service signal, thus rendering it unwatchable. A trap placed on the subscriber's tap port removed this interfering signal, making the service viewable. Several channels belonging to the same service package would be scrambled by the interfering signal. Several packages could also be scrambled by several types of interfering signals where a trap was needed for each tier of programming. This trap method was known as *positive trapping*, and the traps were often referred to as *tiering filters*. Negative traps were also constructed to remove bands of signals. Service tiers could be collected in groups of consecutive channels and a trap removing (negative) the group provided the signal security. Both trapping methods helped to limit the number of traps hanging from a tap. It should be remembered that the negative trap removed service from nonsubscribers and the positive trap provided the premium service to the subscriber. Therefore, a system with few premium subscribers would have many negative traps or few positive traps. Thus one can see where the cost factors lie.

With the increased number of channels, some means for television sets to tune to these channels was needed. Still many television sets only tuned to channels 2 to 13 and UHF (14 to 83). Newer sets, often called *cable-ready*, were able to tune to some of the new channels. The signal converter solved this problem; it received the cable signal and converted the selected channel to a fixed channel, often channel 2, 3, or 4, and fed this channel to the television receiver. Fig. 1.4 illustrates several methods of connecting the converter.

In the early 1980s the synchronizing suppression method of scrambling was developed. This method used video processing circuits at the head end to remove or suppress the horizontal synchronizing pulses from the video signal. No television receiver could view a service with the horizontal holding signals removed. The picture would look as if the horizontal hold control on the television set were misadjusted. The signal restoring circuits were embedded in the subscriber converter circuits. Subscribers desiring the service had their converter pre-programmed with a programmed read-only memory (PROM) chip in-

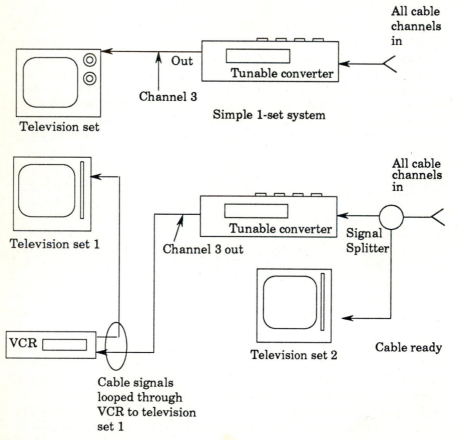

Figure 1.4

tegrated circuit to activate these circuits to restore the suppressed horizontal synchronizing signals.

Later in the 1980s with the development of personal computers, the tunable subscriber converter with the built-in descrambler was made addressable (the computer "talks" to the converter; each converter has an address). Circuits added to the subscriber converter processed data carried by a downstream data carrier. This data carrier was in most cases at the top of the FM radio band (106.250 MHz) and modulated with the addressable data stream. When the converter saw its address appear on the data stream, it extracted the digitally encoded instruction. These instructions essentially replaced the original PROM-provided information. Operationally a subscriber would contact the cable company, for example, for a change of service. The subscriber clerk at the cable company would enter the changes in the computer and through a computer-to-cable modem instruct the converter to respond to the subscriber's orders. No visit to the subscriber's home, either inside or outside, was necessary to effect the change of service.

All of the methods for signal security at one time or another were able to be defeated. Traps were removed from the poles or burned out and replaced. Dishonest persons made or purchased illicit positive traps and placed them in their home out of sight. Signal pirates, as they were often called, sold and installed illicit PROM chips to defeat the descrambling converter. External so-called black boxes were sold to subscribers, who placed them between their addressable converter and the television set. Most cable companies employed various programs to handle their signal security problem.

1.3   The Blue Skies System

The band of frequencies below channel 2 where no television signals exist was unused by cable systems for a long time. Some two-way or reverse capability was desired and suggested by cable operators and manufacturers. The band from 5 to 30 MHz in the upstream direction could provide four 6-MHz channels plus one half-channel. The band between 30 and 55 MHz was reserved as a guard band to separate the upstream (5 to 30 MHz) from the downstream (55 to 550 MHz) service by a diplex filter installed in the amplifiers. This was known as the *subsplit reverse system*. Programming on any or all of the four upstream channels could be picked up from four remote points in a cable system and sent to the head end (often the system hub) to be reprocessed on a downstream channel for subscriber viewing. Since cable loss is a function of frequency, the loss at 5 to 30 MHz was less than the forward path. Therefore, every amplifier location did not need a reverse amplifier unit or low-gain units were installed. The cable system can be thought of as a tree (head end to trunk to limbs to branches) or a wheel (head end to spokes to the rim edge). The branches downstream were separate from the hub and served its separate area. In the reverse path all the spokes, or branches, fed the hub head end. The result of the reverse network topology was a noise buildup in the 5- to 30- MHz band. This in some cases limited the usability of the upstream channels. Cable operators, in many cases, elected to install the reverse amplifiers in selected trunk runs only and activate feeder as necessary to receive the desired upstream services. Addressable feeder switches installed in the reverse path at the bridger amplifiers could connect only the feeder cable where a program pickup location was originating, thus limiting the noise buildup. This technique was known as *bridger leg switching*.

Cable companies in large metropolitan areas often needed more than the 4½ upstream channels. To obtain the needed upstream channel capacity, the split frequency was moved above the FM band. This was known as a *midsplit*, in which, for example, the upstream was in the frequency band of 5 to 112 MHz, giving approximately 17 television channels. The downstream frequency band was 150 to 450 MHz, giving 35 television channels, yielding a 2:1 split. The band between 112 and 150 MHz was the guard band provided by the diplex amplifier filters. The high-split system was also developed and used by some systems; it had an upstream capacity of about 28 television channels (5 to 174 MHz) and a downstream capacity of 36 channels (234 to 450 MHz). For a system with an upper frequency limit of 550 MHz, the downstream capacity was increased to 52 channels. Special converters were used in which an upstream subscriber response could be initiated, for example, to make a change in the service or a so-called pay-per-view selection. It was hoped that uses for the upstream system would add revenue for the cable operator. In many cases the upstream system was never utilized to any great extent and maintaining the system was expensive compared to any advantages. Most systems today are of the subsplit type.

In order to provide cable service more reliably and at a lower cost, many alternatives are appearing in the marketplace. For example, a low-power microwave system could drop the signal into so-called minihubs, eliminating long trunk runs. Fiber-optic trunking is now used by more and more cable operators. Such systems can carry many television channels (20 to date) a distance of 25 km with no electronics between on a single fiber. The costs of the fiber-optic cable and amplifiers are still decreasing, producing more and more interest by cable operators considering restructuring or rebuilding their systems.

Chapter 2

Basic Electric Circuits and Principles

Most people working in the technical area of the cable television business have had some basic electrical training in high school, technical school, or one of the military service schools. Since there are many good books and materials on the basic electrical subjects this work will review some of the fundamental elements pertaining to cable television systems.

Essentially the cable system may be considered as a signal generator system driving a cable system delivering the signals to a load at the subscriber terminal. This system is actually made up of many systems all assembled together. Many of the circuits are simple when treated alone, and so a review of some simple electrical systems will follow.

2.1 Basic Direct Current Circuits

In *direct current* electric current caused by an electromotive force flows through a conductor in one direction. The unit of electric current is the ampere (A) and the electromotive force the volt. Of course, the volt and ampere appear in smaller or larger units as given in Table 2.1.

TABLE 2.1 Common Electrical Quantity Equivalents

1000 volts	1 kilovolt
1 volt	1 volt
1/1000 volt	1 millivolt
1/1,000,000 volt	1 microvolt
1/1000 ampere	1 milliampere
1/1,000,000 ampere	1 microampere

The common current carrying conductor is the metallic wire. Most of us know that copper is a good conductor and is better than aluminum, which is better than iron. Gold and silver are better conductors than copper but because of their scarcity are much more expensive. The better the conductor the lower the resistance to current flow.

2.1.1 Ohm's law

Ohm's law is the basic law of current flow; it states that the resistance in ohms is directly proportional to the voltage and inversely proportional to the current. Mathematically Ohm's law is written in the form of basic equations.

$$R = \frac{V}{I} \quad V = IR \quad \text{and} \quad I = \frac{V}{R}$$

If one of the equations is memorized, the others can be derived by algebra. Another concept is that of power consumed by an electrical resistance. Mathematically

$$P = I^2 R = \frac{V^2}{R} = IV$$

where V = voltage (V)
I = current (A)
R = resistance (Ω)
P = power (W)

If this is difficult to remember, Fig. 2.1 will aid the memory.

The most basic circuit is the battery, which provides the electromotive force and the flow of current connected to a resistor, as shown in the schematic diagram of the circuit in Fig. 2.2.

The current I flows, as a result of V, in the direction shown. By Ohm's law $V = IR$ volts. The voltage across resistance R is equal to the battery voltage.

2.1.2 Kirchoff's laws

Extensions of this elementary concept are Kirchoff's voltage and current laws: Kirchoff's voltage law states that the sum of the voltage drops ($I \times R$) is equal to the sum of the voltage sources (V). Kirchoff's current law states that the current entering a node (junction of conductors) equals the current leaving the node. Fig. 2.3a and b illustrates these laws.

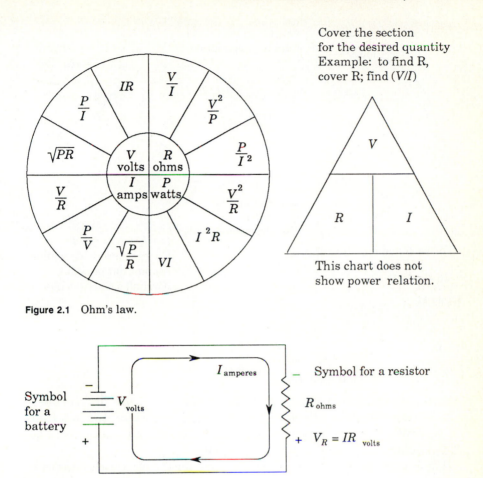

Figure 2.1 Ohm's law.

Cover the section
for the desired quantity
Example: to find R,
cover R; find (V/I)

This chart does not
show power relation.

Figure 2.2 Simple direct current circuit.

2.1.3 Effects of temperature

Heat generated in a resistor (I^2R) changes the amount of resistance of the resistor. The connecting wire carrying the current also can become hot, causing its resistance to increase. Resistance of metals increases with increase in temperature. Therefore, if a resistor is metallic and its resistance increases, the current is decreased; thus the effects are self-limiting. In some electric circuits resistors are made of carbon material, which exhibits a negative temperature coefficient; i.e., the resistance decreases with increase in temperature. More information on resistors' and conductors' characteristics with temperature change can be found in some of the references given at the end of this chapter.

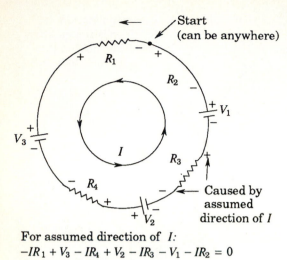

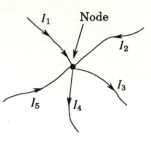

For assumed direction of *I*:

$-IR_1 + V_3 - IR_4 + V_2 - IR_3 - V_1 - IR_2 = 0$

Kirchoff's Voltage Law

$I_1 + I_2 + I_5 = I_3 + I_4$
Current into node equals
current out of node
Kirchoff's Current Law

Figure 2.3

2.1.4 Types of electric circuits

Resistor and batteries appear in what is known as series- and parallel-type circuits. In *series circuits* all the elements are connected in a loop; in *parallel circuits* all elements are connected as in parallel lines. Fig. 2.4 summarizes these two types of circuits.

For two resistors in parallel the equation for the circuit resistance is

$$R = \frac{1}{(1/R_1) + (1/R_2)} \times \frac{R_1 R_2}{R_1 R_2}$$

The last term is simply equal to 1; therefore, it does not change anything.

$$R = \frac{R_1 R_2}{R_1 + R_2}$$

This states that for two resistors in parallel the equivalent circuit resistance R is the product over the sum of the resistors. Now if a third resistor is placed in parallel, the formula can be based on two resistors at a time and calculating the values. Fig. 2.5 illustrates the technique.

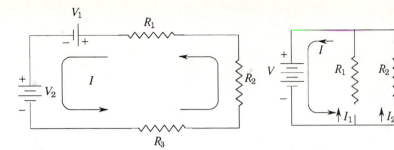

Series circuit
The current may be found by
$I(R_1 + R_2 + R_3) = V_1 + V_2$
This indicates that the total
series resistance is the sum
of the individual resistors.

Parallel circuit
$$I_1 = \frac{V}{R_1}, \qquad I_2 = \frac{V}{R_2}, \qquad I_3 = \frac{V}{R_3},$$
$$I = I_1 + I_2 + I_3 \qquad \text{so}$$
$$I = \frac{V}{R_1} + \frac{V}{R_2} + \frac{V}{R_3} = V\left(\frac{1}{R_1} + \frac{1}{R_2} + \frac{1}{R_3}\right)$$
The total circuit resistance across
the battery is
$$\frac{V}{I} = R = \frac{1}{\frac{1}{R_1} + \frac{1}{R_2} + \frac{1}{R_3}}$$

Figure 2.4

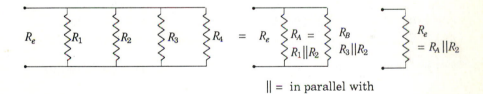

$\| =$ in parallel with

Example   If $R_1 = R_2 = 10\ \Omega$ $R_3 = 15\ \Omega$ and $R_4 = 20\ \Omega$

$R_A = \dfrac{10 \times 10}{10 + 10} = \dfrac{100}{20} = 5\ \Omega$ $5\ \Omega = 1/2$ of $10\ \Omega$; therefore, if

two resistors are equal and placed in parallel, the circuit
equivalent resistance is 1/2 of either resistor

and $R_B = \dfrac{15 \times 20}{15 + 20} = \dfrac{300}{35} = 8.75\ \Omega$ so $R_e = \dfrac{8.57 \times 5}{8.57 + 5} = \dfrac{42.85}{13.57}$

$R_e = 3.16\ \Omega$. Notice that the equivalent circuit resistance
R_e is less than that of any of the resistors in the parallel circuit.

Figure 2.5   Calculation of resistors in parallel.

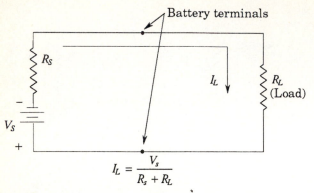

$$I_L = \frac{V_s}{R_s + R_L}$$

V_S is the source voltage (a battery).
R_S is the battery's internal resistance.
R_L is the load resistance.
I_L is the load current.

If $V_s = 10\,\text{V}$ and $R_s = 1\,\Omega$, calculate I_L and P_L for following values of R_L

Case 1: $R_L = 5\,\Omega$ $I_L = \frac{10}{6} = 1.67\,\text{A}$ $P_L = I_L^2\,R_L = [1.67]^2 \times 5 = 13.94\,\text{W}$

Case 2: $R_L = 10\,\Omega$ $I_L = \frac{10}{11} = 0.91\,\text{A}$ $P_L = [0.91]^2 \times 10 = 8.26\,\text{W}$

Case 3: $R_L = 4\,\Omega$ $I_L = \frac{10}{5} = 2.0\,\text{A}$ $P_L = [2]^2 \times 4 = 16\,\text{W}$

Case 4: $R_L = 2\,\Omega$ $I_L = \frac{10}{3} = 3.33\,\text{A}$ $P_L = [3.33]^2 \times 2 = 22.2\,\text{W}$

Case 5: $R_L = 1\,\Omega$ $I_L = \frac{10}{2} = 5.0\,\text{A}$ $P_L = [5]^2 \times 1 = 25\,\text{W}$

Case 6: $R_L = 0.5\,\Omega$ $I_L = \frac{10}{1.5} = 6.67\,\text{A}$ $P_L = [6.67]^2 \times 0.5 = 22.23\,\text{W}$

Figure 2.6 Load resistance matched to source resistance.

2.1.5 Internal resistance matching maximum power transfer

To illustrate a common concept of an electric circuit, consider the diagram in Fig. 2.6. Examining the power taken by the load indicates that maximum power occurs when the load resistance matches the source resistance. A plot on graph paper illustrates this important concept, shown in Fig. 2.7.

2.1.6 Applications of direct current systems

Direct current (dc) at one time was used commercially to distribute electric power to homes and manufacturing plants. Transporting large amounts of direct current became a problem because of the heat and

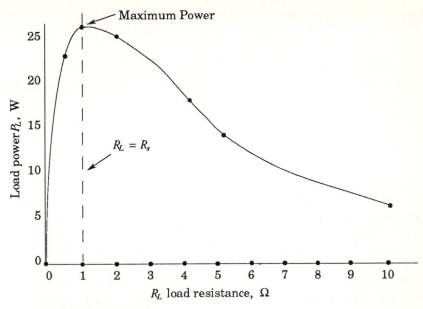

Figure 2.7   Maximum power.

energy loss due to the wires. Alternating current (ac) became the commercial choice mainly because the induction transformer made the control of voltage, current, and power economically feasible. Many electronic devices used today, such as transistors, integrated circuits, and vacuum tubes, have to use sources of direct current to make them work. These devices generally have to use rock-stable (nonvarying) sources of direct current, often necessitating extra circuitry just to hold the sources of direct current stable. These circuits are often referred to as *voltage regulators*. Most sources of direct current used today start with a commercial alternating current (115- V_{ac} line) source, where the voltage level is adjusted by a transformer and then converted to direct current by a rectifier and smoothed to high-quality direct current by a filter.

2.2   Alternating Current Principles

In *alternating current* the current alternates between positive and negative at some rate. The current does not flow only one way but continually reverses direction. The amount of variation between positive and negative per second is called the *frequency*. The common 120- V_{ac} commercial power alternates between positive and negative at 60 times per second. The standard unit for the frequency is the hertz

(Hz), named for Rudolph Heinrich Hertz, a famous German scientist. The old standard unit was the cycle per second (cps).

2.2.1 Sinusoidal alternating currents

The positive to negative alternation for commercial power is called the *sine wave*, where the changes from negative to positive and back to positive, etc., are smooth and cyclical. The word *sine* comes from the mathematics of trigonometry. The *sine function* of an angle is the ratio of the side opposite the angle of a right triangle to the side opposite the right (90°) angle. As the angle varies from 0 to 360°, the sine ratio varies as shown in Fig. 2.8. A full circle of rotation traces out the sine wave. A right triangle contains one right angle, and the side opposite this right angle is called the *hypotenuse*. The sine is the ratio of the length of the side opposite the angle to the length of the hypotenuse. In like manner the ratio of the adjacent side length to the hypotenuse length is called the *cosine* of the angle. The ratio of the length of the opposite side to the length of the adjacent side is called the *tangent* of the angle. The inverse of these three functions of an angle are the *cosecant* (inverse of the sine), *secant* (inverse of the cosine), and *cotangent* (inverse of the tangent). Fig. 2.9 summarizes the functions of an angle in a right triangle.

The plot of instantaneous voltage as a function of time for a sine wave is that shown in Fig. 2.8. The cosine waveform is just like this,

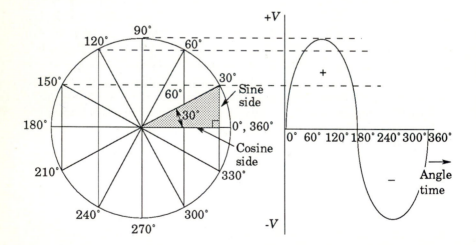

Radius of circle is the hypotenuse of the right triangles.

Figure 2.8 Generation of a sine wave.

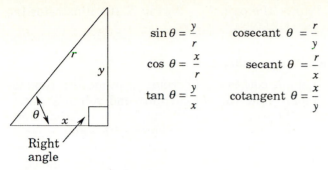

Figure 2.9 Functions of angle θ.

except that at 0° the cosine is at a maximum and at 90° at a minimum. Therefore, it may be said that the cosine wave is 90° out of phase with the sine wave. When two waveforms are at the same frequency and in phase, they are said to be "in step" and are at a maximum and a minimum at the same moment.

2.2.2 Square wave alternating current

Another interesting alternating current waveform is the square wave, as shown in Fig. 2.10a.

A variation is the pulsed alternating current (ac) waveform where,

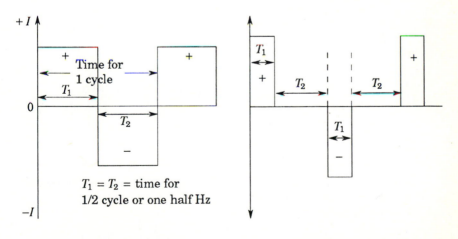

(a) The square wave

(b) Positive-negative pulses

Figure 2.10

T_1 does not equal T_2. For the case $T_1 < T_2$ Fig. 2.10*b* illustrates the pulsed waveform.

2.2.3   Other time-varying power sources

Most of the electric signals conveying information in modern communications systems employ time-variant currents and voltages. Alternating currents and voltages are time-varying; however, other time-varying power sources do not alternate between positive and negative and hence are regarded as a sort of pulsating direct current. Some examples of these waveforms of voltages and/or currents are shown in Fig. 2.11.

2.2.4   Effective value of waveform voltages and currents

Some means is needed to relate the waveform voltages and currents to some equivalent value. The heating of load on the electrical systems is

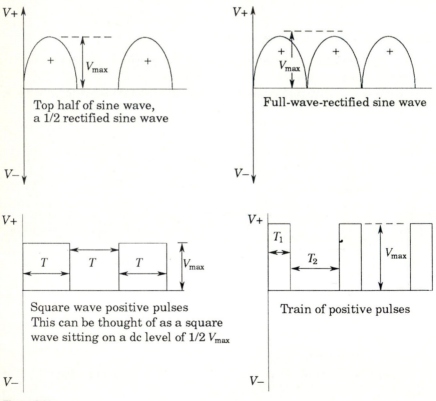

Top half of sine wave, a 1/2 rectified sine wave

Full-wave-rectified sine wave

Square wave positive pulses
This can be thought of as a square wave sitting on a dc level of 1/2 V_{max}

Train of positive pulses

Figure 2.11

the basis of the equivalency. If two different waveforms are able to heat a load to the same temperature, then the electrical values of the power are equivalent. The effective value of a waveform may be found by squaring the amplitudes, finding the mean of the square values, and taking the square root of the mean squared values. Thus the effective value is referred to as the *root mean square* (rms) value, which is the reverse of the procedure. Finding the rms value of a sine wave or one-half or full-wave-rectified sine wave involves use of the calculus, which is beyond the scope of this book. However, it is common knowledge that for a sine wave the effective or rms value is $0.707 \times V_{max}$. For a V_{max} of 162.5 volts (V) the rms value is approximately 115 V_{rms}. Recall that domestic household voltage is 115 V_{rms} at 60 Hz. The rms values of some common signal waveforms are shown in Fig. 2.12.

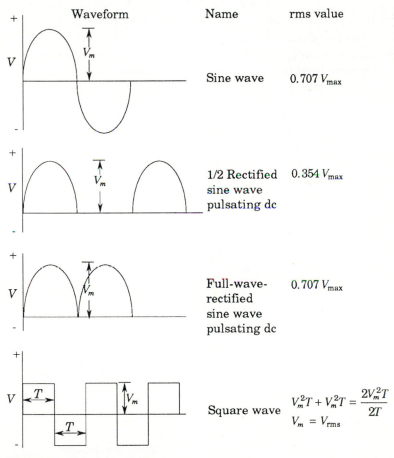

Figure 2.12 Common waveforms.

2.3   Transformer Principles

Previously it was mentioned that a device known as a transformer is often used in a power supply to convert the domestic alternating power source to direct current. Such a power supply is used in head-end processors, modulators, earth station receivers, and most pieces of head-end electronic equipment. The system amplifiers in the cable plant also all use power transformers to adjust the 60-V cable power to the proper value before converting (rectifying) to direct current for the amplifying devices. The 60-V cable power supply is essentially a special step-down regulating transformer device. A condensed review of transformers will be given here; a more complete treatment of the subject will be found in some of the references listed in the Selected Bibliography at the end of this chapter.

Since the transformer is an inductive device, a brief review of inductors and induction is needed. Consider the schematic diagram in Fig. 2.13 for an inductor driven by an alternating current generator where R is the resistance of the wire coil making up the inductor.

Notice the counterelectromotive force (emf) voltage caused by the varying magnetic field buildup and collapse caused by V_s opposes V_s by 180°; hence we have the term *back* or *counter emf*. Energy stored in the magnetic field is given up on alternate cycles of the applied voltage. Notice that back emf is at a negative maximum at the same time the applied voltage is at a positive maximum; when they are equal, zero current flows. The back emf voltage essentially opposes current flow. This opposing action is termed *inductive reactance* X_L, which is in ohms for a sine wave.

Also, for a sine wave X_L may be calculated by using the formula

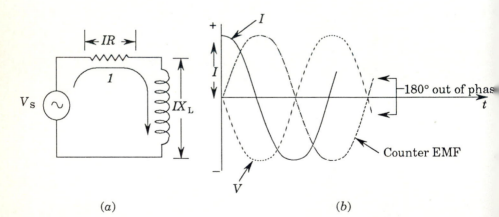

(a)   (b)

The inductive circuit   Circuit waveforms

Figure 2.13

$X_L = 2\pi f L$, where $\pi = 3.14$, $f =$ frequency of the ac in Hz, and $L =$ inductance in henrys (H). Since it is cumbersome to make calculations in the time domain as in Fig. 2.13b, a better method is to stop the alternations at some angle and make the calculations. Such a method is shown in the vector diagram in Fig. 2.14 using the current I as a reference at $0°$.

The diagrams tell us that the voltage across the resistance is in phase with the current through it and lags the voltage across the inductive reactance IX_L by $90°$. Vector addition has to be used to add IX_L and IR to get V_s. The symbol θ is the phase angle that indicates how much the current lags the applied voltage V_s. The time-phase diagram in Fig. 2.13b indicates that $R = 0$ and I lags V_s by $90°$. It should be evident that one cannot have a coil of wire with no resistance. Therefore, the I^2R term in watts is the power loss by the inductor caused by copper winding losses. Recall that $X_L = 2\pi f L$ and the higher the frequency the more inductive reactance results (an *open circuit*). Most power inductors operating in power supply circuits at 60 Hz are of the iron core type as shown in Fig. 2.15. The iron laminations concentrate the magnetic lines of force, thus raising the inductance L. A transformer is really two inductor windings on the same iron-laminated core, where the coefficient of coupling K is 100 percent. This means that all the magnetic lines of force flow through both coils. The coil winding that is connected to a power source is called the *primary*, and that connected to the load side the *secondary*. The schematic diagram for a transformer is shown in Fig. 2.16. With the secondary terminals open V_c (coil voltage) is nearly equal to V_s; thus I_p (primary current) is very small. When R_L is connected across the secondary, I_s flows, caus-

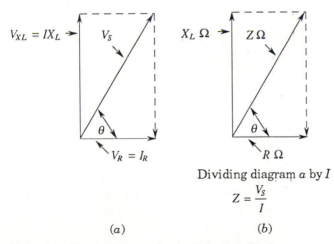

$$Z = \frac{V_s}{I}$$

Dividing diagram a by I

(a) (b)

Figure 2.14 Vector diagrams for the inductive circuit.

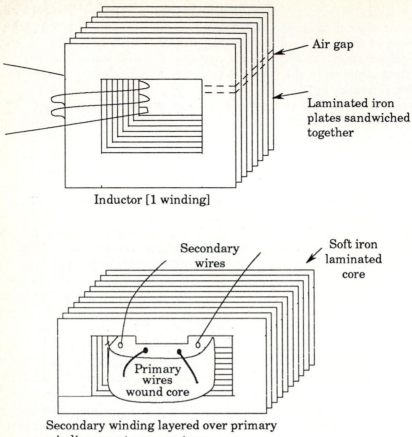

Inductor [1 winding]

Secondary winding layered over primary
windings most common type

Transformer [2 windings]

Figure 2.15

ing the counter-countervoltage (V_{cc}), which opposes V_c, thus aiding V_s and allowing I_p to increase.

2.3.1 Transformer turns ratio

The governing factor for the secondary voltage and current is the turns ratio of the transformer. Simply stated, the ratio of the number of primary turns to the number of secondary turns is the same as the ratio of the primary to secondary voltage. Mathematically:

$$\frac{T_p}{T_s} = \frac{V_p}{V_s}$$

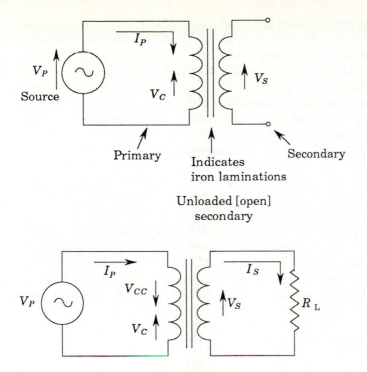

Figure 2.16

where T_p = Number of primary turns
T_s = Number of secondary turns
V_p = Primary voltage
V_s = Secondary voltage

Example

$$T_p = 500 \text{ turns} \qquad T_s = 1000 \text{ turns}$$

$$V_p = 110 \text{ V}_{ac} \qquad V_s = ?$$

Substituting in the above equation and multiplying both sides by $V_s(1000)/500$,

$$\frac{500}{1000} = \frac{110}{V_s} \qquad \frac{V_s(1000)}{500} \times \frac{500}{1000} = \frac{110}{\cancel{V_s}} \frac{\cancel{V_s}(1000)}{500}$$

$$V_s = \frac{110 \times 1000}{500} = 220 \text{ V}_{ac}$$

This transformer is called a *1:2 step-up transformer*, meaning that the primary voltage has been stepped up to 220 V. Most of the transformers used in cable line amplifiers are of the 2:1 step-down type. Here the 60-V cable voltage is stepped down to 30 V_{ac} then rectified and smoothed to +24 V_{dc}. The 60-V cable power supply is a special 2:1 step-down transformer that steps down the 120 V_{ac} to 60 V_{ac}. The special high-voltage winding causes the output wave shape to be a distorted (nearly square wave) sine wave. This does not cause a problem with the line amplifier power transformers using this power.

Stepping up the secondary voltage seems to be a form of free amplification, but it is not because the power taken by the secondary circuit is the same as the primary applied power less the small amount of the usual transformer losses. There is no power gain. Therefore, if the secondary voltage is greater than the primary voltage, the allowable secondary current will be limited to a stepped-down value. An example should clarify the relationship. Assuming a perfect (no loss) 500-watt (W) transformer with a 110-V primary winding has a turns ratio of 4:1, find the allowable primary and secondary currents. Lossless case means $P_p = P_s = 500$ W.

$$P_p = I_p V_p, \quad \text{so} \quad I_p = \frac{P_p}{V_p} = \frac{500}{110} = 4.55 \quad \text{amperes}$$

$$\text{Since } V_p = 110 \text{ V}, \quad V_s = V_p \left(\frac{1}{4}\right) = \frac{110}{4} = 27.5 \ V_{ac}$$

$$P_s = 500 = V_s I_s, \quad \text{so} \quad I_s = \frac{500}{27.5} = 18.2 \ A$$

It should be evident that if the voltage from primary to secondary is stepped down, the current is stepped up by the turns ratio. Therefore, it should be evident that

$$\frac{T_p}{T_s} = \frac{I_s}{I_p} = \frac{V_p}{V_s}, \quad \text{so} \quad V_s I_s = V_p I_p \quad \text{and} \quad P_p = P_s \qquad (2.1)$$

2.3.2 Primary impedance (Z_p) and turns ratio

Recall that to find the secondary voltage

$$V_s = \frac{T_s}{T_p} V_p \quad \text{and} \quad I_s = \frac{T_p}{T_s} I_p$$

The secondary impedance

$$Z_s = \frac{V_s}{I_s} \qquad (2.2)$$

Substituting Eq. (2.1) in (2.2)

$$Z_s = \frac{(T_s / T_p) V_p}{(T_p / T_s) I_p} = \frac{T_s}{T_p} \times \frac{T_s}{T_p} \times Z_p$$

$$= \left(\frac{T_s}{T_p}\right)^2 Z_p$$

For the 4:1 case as above and if the secondary is loaded with a 20-ohm (Ω) resistor, what impedance will the primary see?

$$Z_s = 20\ \Omega \qquad Z_p = \left(\frac{T_p}{T_s}\right)^2 Z_s$$

$$Z_p = \left(\frac{4}{1}\right)^2 (20) = 16 \times 20 = 320\ \Omega$$

Transformers used in this manner can match impedances. Such is the case for the 75-Ω to 300-Ω matching transformer. The frequency is of course in the band of 54 to 450 megahertz (MHz) so very few turns (copper loss) and very little ferrite (iron) are used in such transformers. A metallic shield is used to maintain the fields within the device. It is also important that the wire sizes used in the primary and secondary windings be large enough to carry the design currents.

2.3.3 Transformer characteristics

In the theoretical discussion of transformers, the devices were regarded as perfect, i.e., lossless. Although the efficiency is high (in the range of 90 to 98 percent), transformers do have losses. These losses include copper winding losses and core losses. The core losses are made up of hysteresis and eddy current losses, which cause heating of the core. It is important to use a transformer that is designed for a particular application. A transformer designed for audio frequency work will not function in the cable television frequency bands even though the turns ratios may be correct. Proper size fuses should be used in the primary circuit to protect the power transformer from overload.

2.4 Capacitance and Capacitors

The capacitor acts as the inverse of the inductor. The inductor is a magnetic device; the capacitor is an electrostatic device. The capacitor in its simplest form is constructed of two parallel metallic plates separated by an insulating material. If a source of electromotive force is placed across the two parallel plates, an electric field will be stored in the insulating material. The capacitor can be said to be charged to the applied source of voltage. The capacitor will retain its charge when the source is removed; if the capacitor is perfect, the charge will remain. However, since capacitors are not perfect, the charge will leak off through the insulator. If a charged capacitor is short-circuited, the capacitor will be discharged immediately. The unit of **capacitance** is the farad (F), which for practical purposes is too large. Practical values used in electronic circuitry today are the microfarad (μF = 1×10^{-6}F) and the micromicrofarad $(1 \times 10^{-6} \times 10^{-6}$ F = 1×10^{-12} F), often called the *picofarad* (pF).

2.4.1 The time constant

The time a capacitor needs to attain 63 percent of a full charge when a source of dc voltage is placed across the capacitor is called the *time constant*. The time constant in seconds is directly proportional to the capacitance in farads and any series resistance in ohms. Mathematically $\tau = RC$.

Example.   A 5-μF capacitor is in series with a 1-megaohm (MΩ) resistor. Find the time constant of the circuit

$$\tau = 5 \times 10^{-6} \times 1 \times 10^{6} = 5 \text{ seconds (s)}$$

If this circuit is placed across a 100-V battery, at 5 s the capacitor will have charged to 63 V.

2.4.2 Calculation of capacitance

The capacitance C (in picofarads) of a parallel-plate capacitor can be found by using the formula

$$C = \frac{0.225\,KA}{S}$$

where K = dielectric constant
A = area of one plate (inches squared)
S = plate separation (inches)
0.225 = constant of proportionality

According to this formula, the capacitance is directly proportional to the dielectric constant and the area of the plates and inversely propor-

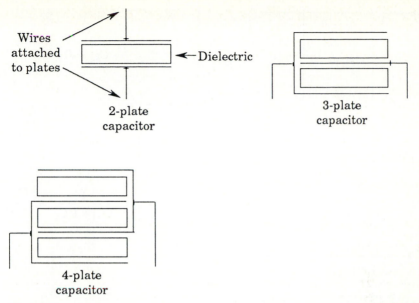

Figure 2.17

tional to the spacing between the plates. The dielectric constant is a measure of the efficiency of the insulator used between the plates. Glass has a dielectric constant in the range of 4 to 7, quartz of 4, and a vacuum of 1. Most practical capacitors use more than one pair of plates. The capacitance of multiple-plate capacitors may be found by multiplying the one pair plate formula by $N - 1$, where N is the number of plates. Fig. 2.17 illustrates the concept of multiple plates. The symbols used in electrical and electronic schematic diagrams are shown in Fig. 2.18.

Capacitance of individual capacitors connected in parallel is simply the sum of the capacitances. The plates may be thought of as increasing the area, thus increasing the capacitance. Capacitors placed in series act by increasing the plate separation and thus decreasing the capacitance. Fig. 2.19 illustrates the concept.

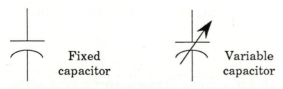

Figure 2.18   Schematic diagram symbols for a capacitor.

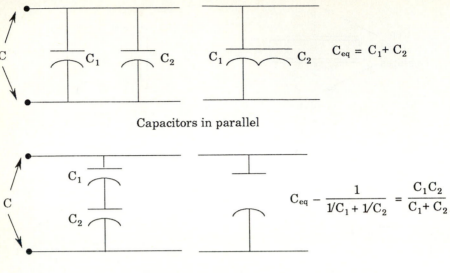

Capacitors in parallel

Capacitors in series

Figure 2.19

2.4.3 Sine wave capacitive circuits

Charging and discharging capacitors oppose current flow in much the same way inductors do. This opposing action is called *capacitive reactance* (in ohms). Mathematically,

$$X_C = \frac{1}{2\pi fC}$$

where $2\pi = 6.28$
f = frequency (Hz)
C = capacitance (F)

This formula states that the capacitive reactance is inversely proportional to the frequency and the capacitance. For a fixed capacitor of 1 μF the reactance at 60 Hz is

$$X_C = \frac{1}{(6.28)(60) \times 1 \times 10^{-6}} = \frac{1}{376.8 \times 10^{-6}}$$

$$= 0.00265 \times 10^6 = 2650\ \Omega \text{ at } 60 \text{ Hz}$$

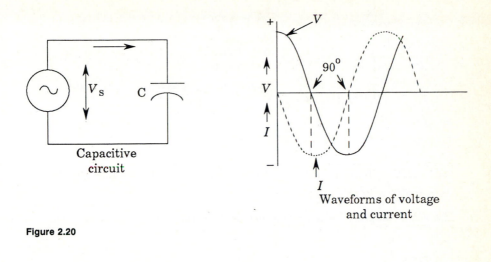

Capacitive
circuit

Waveforms of voltage
and current

Figure 2.20

At 100 MHz the reactance is

$$\frac{1}{(6.28)(1 \times 10^8)(1 \times 10^{-6})} = \frac{1}{6.28 \times 1 \times 10^2} = 0.159 \times 10^{-2}$$

$$= 0.00159 \ \Omega \qquad \text{Nearly a short circuit}$$

If a capacitor is placed across a source of sine wave alternating current, the current leads the voltage by 90° as shown in Fig. 2.20. It may also be said that the voltage lags the current by 90°. It should be clear that the effects of capacitance and inductance are the inverse of each other.

2.5 *RLC* Circuits

To summarize the effects of inductance and capacitance on steady-state alternating current circuits consider the circuit in Fig. 2.21.

2.5.1 Series *RLC* circuit

Since this is a series circuit, the common electrical quantity is the current, which is at 0°; the voltage across the resistor IR is also at 0° (in phase), the voltage across the inductor IX_L is ahead 90°, and the voltage across the capacitor IX_L lags by 90°. The circuit is more capacitive than inductive but is mostly resistive.

2.5.2 Parallel *RLC* circuit

The parallel circuit of an inductor capacitor and resistor is the inverse of the series circuit. The common electrical parameter is the voltage

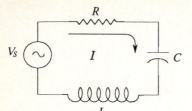

V_S is a source of sine wave voltage
at 100 V$_{\text{rms}}$ at 100 Hz
$R = 10\,\Omega$, $L = 1\,\text{mH}$
$C = 15\,\mu\text{F}$
Find the voltages across $R, L,$ and C.

Solution

$$X_L = 2\pi fL = 6.28 \times 1 \times 10^3 \times 1 \times 10^{-3} = 6.28\,\Omega$$

$$X_C = \frac{1}{2\pi fC} = \frac{1}{(6.28)(1 \times 10^3)(15 \times 10^{-6})} = \frac{1}{6.28 \times 15 \times 10^{-3}} = \frac{1}{94.2 \times 10^{-3}}$$

$$= 0.0106 \times 10^3 = 10.6\,\Omega$$

The vector diagram for the resistance and the reactances.

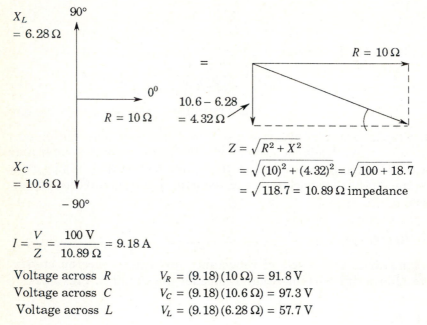

$$I = \frac{V}{Z} = \frac{100\,\text{V}}{10.89\,\Omega} = 9.18\,\text{A}$$

Voltage across R	$V_R = (9.18)(10\,\Omega) = 91.8\,\text{V}$
Voltage across C	$V_C = (9.18)(10.6\,\Omega) = 97.3\,\text{V}$
Voltage across L	$V_L = (9.18)(6.28\,\Omega) = 57.7\,\text{V}$

Figure 2.21 RLC circuit.

that appears across each element. To demonstrate this condition consider the example shown in Figure 2.22.

2.6 Resonance and Tuning

The condition when $X_L = X_C$ results in a circuit that is said to be in *resonance*. For the series resonant circuit the effects of the inductor and capacitor are canceled and the resulting source voltage appears

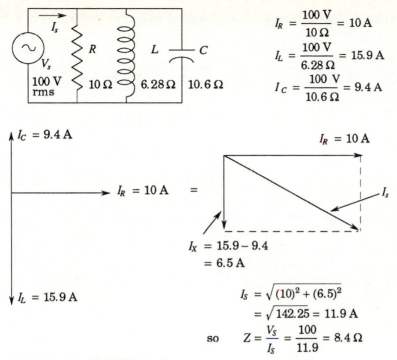

$$I_R = \frac{100\ V}{10\ \Omega} = 10\ A$$

$$I_L = \frac{100\ V}{6.28\ \Omega} = 15.9\ A$$

$$I_C = \frac{100\ V}{10.6\ \Omega} = 9.4\ A$$

$$I_X = 15.9 - 9.4$$
$$= 6.5\ A$$

$$I_S = \sqrt{(10)^2 + (6.5)^2}$$
$$= \sqrt{142.25} = 11.9\ A$$

so $\quad Z = \dfrac{V_S}{I_S} = \dfrac{100}{11.9} = 8.4\ \Omega$

Figure 2.22 Parallel *RLC* circuit.

across R. If $R = 0\ \Omega$, then the circuit constitutes a short circuit. The parallel resonant circuit presents an open circuit to the source voltage, allowing nearly no current to flow. The condition of resonance occurs for one particular frequency. Mathematically for $X_L = X_C$ then

$$2\pi fL = \frac{1}{2\pi fC}$$

Multiplying both sides by $2\pi fC$ gives $(2\pi fL)(2\pi fC) = 1$. Then

$$f^2(4\pi^2 LC) = 1 \qquad f^2 = \frac{1}{4\pi^2 LC} \qquad \text{and} \qquad f = \frac{1}{2\pi\sqrt{LC}}$$

where f is the resonant frequency in hertz when L is in henrys and C is in farads.

The following example will illustrate the principle for $R = 100\ \Omega$, $C = 10$ pF, $L = 10$ microhenrys (μH).

$$f = \frac{1}{2\pi\sqrt{10 \times 10^{-12} \times 10 \times 10^{-6}}} = \frac{1}{2\pi\sqrt{100 \times 10^{-18}}}$$

$$= \frac{1}{2\pi\sqrt{1 \times 10^{-16}}} = \frac{1}{2\pi \times 10^{-8}}$$

$$= \frac{1 \times 10^8}{6.28} = 0.159 \times 10^8 = 15.9 \times 10^6 = 15.9 \text{ MHz}$$

$$X_C = \frac{1}{2\pi fC} = \frac{1}{(6.28)(15.9 \times 10^6)(10 \times 10^{-12})} = \frac{1}{100 \times 10 \times 10^{-6}}$$

$$= \frac{1}{1 \times 10^{-3}} = 1 \times 10^3 = 1000 \ \Omega \qquad \text{Also } X_C = X_L = 1000 \ \Omega$$

Fig. 2.23a is for the series case and 2.23b for the parallel case.

The conclusions are drawn for the series case because at resonance the source voltage V_s appears across R, the series resistance, since the voltage across points A to B is zero. However, the voltage across L and C can be very large where the reactances are large. If R is 0 Ω or very small, the voltage V_s is practically short-circuited although the individual voltages across L and C can be large. For the parallel case, at resonance most of the current drawn from the source is taken by R. If R is removed (infinite ohms), then the current I_s will be very small.

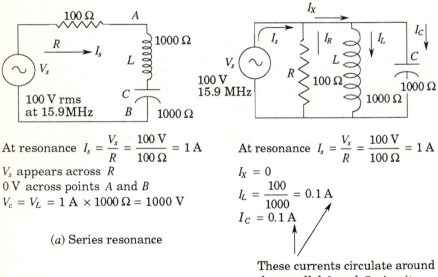

At resonance $I_s = \dfrac{V_s}{R} = \dfrac{100 \text{ V}}{100 \ \Omega} = 1 \text{ A}$

V_s appears across R

0 V across points A and B

$V_c = V_L = 1 \text{ A} \times 1000 \ \Omega = 1000 \text{ V}$

(a) Series resonance

At resonance $I_s = \dfrac{V_s}{R} = \dfrac{100 \text{ V}}{100 \ \Omega} = 1 \text{ A}$

$I_X = 0$

$I_L = \dfrac{100}{1000} = 0.1 \text{ A}$

$I_C = 0.1 \text{ A}$

These currents circulate around the parallel L and C circuit not contributing to I_s

(b) Parallel resonance

Figure 2.23

However, the circulating currents I_C and I_L can be large and usually are greater than the total current drawn from the source. A series resonant circuit provides a very low impedance, and a parallel resonant circuit provides a very high impedance at the resonant frequency. These circuits when placed in series or parallel combinations constitute frequency selectivity, or what is simply known as *tuning*. Such circuits are used in television and radio transmitters and receivers as well as in frequency selective filters which are used in a variety of applications.

2.7 Basic Power Supply Circuits

Since transformer principles and basic inductive, capacitive, and resistive circuits have been reviewed, power supply systems used in cable television systems will now be discussed. The power supplies used in head-end electronic equipment usually operate from 115-V_{ac} line power, and the line amplifiers in the cable distribution plant from the quasi-square-wave 60-V_{ac} power provided on the cable. The 60-V cable power supply is basically a ferroresonant transformer driven from the 115-V_{ac} commercial power source, which provides a regulated (nearly constant) 60 V to the power inserter. This power supply is discussed in more detail in the following chapters.

The power supply systems for the electronic equipment require a direct current. Hence the power supplies have to adjust the ac voltage level from the input power, rectify this ac voltage (pulsating dc), and then filter and regulate the resulting direct current. Examples of the circuit configurations are shown in Fig. 2.24.

The rectifier circuit usually consists of solid-state diodes (two-terminal devices) which pass current in only one direction. An elementary half-wave rectifier is shown in Fig. 2.25.

By reversing the diode the negative half of the 27-V sine wave will be passed through. By using a center tapped secondary for the transformer and using two diodes, both halves of the sine wave can be used. Such a rectifier circuit is called a *full-wave rectifier* and is shown in Fig. 2.26.

A popular type of rectifier circuit used in many systems is a *bridge-type rectifier*, which also is a full-wave rectifier. This circuit does not require a center tapped transformer but uses four diodes. Many diode manufacturers package the four diodes in one case. The bridge rectifier circuit is shown in Fig. 2.27.

The rectified waveform needs some low-pass filtering circuits to smooth the pulsating direct current, followed by solid state regulating circuits which prevent the smooth direct current output from varying as a result of power line fluctuations or changes in load requirements. A low-pass filter that passes only direct current (zero frequency) will,

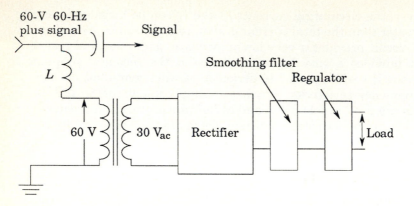

Cable amplifier power supply

Head-end equipment power supply

Figure 2.24

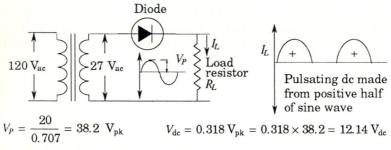

$$V_P = \frac{20}{0.707} = 38.2 \ V_{pk} \qquad V_{dc} = 0.318 \ V_{pk} = 0.318 \times 38.2 = 12.14 \ V_{dc}$$

Figure 2.25   Simple half-wave rectifier.

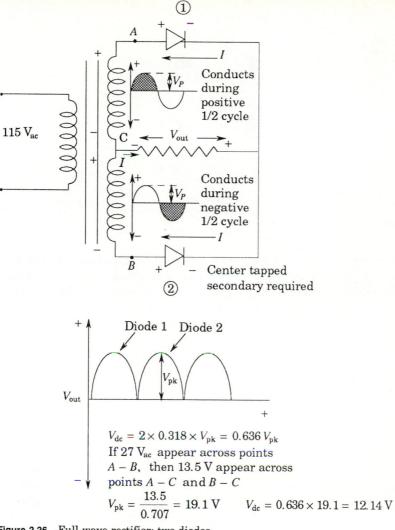

$$V_{dc} = 2 \times 0.318 \times V_{pk} = 0.636\, V_{pk}$$

If 27 V_{ac} appear across points
$A - B$, then 13.5 V appear across
points $A - C$ and $B - C$

$$V_{pk} = \frac{13.5}{0.707} = 19.1 \text{ V} \qquad V_{dc} = 0.636 \times 19.1 = 12.14 \text{ V}$$

Figure 2.26   Full-wave rectifier: two diodes.

of course, do the job; however, practical filters made from resistors, capacitors, and inductors where attenuation is high at the 60-Hz frequency perform adequately.

2.8 Filters and Signal Filters

Filters are used in the selection of signals in a band of frequencies and are classified as *low-pass* (pass the low frequencies), *high-pass* (pass the high-frequency band), *bandpass* (pass only a band of frequencies),

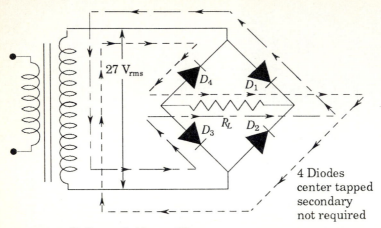

Figure 2.27   Full-wave bridge rectifier.

and *bandstop* (reject a band of frequencies). The usual plots of the frequency response characteristics of filters are attenuation versus frequency with the attenuation in decibels (dB) plotted on a linear scale and the frequency plotted on a logarithmic scale. Examples of the idealized plots for the four types of filters are shown in Fig. 2.28.

Filters with the same input/output impedance may be coupled together to control the overall bandpass/reject characteristics. For example, a low-pass filter may be coupled to a high-pass filter to make up a bandpass filter, as illustrated in Fig. 2.29. Notice that the cutoff frequency for the high-pass filter is below the cutoff frequency of the low-pass filter.

By coupling in a different manner a band reject filter can be made. Coupling together by means of two hybrid splitters and spreading the cutoff frequencies apart will accomplish the task. Fig. 2.30 illustrates this case. Of course, the low-pass and high-pass filters with the overlapping cutoff frequencies can be connected in this manner to make up the bandpass case instead of coupling end to end. The diagrams of the frequency response characteristics of the four types of filters so far has been idealized, e.g., by the square sides at the cutoff frequency. By now we should realize that nothing is perfect and that the drop off in attenuation at the cutoff frequency is sloped according to the following diagram (Fig. 2.31). The cutoff frequency is now defined as occurring when the attenuation has dropped 3 dB (the half-power point).

2.8.1   Frequency response

Since the filters are named low-pass, high-pass, etc., it is often more convenient to show the frequency response characteristics not in

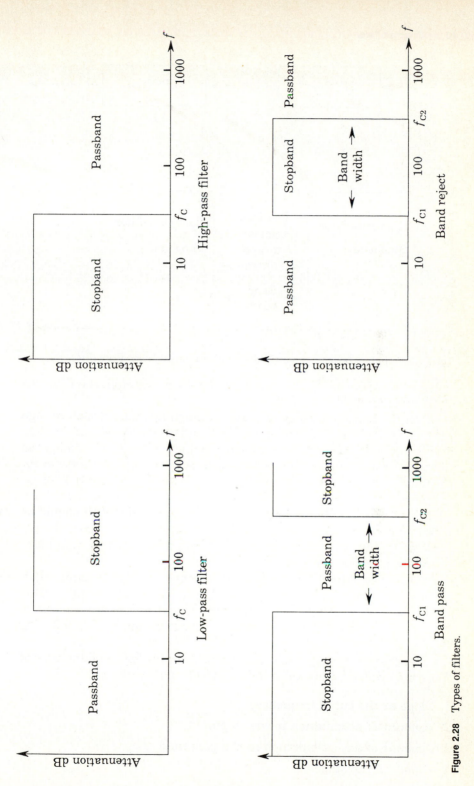

Figure 2.28 Types of filters.

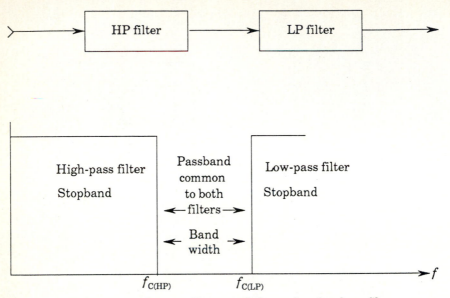

Figure 2.29   High-pass and low-pass filters coupled to make a bandpass filter.

terms of attenuation but of signal level passing through the filter, as shown in Fig. 2.31.

Often the filter frequency response is given as input/output voltage ratio in decibels for the case of an input and output impedence match, producing a signal level response curve which is the mirror image, so to speak, of the attenuation versus frequency plots. The diagram for the bandpass case should clarify the relation, as shown in Fig. 2.32. f_{C_1} is known as the *lower cutoff frequency* and f_{C_2} as the *upper cutoff frequency*. When studying filter passband and stopband plots it should be noted immediately whether attenuation or signal amplitude is the parameter. Many textbooks use the preceding notation, i.e., signal amplitude versus frequency.

Recall that signal amplitude, in decibels, is for a gain dB = 10 log (P_{out}/P_{in}) and for matched filters (impedance) gain dB = 20 log (V_{out}/V_{in}). Filters have attenuation and hence loss; therefore, V_{out} is less than V_{in} so the gain is negative gain dB = −20 log (V_{in}/V_{out}).

Filter frequency response characteristic curves for the same type of filter may differ by the following important parameters:

1. Slope at the cutoff frequency

2. Amount of attenuation in the stopband

3. Amount of any attenuation in the passband

4. Ripple in the stopband

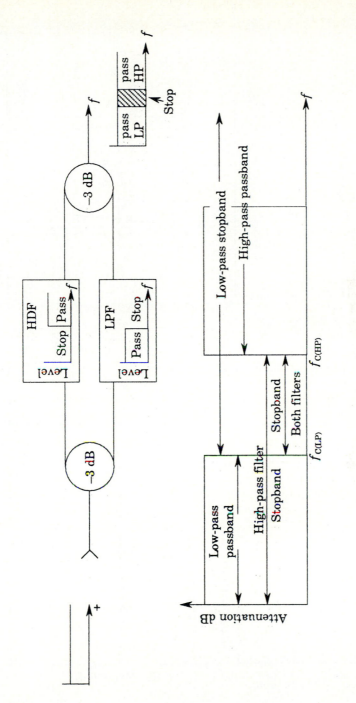

Figure 2.30   High-pass–low-pass coupling to make a band reject filter.

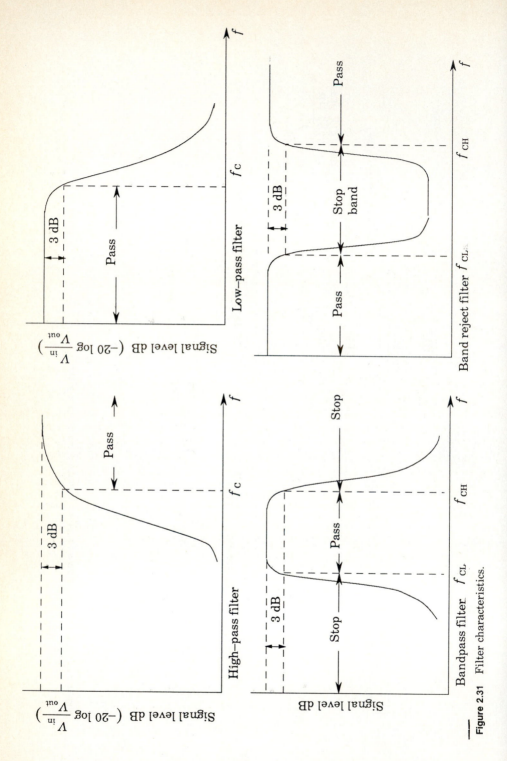

Figure 2.31 Filter characteristics.

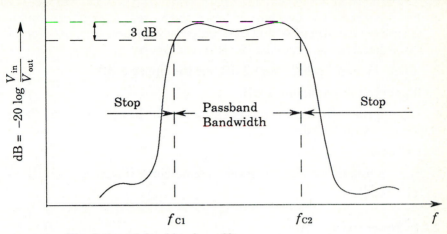

Figure 2.32   Signal through level bandpass filter.

5. Ripple in the passband
6. Signal peaking near the cutoff frequency
7. Passband (bandwidth)
8. Center frequency

The annotated curve in Fig. 2.33 illustrates these eight parameters:

1. Slope at cutoff 18 dB/octave (remember from music that an octave occurs every time we double the frequency).

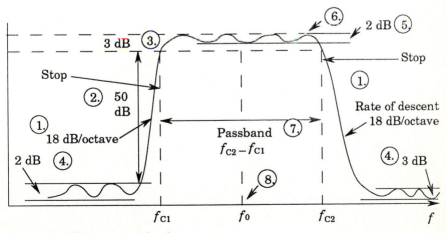

Figure 2.33   Filter response bandpass case.

2. Attenuation in the stopband: 50 dB below that of passband.

3. Attenuation in the passband: shown as the reference (0 dB) (manufacturers usually give this parameter as insertion loss, usually on the order of a few decibels, (2 dB in the figure).

4. Ripple in stopband: (lower) 2 dB, for the upper 3 dB.

5. Ripple in the passband: 2 dB.

6. Peaking: at 2 dB (sometimes peaking may be lumped with ripple in the passband to 3 dB).

7. Passband: $f_{C_2} - f_{C_1}$.

8. Center frequency: f_o (for good symmetrical filters f_o usually is $f_{C_1} - f_{C_2}/2$).

2.8.2 Power supply filters

Probably the most basic filter, studied in the first part of this book, is the simple low-pass resistor-capacitor (RC) filter used to filter the power from a rectifier circuit. The diagram in Fig. 2.34 should help us to recall it.

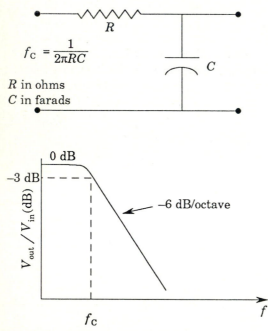

Figure 2.34 RC power supply filter.

Example

$$R = 100 \ \Omega$$

$$C = 100 \ \mu F = 1 \times 10^2 \times 10^{-6} = 1 \times 10^{-4} \, F$$

$$RC = 1 \times 10^2 \times 1 \times 10^{-4} = 1 \times 10^{-2}$$

$$f_C = \frac{1}{2\pi(10^{-2})} = \frac{1}{6.28 \times 10^{-2}} = \frac{100}{6.28} = 16 \text{ Hz}$$

which is very much less than the 60- or 120-Hz components trying to get through the filter. The ripple in this case is the amount of 60- or 120-Hz signal appearing in the output voltage, not the response of the filter. This type of filter ideally should pass only direct current (dc), which means a cutoff frequency of 0 Hz. This, of course, would make C very large and R small. The resistor R essentially limits the charging current of the capacitor at turn-on, so for large capacitors we will have large charging currents not limited very much by a low value of R and possibly enough to exceed the allowable diode rectifier currents, thus causing them to burn out.

Another slightly more costly but superior power supply low-pass filter replaces R with an appropriate value of inductor as shown in Fig. 2.35.

The simple RC-type filter has only one energy storage element (reactive), the capacitor, which is referred to as a *first-order filter*. The LC-type filter exhibits a much steeper response, contains two reactive elements, and is referred to as a *second-order filter*; its slope is twice that of the first-order filter. Naturally a third-order filter will contain three reactive elements and will exhibit an 18-dB/octave slope.

The elements of filter construction, as studied in the previous material, are made up of resistors, capacitors, and inductors by them-

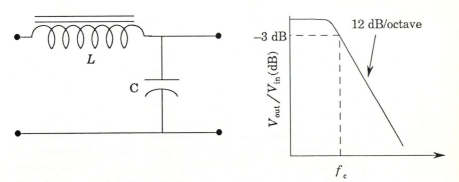

Figure 2.35 LC power supply filter.

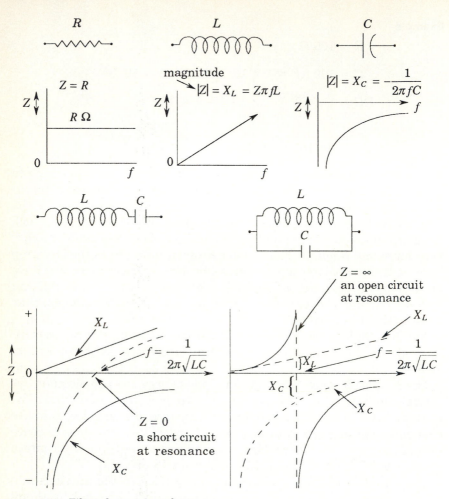

Figure 2.36   Filter element impedances.

selves or in series or parallel combinations. Recall the impedances or reactances for the following cases in Fig. 2.36.

Filters are made from these elements placed either in series or across the network terminals. If a parallel-tuned circuit is placed in the series arm, a series-tuned circuit, is placed across the terminals, and, they are tuned to the same frequency, this signal will be eliminated in the output. The diagram in Fig. 2.37 illustrates the point. When L_1 and C_1 are at resonance, the impedance between A and B is very large, thus attenuating the signal level at the tuned frequency. If L_2 and C_2 are tuned to the same frequency, the signal will be short-circuited from B to C. Such a filter is a very narrow band notch filter

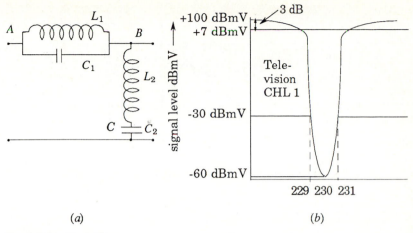

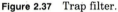

(a)   (b)

Figure 2.37   Trap filter.

that attenuates a huge amount around a narrow band of frequencies. This type of filter is often referred to as a *trap* because the signal is essentially trapped out from the output. Signal traps are used widely in cable television applications to delete a channel or band of channels from a subscriber. Such traps can be stacked or placed in series, causing channels to be deleted. A sample of a trap response is shown in Fig. 2.37*b*.

2.8.3   Transmission line filters

A study of cable as a transmission line will show that short sections of cable can act as resistance, capacitance, and inductance. Consider the diagram of a short-circuited piece of coaxial cable in Fig. 2.38. As we move away from the short circuit (low pure resistance) the circuit gets increasingly more inductive until the one-quarter wavelength, where it gets very high resistance. Past the one-quarter wavelength point the capacitance increases until a very low resistance point is achieved at the one-half wavelength point. Essentially the short circuit is repeated every half wavelength, thus causing zero voltage and high current nodes in the standing wave pattern. Also an open circuit appears at every quarter wavelength point. As one might expect, the mirror image phenomenon occurs for the open-circuited line. Figure 2.39 shows this condition. Now as we move away from the very high resistance (open circuit) the cable becomes more capacitive when at one-eighth wavelength the capacitive reactance equals Z_0. At one-quarter wavelength the cable has a very low resistance point (a short circuit). Then the line becomes increasingly more inductive until we arrive at

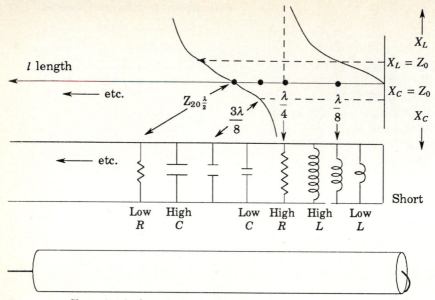

Figure 2.38 Short-circuited transmission line.

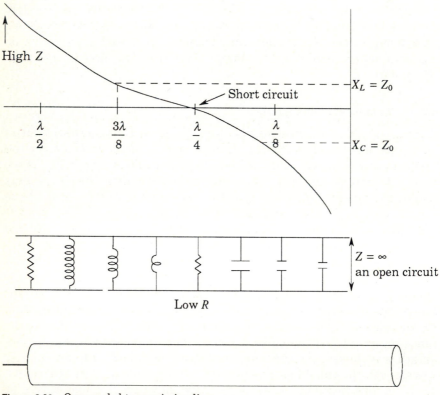

Figure 2.39 Open-ended transmission line.

an open circuit at one-half wavelength. At three-eighths wavelength the line inductive reactance equals Z_0.

By now the question may be, How can we use this information to work for us? The answer is simple: we can construct some very sharp and deep notch filters by using transmission lines. The best way to illustrate the point is to take an example. Suppose we needed a sharp trap at say 200 MHz. The first step is to calculate the wavelength for a frequency of 200 MHz. In feet

$$\lambda_{(air)} = \frac{984}{f_{(MHz)}} = \frac{984}{200} = 4.92 \text{ ft}$$

Since this is for a full wavelength and we would like to use the shortest length of cable, only a quarter wavelength is needed. Therefore, $(\lambda/4) = (4.92/4) = 1.23$ ft. For a more accurate measure we will convert 1.23 ft to inches. So

$$0.23 \text{ ft} \times \frac{(12 \text{ in})}{\text{ft}} = 2.76 \text{ in}$$

Since we do not know the precise cable characteristics we will allow ourselves some extra length. To construct a 75-Ω trap we will cut off 1 ft and say another 5 in of RG6 or RG59 drop cable. On one end fasten an F male connector. Now take a two-way hybrid (splitter of the blister can variety) and carefully remove the rear cover. Remove the contents and wire the center conductors of the F connectors together by soldering in some bare tinned copper wire. The splitter now is a T connector with F female connectors. The rear cover will now be replaced and soldered back on. Now we can use a signal generator and signal level meter to make the final adjustments. Fig. 2.40 illustrates the procedure.

Once this setup has been connected and the equipment turned on long enough to stabilize, we can proceed. Adjust the signal level meter attenuator for a ¾-meter (m) scale indication.

With a sharp pair of cutting pliers start to cut off ¼-in sections until the meter indicator starts to decrease. Keep lowering the attenuator control on the signal level meter for a ½- to ¾-m reading. Keep trimming until the lowest reading is obtainable. The generator should be able to provide at least 40 decibel-millivolts (dBmV) of signal level. Most signal level meters are only able to handle signals as low as -30 dBmV. Traps of this type can usually provide notches as deep as 50 to 70 dB down if care is taken. It must be remembered that if a trap is not cut precisely to the desired frequency it will provide a deep notch at some other frequency slightly off the desired one. After this trap has been adjusted and the modified splitter made into a T connector it

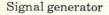

Signal generator

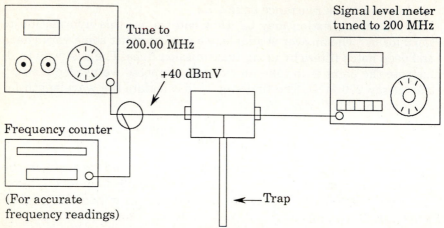

Tune to
200.00 MHz

Signal level meter
tuned to 200 MHz

+40 dBmV

Frequency counter

(For accurate
frequency readings)

←— Trap

Figure 2.40   Transmission line trap tuning procedure.

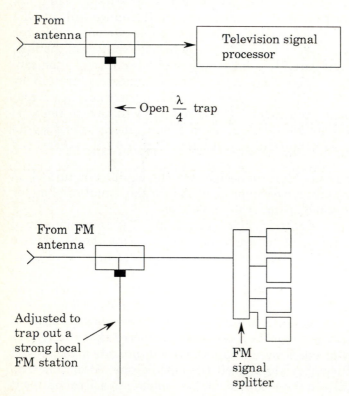

From
antenna

Television signal
processor

←— Open $\frac{\lambda}{4}$ trap

From FM
antenna

Adjusted to
trap out a
strong local
FM station

FM
signal
splitter

Figure 2.41   Trap applications.

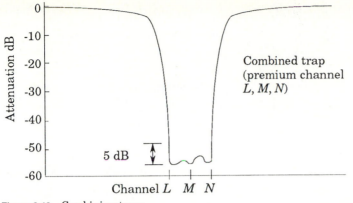

Figure 2.42   Combining traps.

can be removed and put into service, possibly in the manner shown in Fig. 2.41.

The same type of trap can be constructed at this same frequency by cutting the stub to a one-half wavelength and short-circuiting the end. There are pros and cons to using half wavelength stub traps. The advantage is that the end is short-circuited by the F connector which shields the ends against detuning. The disadvantages are that for lower frequencies the trap can become quite long and stepping on the cable can distort the cable and cause detuning. Detuning can cause some desired frequency to become attenuated (trapped out).

Now we have learned that short pieces of transmission lines can be used as narrow-band notch filters. Like many lumped constant filters (made from R, L, and C) this type of filter can be coupled together to extend the bandwidth. The diagram in Fig. 2.42 illustrates three consecutive channel traps placed together. The LC lumped trapped has a similar response characteristic. The transmission line traps when used as subscriber traps look very much like excess cable ty-wrapped to the strand, whereas the cylindrical silver bullet trap is easy to spot. The transmission line traps are usually quite sharp and hence do not significantly hurt the adjacent channel signals.

2.8.4   Testing filters

Filters and traps of either type can be tested for impedance and frequency response as transmission lines are. Essentially the filter should have an impedance in and out of the bandwidth that is equal to the input driving source impedance and the load impedance. To test for impedance the open-circuit, short-circuit test can be conducted. For more understanding of the test consider the circuit diagram for a filter

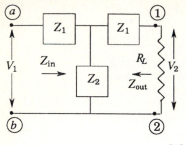

For a correct impedance match $Z_{\text{out}} = R_L$ and $Z_{\text{in}} = Z_{\text{out}}$

For terminals a and b, we may write

$$Z = Z_{\text{in}} = Z_1 + \frac{Z_2(Z_1 + R_L)}{Z_1 + Z_2 + R_L}$$

Solving for R_L by multiplying out

$$Z_1 R_L + Z_1^2 + Z_1 Z_2 + Z_1 Z_2 + Z_2 R_L = R_L^2 + R_L Z_1 + R_L Z_2$$

$$R_L^2 = Z_1^2 + 2Z_1 Z_2 \qquad R_L = \sqrt{Z_1^2 + 2Z_1 Z_2} = R_0$$

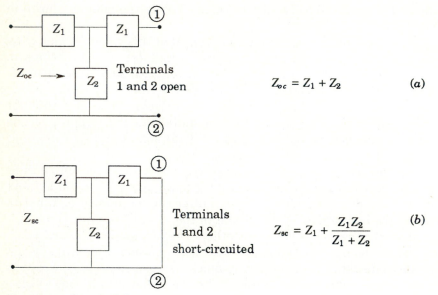

Terminals 1 and 2 open

$$Z_{oc} = Z_1 + Z_2 \qquad (a)$$

Terminals 1 and 2 short-circuited

$$Z_{sc} = Z_1 + \frac{Z_1 Z_2}{Z_1 + Z_2} \qquad (b)$$

Figure 2.43   Open and short test.

or transmission line in Fig. 2.43. The circuit is for a balanced T configuration.

The equations of Fig. 2.43 indicate that we can simplify matters if we multiply the equations together.

$$Z_{OC} Z_{SC} = (Z_1 + Z_2)\left(Z_1 + \frac{Z_1 Z_2}{Z_1 + Z_2}\right)$$

$$= Z_1^2 + Z_1 Z_2 + Z_1 Z_2 = Z_1^2 + 2Z_1 Z_2 = R_O^2$$

so $\qquad R_O{}^2 = Z_{OC}Z_{SC} \qquad$ and $\qquad R_O = Z_O = \sqrt{Z_{OC}Z_{SC}}$

What we have proved so far is that R_O is the geometric mean between the open and short circuit impedances. The geometric mean between two numbers is the square root of the product of the two numbers.

Transmission line, or coaxial cable as we know it, can act as a notch filter because of its distributed resistance, capacitance, and inductance properties. These parameters are shown as an equivalent lumped circuit consisting of R, L, and C in Fig. 2.44 for a given section Δd or line distance.

For the coaxial cable example $R_1\Delta d$ is the resistance of the center conductor and shield resistance per unit distance. Usually providing low resistance for good cable, $L\,\Delta d$ is the inductance per unit distance. $C\,\Delta d$ is the capacitance per unit distance. $R_2\,\Delta d$ is the dielectric leakage resistance per unit distance, which is usually high for good cable.

The per unit distance is the foot for the English system and the meter for the (metric) system. Manufacturers usually give the parameters, e.g., capacitance in picofarads per foot and the inductance in microhenrys per foot. At high frequencies the characteristic impedance can be approximated as $Z_O = \sqrt{L/C}$, where C is in farads and L is in henrys.

The approximate characteristic impedance of a section of coaxial cable can be found by using one of the new digital-type instruments capable of measuring small amounts of L and C. These instruments are known as L, C, and/or Z meters. If the instrument cannot measure capacitances or inductances on the order of a few microhenrys or picofarads, then select a long enough cable sample, for example, 100 ft, make the measurement according to the instrument instructions, and calculate Z. For example, assume that a manufacturer's catalogue gives for RG6 drop-type coaxial cable a capacitance per foot of 16 pF and a velocity of propagation of 0.82. This velocity of propagation is

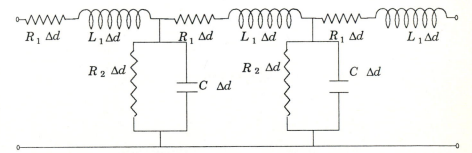

Figure 2.44 Equivalent circuit of a transmission line section.

given as a percentage of the velocity of free space at 3×10^8 m/s; since there is 3.28 ft per meter, the velocity in space at $3.28 \times 3 \times 10^8 = 9.84 \times 10^8$ ft/s.

The velocity in the cable is 0.82 times this number and the velocity is given as $v = (1/\sqrt{LC})$. Now substituting and solving for L,

$$(0.82)(9.84 \times 10^8) = \frac{1}{\sqrt{L \times 16 \times 10^{-12}}}$$

$$8.07 \times 10^8 = \frac{1}{\sqrt{L \times 16 \times 10^{-12}}}$$

$$\sqrt{L \times 16 \times 10^{-12}} = \frac{1}{8.07 \times 10^8}$$

$$L \times 16 \times 10^{-12} = \frac{1}{65.11 \times 10^{16}}$$

$$L = \frac{0.0154 \times 10^{-16}}{16 \times 10^{-12}} = 0.963 \times 10^{-7} \text{ H} = 0.0963 \ \mu\text{H}$$

Substituting $Z_O = \sqrt{L/C}$

$$Z_O = \sqrt{\frac{9.63 \times 10^{-8}}{16 \times 10^{-12}}} = \sqrt{0.60 \times 10^{11}} = 77.6 \ \Omega$$

This value is 2.6 Ω; too high, however, since we may have rounded some numbers, they could account for the difference. The point is that we can calculate the characteristic impedance quite well by measuring the capacitance and inductance of the cable.

Another method is to use the noise generator bridge instrument and measure the return loss for a properly terminated piece of cable. If the termination is exactly equal to the cable, a very high value of return loss will be measured. Recall the formula for return loss for a piece of cable:

$$RL = 20 \log \frac{|Z_R - Z_O|}{|Z_R + Z_O|}$$

When the cable is properly terminated,

$$Z_R = Z_O$$

Then

$$RL = 20 \log \frac{0}{2 Z_O} \qquad RL = \infty \text{ (very large)}$$

For example, suppose we measured our 77.6-Ω cable as above with our 75-Ω terminator.

$$RL = 20 \log \frac{75 - 77.6}{75 + 77.6}$$

Notice that we have to change the equation to

$$RL = 20 \log \frac{77.6 - 75}{77.6 + 75}$$

$$= 20 \log \frac{2.6}{152.6} = -35.4 \text{ dB}$$

To zero in on the above cable, we can place a variable attenuator for the termination, maximize the return loss, and take the impedance reading from the dial.

Now we should be able to understand more fully why short-circuited or open-circuited cable sections can act as inductors, capacitors, and filter/trap sections.

Lumped element filters are made up of L and C as tuned circuits used to pass or deny signals to appear at the output. The low-pass and high-pass filters are made up of L and C sections; the bandpass and bandstop filters are made of various combinations of series and parallel tuned circuits. A summary of such filters and sample configurations is shown in Fig. 2.45.

2.8.5 Measurement of filter frequency response

As for any device or network where the frequency response characteristics are desired, the procedure is to use a sweep signal generator and a detector or sweep receiver. In order to measure large amounts of attenuation either a very high signal level or a very sensitive detector/receiver or a compromise of both is needed to produce good results. If a sweep setup is not available, a manually tuned signal generator and a signal level meter will take a little longer but produce very accurate measurements. The manual system is shown in Fig. 2.46.

Most frequency counters that operate in the television band need 25 mV of signal level, which corresponds to about + 28 dBmV according to our chart or calculated as follows:

$$\text{dBmV} = 20 \log \frac{25 \text{ mV}}{1 \text{ mV}} = 20 \log 25 = +27.958 \approx 28 \text{ dBmV}$$

If the frequency counter needs more signal, either the generator output level can be increased by an appropriate amount or the value of the directional coupler can be changed.

Erratic operation of the frequency counter often indicates lack of

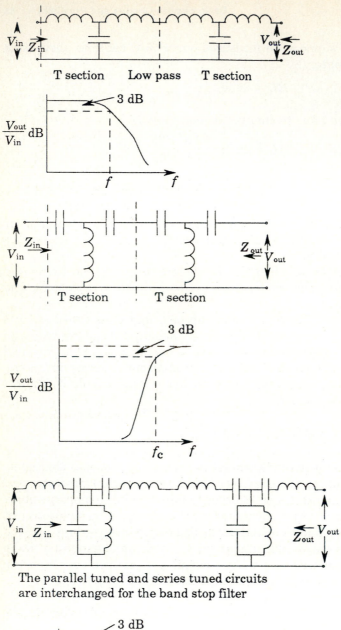

T section   Low pass   T section

$\dfrac{V_{\text{out}}}{V_{\text{in}}}$ dB

3 dB

f f

T section   T section

$\dfrac{V_{\text{out}}}{V_{\text{in}}}$ dB

3 dB

f_{c} f

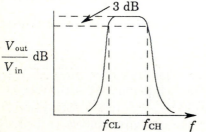

The parallel tuned and series tuned circuits are interchanged for the band stop filter

$\dfrac{V_{\text{out}}}{V_{\text{in}}}$ dB

3 dB

f_{CL} f_{CH} f

Figure 2.45   Lumped element filters.

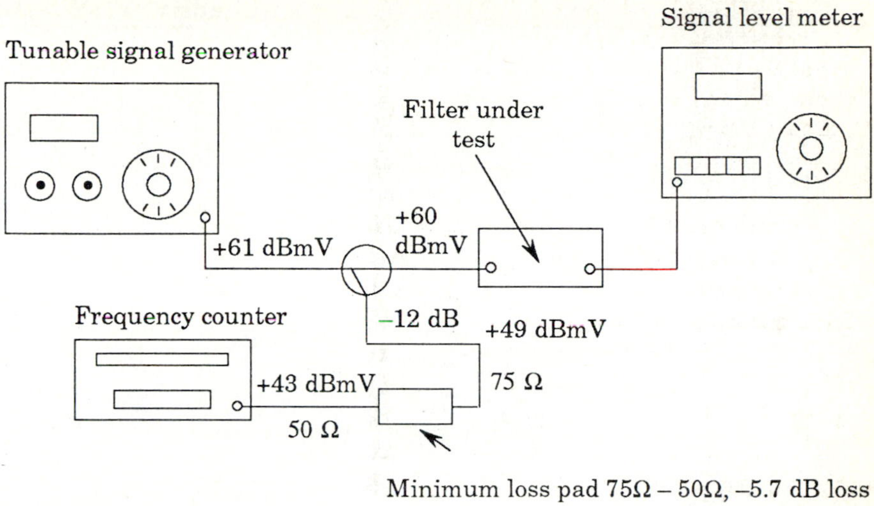

Signal level meter

Tunable signal generator

Filter under
test

+60
dBmV

+61 dBmV

Frequency counter

−12 dB   +49 dBmV

+43 dBmV   75 Ω

50 Ω

Minimum loss pad 75Ω − 50Ω, −5.7 dB loss

Figure 2.46   Filter response test.

proper input signal level. The frequency counter is used to provide an accurate measure of the test frequency. The procedure is to select the test frequency on the generator dial, tune the signal level meter to this frequency, and record the level. Small steps in frequency around the cutoff frequencies should be made in order to measure the roll-off filter characteristics. When enough points are taken, the data can be plotted on graph paper. An example is shown in Fig. 2.47.

Most manufacturers use computer operated test equipment and automatic plotter-recorders to yield very precise plots easily and quickly. However, equally accurate results can be obtained by a competent cable technician taking careful measurements and hand plotting the graph.

An automatic sweep system can also produce good results. A diagram of a sweep recovery system for a filter test is shown in Fig. 2.48.

Caution should be observed for very deep (high-attenuation) filters because the sweep receiver range of attenuation may be limited; if it is, several steps and readings may have to be made. An instant scope camera can record the results, making the job easier. The equipment operator's manual should be consulted to aid in making the filter response test accurately.

Two filters of particular interest are the 4.2-MHz video low-pass filter, which is used in making baseband video signal-to-noise ratio tests, and the weighting filter, which allows the video noise to be weighted (scaled) in terms of degree of picture impairment. Both filters are shown in the schematic diagrams in Fig. 2.49. These filters can be constructed on a plastic or phenolic perforated circuit board

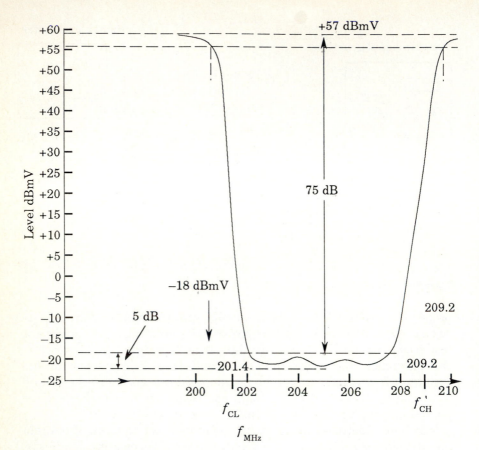

Figure 2.47 Frequency response plot.

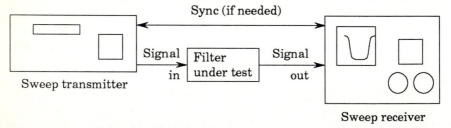

Figure 2.48 Filter automatic sweep test.

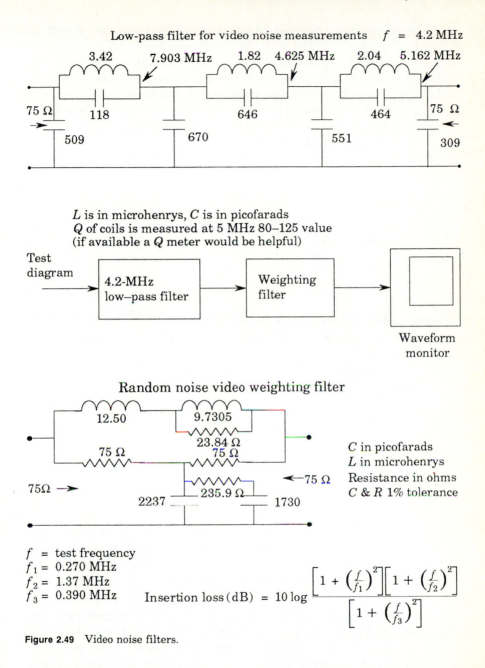

Low-pass filter for video noise measurements $f = 4.2\,\text{MHz}$

L is in microhenrys, C is in picofarads
Q of coils is measured at 5 MHz 80–125 value
(if available a Q meter would be helpful)

Test diagram

Random noise video weighting filter

C in picofarads
L in microhenrys
Resistance in ohms
C & R 1% tolerance

f = test frequency
f_1 = 0.270 MHz
f_2 = 1.37 MHz
f_3 = 0.390 MHz

$$\text{Insertion loss (dB)} = 10\log\frac{\left[1 + \left(\frac{f}{f_1}\right)^2\right]\left[1 + \left(\frac{f}{f_2}\right)^2\right]}{\left[1 + \left(\frac{f}{f_3}\right)^2\right]}$$

Figure 2.49 Video noise filters.

with the components laid out in the symmetrical manner shown in Fig. 2.49. The circuit board with the components soldered in place can then be mounted in an aluminum box with an input/output F-type female connector on both ends. Since most of the circuit elements R, L, and C are not easily obtainable, the resistor and capacitor values will have to be made from series or parallel combinations of standard elements. The coils can be constructed by experimenting with a few turns of 20 to 22 gauge enameled copper wire on pieces of plastic. The number of turns can be estimated by using the following formula for an air core inductor.

$$L = \frac{r^2 N^2}{9r + 10l}$$

where L = (μH) ($\times$ 10^{-6} H)
    r = radius of coil (in)
    N = number of turns
    l = coil length (in)

Example   For a coil value of 9.73 μH for one of the values required by the weighting filter and an 0.5-in-diameter plastic rod 1.5 in long we can calculate the number of turns.

$$9.73 \frac{(0.25)N^2}{(9)(0.25) + (10)(1.5)} = \frac{(0.0625)N^2}{2.25 + 15} = \frac{(0.0625)N^2}{17.25}$$

$$0.0625N^2 = (9.73)(17.25) \qquad N^2 = \frac{(17.25)(9.73)}{0.0625}$$

$$N^2 = 2685.5$$

so

$$N = 51.8 \text{ turns}$$

Since this value coil is the second largest amount of inductance needed, smaller coil forms can be used. To test the inductance value and make any adjustments in inductance—or for that matter in capacitance—it is necessary to have one of the new LCR digital meters.

Filter design is a very specialized area of electrical engineering. There are many mathematical models, kinds of filters, and computer programs available to do the design work. At best, it is a complicated and specialized field. Therefore, for our purposes we need to understand the characteristics, methods of connections, and ways to test them. The foregoing material should aid technicians to work with filters and filter techniques and to select and communicate to vendors the system signal processing needs.

2.8.6 Signal attenuators

Signal pads (attenuators) are essentially four terminal devices, two of which may in some cases be common. These devices are used to decrease the signal level by a precise amount so amplifiers and signal conditioning equipment do not become overdriven (too much signal) and cause distortion. Pads, unlike filters, have a flat frequency response characteristic and merely lower the signal level. The diagram in Fig. 2.50 illustrates the point.

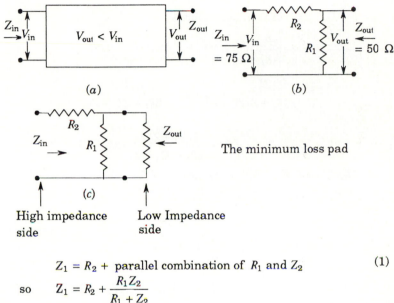

(a)   (b)

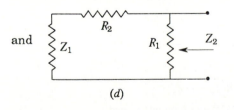

(c)

The minimum loss pad

High impedance   Low Impedance
side   side

$$Z_1 = R_2 + \text{parallel combination of } R_1 \text{ and } Z_2 \qquad (1)$$

so $$Z_1 = R_2 + \frac{R_1 Z_2}{R_1 + Z_2}$$

Recall parallel combination of 2 impedances is product over sum.

and

(d)

$$Z_2 = R_2 + Z_2 \quad \text{in parallel with } R_1 \qquad Z_2 = \frac{R_1 (R_2 + Z_1)}{R_1 + (R_2 + Z_1)}$$

Multiplying out and gathering like terms:

$$Z_2 R_1 + Z_2 R_2 + Z_1 Z_2 = R_1 (R_2 + Z_1)$$

Figure 2.50 Development of the minimum loss pad.

Solving for R_1 :

$$R_1 (R_2 + Z_1) - Z_2 R_1 = Z_2 R_2 + Z_1 Z_2$$

and $\quad R_1 (R_2 + Z_1 - Z_2) = Z_2 R_2 + Z_1 Z_2$

$$R_1 = \frac{Z_2 R_2 + Z_1 Z_2}{R_2 + Z_1 - Z_2} \tag{2}$$

(1) may be written in terms of R_2 as:

$$R_2 = Z_1 - \frac{R_1 Z_2}{R_1 + Z_2} \tag{1a}$$

If we substitute Eq. (2) in (1a) and solve for R_2,

$$R_2 = Z_1 - \left[\frac{Z_2 \left[\frac{Z_2 R_2 + Z_1 Z_2}{R_2 + Z_1 - Z_2} \right]}{\left[\frac{Z_2 R_2 + Z_1 Z_2}{R_2 = Z_1 - Z_2} \right] + Z_2} \right] \frac{(R_2 + Z_1 - Z_2)}{(R_2 + Z_1 - Z_2)}$$

$$R_2 = Z_1 - \frac{Z_2^{\cancel{2}} R_2 + Z_1 Z_2^{\cancel{2}}}{\cancel{Z_2} R_2 + Z_1 \cancel{Z_2} + \cancel{Z_2} R_2 + Z_1 \cancel{Z_2} - Z_2^{\cancel{2}}} = Z_1 - \frac{Z_2 R_2 + Z_1 Z_2}{2R_2 + 2Z_1 - Z_2}$$

Solving for R_2, $(2R_2 + 2Z_1 - Z_2) R_2 = Z_1 - \dfrac{Z_2 R_2 + Z_1 Z_2}{2R_2 + 2Z_1 - Z_2} \, (2R_2 + 2Z_1 - Z_2)$

$$2R_2^2 + \cancel{2Z_1 R_2} - \cancel{R_2 Z_2} = \cancel{2R_2 Z_1} + 2Z_1^2 - Z_1 Z_2 - \cancel{Z_2 R_2} - Z_1 Z_2$$

$$2R_2^2 = 2Z_1^2 - 2Z_1 Z_2 \qquad R_2 = \sqrt{Z_1^2 - Z_1 Z_2}$$

To put in better form $R_2 = \sqrt{Z_1^2 - \left(1 - \dfrac{Z_2}{Z_1} \right)} = Z_1 \sqrt{1 - \dfrac{Z_2}{Z_1}}$

Since now we have solved for R, we will do the same thing for R_1 by substituting Eq. 1a in 2.

$$R_1 = \frac{Z_2 R_2 + Z_1 Z_2}{Z_1 + R_2 - Z_2} \qquad \text{substituting for } R_2 = Z_1 - \frac{R_1 Z_2}{R_1 + Z_2} \tag{2}$$

$$R_1 = \frac{Z_2 \left[Z_1 - \frac{R_1 Z_2}{R_1 + Z_2} \right] + Z_1 Z_2}{Z_1 + \left[Z_1 - \frac{R_1 Z_2}{R_1 + Z_2} \right] - Z_2} \times \frac{(R_1 + Z_1)}{(R + Z_1)}$$

$$= \frac{Z_1 Z_2 R_1 + Z_1 Z_2^2 - R_1 Z_2^2 + R_1 Z_1 Z_2 + Z_1 Z_2^2}{Z_1 R_1 + Z_1 Z_2 + Z_1 R_1 + Z_1 Z_2 - R_1 Z_2 - R_1 Z_2 - Z_2^2}$$

Figure 2.50 (*Continued*) Development of the minimum loss pad.

Combining like terms,

$$R_1 = \frac{2R_1 Z_1 Z_2 - R_1 Z_2^2 + 2Z_1 Z_2^2}{2R_1 Z_1 + 2Z_1 Z_2 - 2R_1 Z_2 - Z_2^2}$$

$$R_1 \left(2R_1 Z_1 + 2Z_1 Z_2 - 2R_1 Z_2 - Z_2^2 \right) = 2R_1 Z_1 Z_2 - R_1 Z_2^2 + 2Z_1 Z_2^2$$

Multiplying and gathering like terms,

$$2R_1^2 Z_1 + 2\cancel{R_1 Z_1 Z_2} - 2R_1^2 Z_2 - \cancel{R_1 Z_2^2} = 2\cancel{R_1 Z_1 Z_2} - \cancel{R_1 Z_2^2} + 2Z_1 Z_2^2$$

$$\cancel{2}R_1^2 Z_1 - \cancel{2}R_1^2 Z_2 = \cancel{2}Z_1 Z_2^2 \qquad R_1^2 (Z_1 - Z_2) = Z_1 Z_2^2$$

so $\quad R_1^2 = \dfrac{Z_1 Z_2^2}{Z_1 - Z_2} \quad$ and $\quad R_2 = Z_2 \sqrt{\dfrac{Z_1}{Z_1 - Z_2}} \times \dfrac{\sqrt{\frac{1}{Z_1}}}{\sqrt{\frac{1}{Z_1}}}$

$$R_1 = Z_2 \sqrt{\frac{1}{1 - \frac{Z_2}{Z_1}}} \qquad \text{Reorganizing} \qquad R_1 = Z_2 \frac{1}{\sqrt{1 - \frac{Z_2}{Z_1}}}$$

Figure 2.50 (*Continued*) Development of the minimum loss pad.

For an impedance matching type of pad, minimum loss is usually desired; hence such pads are often referred to as *minimum loss pads*. A typical use may be to match a 75-Ω cable system to a 50-Ω spectrum analyzer. The simplest type of such a box would be the L-type pad, so called because the circuit resembles the letter L.

Since it may be desirable to use this type of pad for a variety of impedance matching problems, a set of design equations will be generated by using the theory we have learned in the first sections of this book. Therefore, let us consider the basic circuit in Fig. 2.50c.

Now suppose, e.g., we want to construct an attenuator pad for 75 to 50 Ω as shown in Fig. 2.51. To calculate the loss, elementary circuit techniques can provide the information as follows. Recall that the loss as a voltage ratio for $20 \log N$ cannot be used because the impedance ratios are different. Therefore, loss $= -10 \log (P_{in}/P_{out})$. Suppose the 50-$\Omega$ resistor dissipates by 1 mW. Then

$$I^2 R = 1 \text{ mW} \quad \text{so} \quad I^2 = \frac{1 \text{ mW}}{50 \ \Omega} = \frac{1 \times 10^{-3}}{50}$$

$$I = \sqrt{\frac{1 \times 10^{-3}}{5 \times 10^1}} = \sqrt{\frac{1 \times 10^{-4}}{5}} = \sqrt{0.2 \times 10^{-4}} = 10^{-2}\sqrt{0.2}$$

$$= 0.45 \times 10^{-2} = 4.5 \text{ mA}$$

Now the output voltage $IR = 4.5 \times 10^{-3} \times 50 = 0.225$ V.

So 0.225 V also appears across the R_1 at 86.6 Ω; therefore, the current through it is $I = (0.225 \text{ V}/86.6 \ \Omega)$ and $I = 0.0026$ A $= 2.6$ mA.

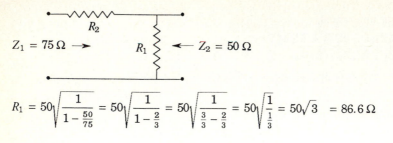

$$R_1 = 50\sqrt{\frac{1}{1-\frac{50}{75}}} = 50\sqrt{\frac{1}{1-\frac{2}{3}}} = 50\sqrt{\frac{1}{\frac{3}{3}-\frac{2}{3}}} = 50\sqrt{\frac{1}{\frac{1}{3}}} = 50\sqrt{3} = 86.6\,\Omega$$

and

$$R_2 = Z_1\sqrt{1-\frac{Z_2}{Z_1}} = 75\sqrt{1-\frac{50}{75}} = 75\sqrt{1-\frac{2}{3}} = 75\sqrt{\frac{1}{3}} = 75\times 0.577 = 43.3\,\Omega$$

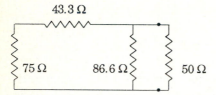

Figure 2.51   75- to 50-Ω minimum loss pad.

Circuit theory tells us that the input current has to be the sum of the branch currents (Kirchoff's current law), as in Fig. 2.52.

The attenuator loss is essentially the power burned up by the 43.3-Ω and 86.6-Ω resistors. To add up the power consumption for the 86.6-Ω resistor

$$86.6\,\Omega\ I^2R = (0.0026)^2\,(86.6) = 0.00059\text{ W}$$
$$43.3\,\Omega\ I^2R = (0.0071)^2\,(43.3) = \underline{0.00218\text{ W}}$$
$$0.00277\text{ W}$$

plus the load current of 1 mW; total power is 3.77 mW. Therefore, the attenuation is

$$10\log\frac{3.77\text{ mW}}{1.0\text{ mW}} = 5.76\text{ dB}$$

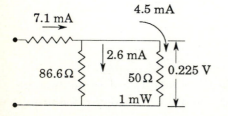

Figure 2.52 Currents through the "L" pad.

As a check, the input voltage between terminals 1 and 2 is the drop across the 43.3-Ω resistor plus the 0.225 V across the parallel combination of the 86.6- and 50-Ω resistors. The resistance across terminals 1 and 2 is 75 Ω, so the input power will be V_{in}^2/R_{in}

$$\frac{V_{in}^2}{R_{in}} = \frac{(0.31\ V + 0.225\ V)^2}{75} = 0.0037\ W$$

$$\text{Loss} = 10 \log \frac{3.7\ mW}{1\ mV} = 5.7\ dB$$

Most of us know that when a 75- to 50-Ω minimum loss pad is purchased, this exact amount of loss is written on the pad.

Now that the technique has been studied the discussion of pads and attenuators can continue. The previous L-type pad is usually used for impedance matching applications, and the loss is usually as low as it can be. The usual signal attenuators in cable television applications are used only to decrease the signal level and not to change the im-

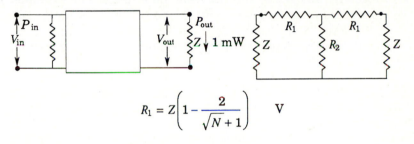

$$R_1 = Z\left(1 - \frac{2}{\sqrt{N}+1}\right) \qquad V$$

and since the voltage ratio is related to the power ratio

$$R_1 = Z\left(\frac{V-1}{V+1}\right)$$

and in like manner

$$R_2 = \frac{2Z}{\sqrt{N} - \frac{1}{\sqrt{N}}} = Z\left[\frac{2V}{(V^2 - 1)}\right]$$

Now for a 6-dB attenuator where $Z_1 = Z_2 = Z = 75\ \Omega$ then $20 \log V = 6$ dB, $\log V = 0.3$ and $V = 2$. Now

$$R_1 = 75\left(\frac{2-1}{2+1}\right) = \overset{25}{\cancel{75}}\left(\frac{1}{\cancel{3}}\right) = 25\ \Omega$$

and $\qquad R_2 = 75\left[\frac{2 \times 2}{4-1}\right] = \overset{25}{\cancel{75}}\left(\frac{4}{\cancel{3}}\right) = 100\ \Omega$

Figure 2.53 "T" attenuator design.

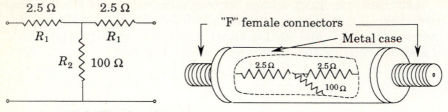

Figure 2.54   75-Ω "T" pad configuration.

pedances. Therefore, $Z_1 = Z_2 = Z$ and $N = (P_{in} - P_{out})/P_{out}$ according to the diagram in Fig. 2.53.

After much algebra and using the T-type network shown in Fig. 2.53, we can derive the equations for the resistors in the box.

Now the pad looks like the diagram in Fig. 2.54. Of course, no one wants to compute the values every time so a chart can be made by calculating the values. One such chart is shown in Table 2.2.

To scale Z up or down, $(Z_{desired}/75)$ multiplied by R_1 and R_2 will give new values for R_1 and R_2 at the different impedance. For example, suppose we wanted to scale down to 50 Ω; then $^{50}/_{75}$ is the multiplier $^2/_3$. For a 50-Ω, 6-dB pad $R_1 = 25(^2/_3) = 16.67$ Ω and $R = 100.5(^2/_3) = 66.7$ Ω. The diagram in Fig. 2.55 shows the configuration.

TABLE 2.2   Attenuation versus Resistance for 75-Ω T Pad

dB	R_1 Ω	R_2 Ω
1	4.3	650
2	8.6	323.0
3	12.8	213.0
4	17.0	157.3
5	21.0	123.4
6	25.0	100.5
7	28.7	83.7
8	32.2	71.0
9	35.7	61.0
10	39.0	52.7
11	42.0	46.0
12	45.0	40.2
13	47.6	35.4
14	50.0	31.2
15	52.3	27.5
16	54.5	24.4
17	56.4	21.5
18	58.2	19.2
19	60.0	17.0
20	62.5	15.2

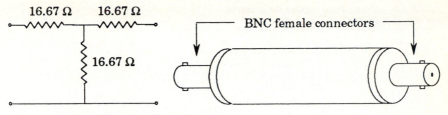

Figure 2.55   50-Ω "T" pad configuration.

This same T pad can be reconfigured to a balanced H-type pad used for balanced 50-Ω or 300-Ω RF transmission line (television flat lead). Since cable television systems do not use this type of cable, the balanced H-type pad will be for balanced 600-Ω audio applications. The H pad parameters are found by using the formulas for R_1 and R_2 used for the T pad as shown in Fig. 2.56.

For audio applications then, we can take a short cut, e.g., scale up the 75 Ω to 600 Ω from the previous chart, or we can make a new chart. To scale up we can proceed as follows: (600/75) = 8, the scale factor. For a 6-dB pad R_1 = 100.5 × 8 = 804 Ω and R_2 = 25 × 8 = 200 Ω.

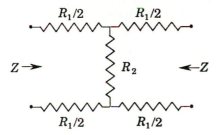

"T" to "H"

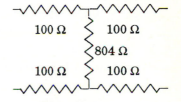

600-Ω 6-dB pad

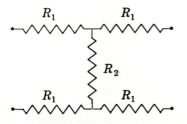

"H" pad 600-Ω value in table 2.2

Figure 2.56   Balanced "H" pad.

TABLE 2.3   Attenuation versus Resistance
for 600-Ω "H" Pad

dB	R_1	R_2
1	17.2	5200.0
3	51.3	1703.0
6	100	804
10	156	420
15	210	220
20	250	120

Common values for 600-Ω balanced audio pads are shown in Table 2.3.

2.9   Signal Level Gains and Losses in Community-Antenna Television Systems

Instruments commonly used in community-antenna television (CATV) systems indicate signal level measurements in decibel-millivolts (dBmV). Instruments used by radio and television transmitting facilities use instruments to measure field strength in microvolts/meter. Engineers measure electrical quantities in volts, amperes, watts, and ohms. The one measure common to electrical systems is power, where the electrical quantity is in watts or milliwatts, microwatts, kilowatts, and megawatts. Gains and losses in electrical systems may be kept track of easily by using the decibel (dB).

2.9.1   Logarithm theory: the decibel

Historically, the bel was a shortened version of the name Bell, after the inventor of the telephone, Alexander Graham Bell. The bel follows a logarithmic scale, since the human ear responds logarithmically to sound stimulus. Therefore, a short lesson on logarithms is in order.

Consider the present numbering method called the decimal system, which is based on the number 10. Most likely, this system was adopted because humans have 10 fingers. The decimal counting or numbering system has 10 symbols, which are placed in rows or columns to indicate system counting. These symbols are 0 through 9; when we count to 10 we shift the symbol 1 one position to the left, and then, starting with the 0, proceed to symbol 9, thus counting to 19. Next the left digit goes to 2 and the right digit starts at 0, and we proceed by counting to 9, thus reaching the number 29. This process is repeated to continue counting. Any number can be interpreted by knowing the counting sequence. Consider the number

Thousands position $\rightarrow$ 5976$\leftarrow$ Units position
Hundreds position $\nearrow$ $\nwarrow$ Tens position

Our numbering system is based on 10, which is termed the *base* or *radix*. Consider the number 1000, in which the units, tens, and hundreds positions are all zeros, and the thousands position is a one. This number can be represented by the number 10 multiplied by itself three times (three zeros). Mathematically this can be written 10^3 where the 3 indicates the number of times the radix or base is multiplied by itself: $10 \times 10 \times 10 = 10^3 = 1000$. The = symbol indicates equivalence or balance and is often referred to as the *equals sign*. The quantity to the left of the sign is equivalent to the quantity to the right of the sign. For example, $a = b$ means a is the same as b or a is equivalent to b. The expression $a = b$ is called an equation.

Going back to the equation $1000 = 10^3 = 10 \times 10 \times 10$, the logarithm of the number 1000 is 3, which is the number of times the base or radix is multiplied by itself. The numbers 10 and 100 have logarithms of 1 and 2, respectively. Numbers falling between 1, 10, 100, etc., have logarithms not equal to single digits and have to be calculated. The logarithm of a number is a decimal number where the number to the left of the decimal point is called the *characteristic* and the numbers to the right of the decimal point are called the *mantissa*. For example, given the logarithm

1.7586
Characteristic $\nearrow$ $\nwarrow$ Mantissa

The characteristic tells us where the decimal point is placed, i.e., whether the number falls between 10 and 100, and the mantissa represents where the number falls in the range. Numbers that are even powers of 10 have logarithms that have only a characteristic and a mantissa of zero. Cumbersome numbers can be represented in simple terms by their logarithms.

One must suspect by now that the use of logarithms must have some advantage since quite a bit of effort may be spent to find the logarithms of various numbers.

Since gains and losses in electrical systems involve extensive multiplication and division, the use of logarithms will reduce the process to simple addition and subtraction of logarithms. Consider the problem of multiplying 1000 by 100, in equation form, $1000 \times 100 = 100,000$. The logarithm of $1000 = 3$ and of $100 = 2$; $3 + 2 = 5$, and the number that has 5 as a logarithm (antilogarithm) is equal to 100,000. It should be clear now that to multiply two numbers together, find the logarithms (logs) of each number, add the two logs together, and find the antilog of the sum to get the answer. To divide two numbers, subtract the logs and find the antilog of the difference. In short, multiplication and division are reduced to addition and subtraction if one can find the logarithms of the numbers.

Logarithms can be found by several methods, most commonly by tables and by scientific pocket calculators. The second method is most used today because such calculators can be purchased almost anywhere and cost little. Most pocket calculators that include the log function are used by simply pressing the keys to enter the number, then pressing the log key with the logarithm displayed. If the calculator has several memories, the logs can be stored in memory, then the contents of the memories can be summed. Activating the inv and log keys will then display the number corresponding to the antilog, which is the answer. The student should refer to the calculator's instruction manual to clarify the logarithm functions.

As stated, the decibel is related to power. Specifically the decibel is equal to 10 times the logarithm of the power ratio. In mathematical terms, number of decibels = dB = $10 \log (P_{in}/P_{out})$ where P_{out} and P_{in} are in the same units, that is milliwatts, microwatts, watts, etc. The prefixes milli-, micro-, kilo-, etc., are shown in Table 2.4.

It should be evident that 1 milliwatt (mW) is one-thousandth of a watt = 1×10^{-3} W.

Expressing numbers in powers of 10 is very easy and can simplify many calculations. Essentially powers of 10 are very closely allied to logarithms.

Consider Table 2.5.

TABLE 2.4   Numerical Prefixes and Equivalent Values

Micro-	$\dfrac{1}{1,000,000}$	1 millionth	10^{-6}
Milli-	$\dfrac{1}{1000}$	1 thousandth	10^{-3}
Kilo-	1000	1 thousand	10^{3}
Mega-	1,000,000	1 million	10^{6}

TABLE 2.5   Powers of 10 and Logarithm Equivalents

1×10	10	10^{1}	10 has one zero	$\log 10 = 1$
1×100	100	10^{2}	100 has two zeros	$\log 100 = 2$
1×1000	1000	10^{3}	1000 has three zeros	$\log 1000 = 3$
1×1000	10,000	10^{4}	10,000 has four zeros	$\log 10,000 = 4$

Consider the following multiplication problem: 1285 × 40 can be expressed as

$$1.285 \times 10^3 \times 4 \times 10^1 = 4 \times 1.285 \times 10^{3+1}$$
$$= 5.140 \times 10^4 = 51,400$$

Move decimal point four places to the right ↗

Using the decibel expression for gain and loss specifically in a cable system, consider the following examples:

1. Given a length of cable where the input is 1 mW and the output is $\frac{1}{10}$ mW, as shown in Fig. 2.57a. This is clearly a loss from input to output. The output is less than the input.

$$dB = -10 \log \frac{P_{in}}{P_{out}} = -10 \log \frac{1 \text{ mW}}{1.10 \text{ mW}} = -10 \log \frac{1}{10^{-1}}$$

$$= -10 \log 10^1 = -10 \log 10 = -10 \times 1 = -10 \text{ dB}$$

2. Given an amplifier with $\frac{1}{10}$-mW input and 1-mW output, shown in Fig. 2.57b.

$$dB = 10 \log \frac{P_{out}}{P_{in}} = 10 \log \frac{1 \text{ mW}}{\frac{1}{10} \text{ mW}} = 10 \log 10$$

$$= 10 \times 1 = 10 \text{ dB} \qquad \text{indicates gain}$$

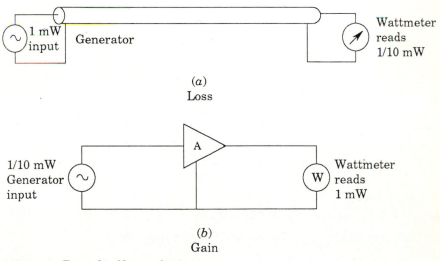

(a)
Loss

(b)
Gain

Figure 2.57 Example of loss and gain.

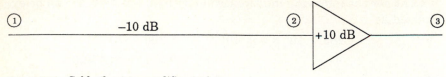

Figure 2.58   Cable (loss) to amplifier (gain).

Cable systems are usually constructed in what are termed *unity gain building blocks*, where the cable loss is compensated completely by the gain of the amplifier.

Consider placing the cable in example 1 and the amplifier in example 2 together, as shown in Fig. 2.58. Between points 1 and 2 there is 10-dB loss (−10 dB). Between points 2 and 3 there is 10-dB gain (10 dB/0 dB). Therefore, between points 1 and 3 there is no loss. So if 1 mW is fed into the cable, then 1 mW comes out of the amplifier and 1 mW in times 1 = 1 mW out; hence, there is unity (1) gain.

A mathematical relationship exists among electrical force, voltage, and electrical circuit resistance to power.

$$P = \frac{(\text{voltage V})^2}{(\text{resistance } \Omega)} = \frac{V^2}{R}$$

A mathematical relationship pertaining to logarithms is expressed $\log N^2 = 2 \log N$.

Now, substituting the expression for power in terms of voltage and resistance with the logarithm expression in the decibel formula gives

$$dB = 10 \log \frac{P_{out}}{P_{in}} = 10 \log \frac{(V_{out}^2/R)}{(V_{in}^2/R)} \times \frac{R}{R} \qquad (\text{multiplying by 1})$$

$$= 10 \log \frac{V_{out}^2}{V_{in}} = 10 \log \left(\frac{V_{out}}{V_{in}}\right)^2 = 2 \times 10 \log \frac{V_{out}}{V_{in}}$$

$$= 20 \log \frac{V_{out}}{V_{in}} \qquad \text{This is still a power ratio since } R_{in} = R_{out}.$$

Table 2.6 and the following examples summarize the procedure.

Example

$P_{in} = P_{out}$ (no gain or loss) dB $= 10 \log \dfrac{P}{P}$

$\qquad = 10 \log 1 = 10 \times 0 = 0$ dB $\qquad\qquad\qquad\qquad\qquad$ (2.3)

$P_{in} = 100\, P_{out}$ (0 loss) dB $= -10 \log 100 = -10 \times 2 = -20$ dB $\qquad$ (2.4)

$P_{in} = 1{,}000{,}000\, P_{out}$ (a loss) dB $= -10 \log 1{,}000{,}000$

$\qquad = -10 \times 6 = -60$ dB $\qquad\qquad\qquad\qquad\qquad\qquad$ (2.5)

TABLE 2.6   Decibels versus Power and Voltage Ratios for $R_{in} = R_{out}$

dB	P_{in}/P_{out}	V_{in}/V_{out}
0	1	1
20	100	10
40	10,000	100
60	1,000,000	1,000
80	100,000,000	10,000
100	10,000,000,000	100,000
120	100,000,000,000	1,000,000

$$V_{in} = V_{out} \; dB = -20 \log \frac{V_{out}}{V_{out}} = -20 \log 1 = 20 \times 0 = 0 \; dB \qquad (2.6)$$

$$V_{in} = 100 \; V_{out} \; dB = -20 \log 100 = -20 \times 2 = -40 \; dB \qquad (2.7)$$

$$V_{in} = 1,000,000 \; V_{out} \; dB = -20 \log 1,000,000$$

$$= -20 \; dB \times 6 = -120 \; dB \qquad (2.8)$$

So far, only ratios of power and voltage have been discussed. Signal level can be discussed if the decibel expression power level is referenced to a fixed level.

2.9.2   The decibel-millivolt

For CATV work the reference level is specified as 1 mV across a 75-Ω standard impedance mathematically.

$$dBmV = 20 \log \frac{\text{voltage (mV)}}{1 \; mV} \qquad \text{(dB referred to mV across 75 } \Omega\text{)}$$

or dBmV = 20 log voltage (mV)

Example   1 V appears across a 75-Ω resistor. What level does this represent in dBmV?

$$dBmV = 20 \frac{1000 \; mV}{1 \; mV} = 20 \log 1000 = 20 \times 3 = 60 \; dBmV$$

Example   10 mV across 75: dBmV = ?

$$dBmV = 20 \log \frac{10}{1} = 20 \times 1 = 20 \; dBmV$$

Example   100 μV across 75 dBmV = ?

$$dBmV = -20 \log \frac{1 \; mV}{0.1 \; mV} = -20 \log \frac{1}{0.1} = -20 \log 10$$

$$= -20 \times 1 = -20 \; dBmV$$

TABLE 2.7   dBmV Values versus Voltage Ratio across $R_{in} = R = 75\ \Omega$

dBmV	Voltage ratio across 75 Ω, rms
+60	1 V (1000 mV)
+40	0.1 V (100 mV)
+20	0.01 V (10 mV)
0	0.001 V (1 mV)
−20	100 μV (0.1 mV)
−40	10 μV (0.01 mV)
−60	1 μV (0.001 mV)

Table 2.7 summarizes the process.

By now it should be evident that to keep track of gains and losses much use will be made of the decibel (dB) and decibel-millivolt (dBmV).

2.9.3   The unity gain building block

As mentioned, CATV cable systems are designed by using the unity gain building-block concept.

Consider a diagram of a piece of a cable system trunk as given in Fig. 2.59. Notice that all amplifiers have the same input levels and the same output levels because the cable section loss is the same. This yields the cable amplifier section with unity gain, which is the building block. The preceding example illustrates only the trunk system of a CATV system. The trunk cable is only the cable to transport signal to outlying areas, similar to a high-pressure water main or high-tension power line. The procedure is essentially to prevent the trunk system signals from being contaminated by the subscriber distribution system. The distribution cable containing the subscriber taps is

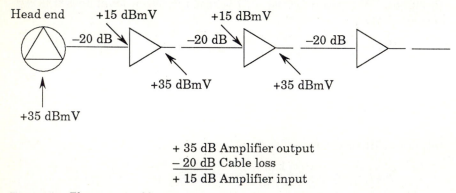

+ 35 dB Amplifier output
− 20 dB Cable loss
+ 15 dB Amplifier input

Figure 2.59   Elementary cable system.

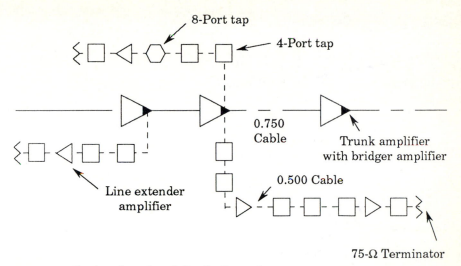

Figure 2.60 Section of trunk and distribution system.

connected to the trunk cable via a bridging amplifier, shown schematically in Fig. 2.60.

Now that gains and losses in decibels (dB) and signal levels as decibel-millivolts (dBmV) have been introduced, some examples should clear up any misunderstanding and illustrate how simple the technique really is. Consider a device with an input signal level of +10 dBmV that has a loss of 3 dB from input to output. One simply subtracts the device loss from the input signal to find the output signal level as shown in schematic form in Fig. 2.61.

In actual practice a two-way splitter has −3.5 dB from input to either output port. The four-way splitter has twice the loss of the two-way device: −3.5 dB − 3.5 dB = −7 dB.

Up until now losses have been referenced to one television video carrier, or a single-frequency alternating current or flat loss across the whole television spectrum was assumed. This is not the case in actual practice. However, most splitters, directional couplers, and signal attenuators (pads) are fairly flat; that is, they have a constant loss across the television spectrum 55 to 550 MHz. Amplifiers provide gain which is quite constant over the television spectrum. With cable this is not the case.

2.10 Signal Waveforms

Most of the electric power circuits studied so far involved either direct current (dc) or alternating current (ac) conditions. Batteries, dc generators, or filtered rectifiers were used as sources of dc power. Com-

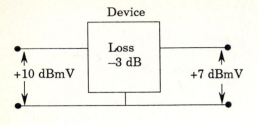

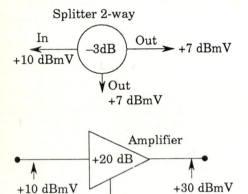

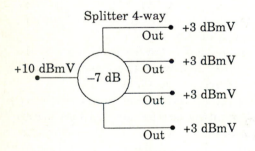

Figure 2.61 Examples of gains and losses using dBmV levels.

mercial single-phase lighting alternating current was the primary source of ac power. Electronic oscillating circuits were used for low-power, high-frequency ac power. In the case of alternating current, the principal waveform was for the sine/cosine case. Other waveforms of alternating currents briefly mentioned were square waves, sawtooth, ramps, and half- or full-wave sine/cosine types.

Mathematics and electrical experimentation have clearly demonstrated that any waveform can be formed by sine/cosine waves of various amplitudes and phases. For example, the sawtooth wave case shown in Fig. 2.62 can be produced by a series of sine wave generators.

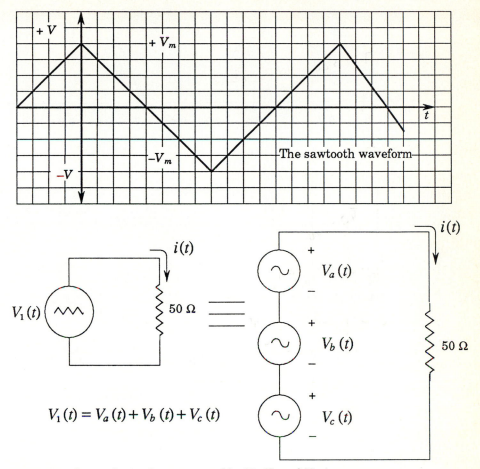

Figure 2.62 Sawtooth waveform generated by V_a, V_b, and V_c sine waves.

For the square wave condition, like that of the sawtooth, a series of odd harmonics can do the job; however, different amplitudes and phases of the signal component will be required. Odd harmonics in various amplitudes and phases can generate a ramp waveform. A series of pulses, e.g., a clipped square wave, can be formed by a series of even harmonics: 0, 2, 4, 6, 8, etc.

Consider the rectified sine wave case where the period $t = 0.4$ s is shown in Fig. 2.63. For $t = 0.4$ s then,

$$f = (1/t) = (1/0.4) = 2.5 \text{ Hz}$$

and we have a half-rectified sine wave.

The purpose of this discussion is to make emphatically clear that if no sinusoidal-type waveforms are passed through filter circuits or

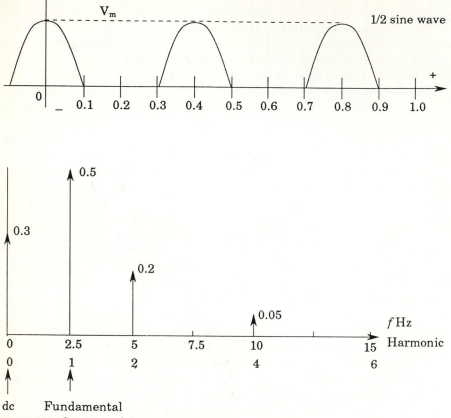

Figure 2.63 Spectrum of sine wave components.

frequency-limiting circuits, severe distortion of the resulting wave-form can occur. The spectrum of the half-wave sine wave has a dc term at zero frequency. This we know. Some fundamental frequency at 0.5 V_m (V_m is maximum voltage amplitude) is present as well as second harmonic (twice the fundamental frequency) at 0.2 V_m and fourth harmonic at about 0.05 V_m, and a very small amount of sixth harmonic. To preserve the wave shape a filter that passes fourth harmonic four times the fundamental frequency should be adequate. Now if we pass this signal through a filter say 0.1 times the fundamental, then the resulting waveform will have everything above zero frequency filtered out and all we will have left is 0.3-V_m direct current, which is what we want for direct current.

By now it should be evident that when a series of pulses amplitude modulate a sine wave high-frequency carrier, then the rich, even harmonic content of the pulse train (horizontal synchronization pulses) produces several harmonic sidebands. The video waveform for National Television System Committee (NTSC) television is indeed complicated. The audio waveforms produced by orchestral music and voice are also extremely complicated. The electronic amplifiers and circuits that work with these waveforms need adequate bandwidth either to pass undistorted signals or provide calculated amounts of distortion to these signals. Knowledge of the spectral content of these complicated waveforms is helpful in analyzing, designing, or understanding electronic circuits using or measuring these signals.

The voltage or current waveforms of video and audio are known as *baseband signals*. To transmit these signals over the air via a radio transmission carrier technique the principle of modulation is applied. Essentially the method of piggybacking the baseband signals on a high-frequency sine wave carrier is modulation and demodulation. For the NTSC type of television signals the video waveform (baseband) amplitude modulates a carrier (video carrier); the audio baseband signal frequency modulates a separate related carrier located 4.5 MHz above the amplitude-modulated (AM) video carrier. The different kinds of modulation along with the 4.5-MHz-frequency separation minimize or eliminate mixing of the two signals. When audio signals are mixed with video signals the result is a picture that seems to depict worms moving across the screen at the audio rate heard on the speaker. Decreasing the amplitude of the audio carrier 15 dB below the picture carrier, in the cable television case, where adjacent channels are used, clears the picture. It will become evident when studying modulation techniques that audio loudness for frequency modulation is a function of frequency deviation, not amplitude.

2.10.1 Television video baseband signals

A television picture is made by a television receiver with the picture tube (cathode-ray tube [CRT]) scanning or drawing lines from the upper left-hand corner across the screen painting a line. The intensity or brightness of the line varies according to the scene that is scanned by the camera. It should be obvious that the receiver has to be synchronized to the camera for the receiver to show what the camera is looking at. On the right-hand side of the screen the beam is blanked off where it returns to the left-hand side to start another line. This procedure is repeated until the beam end is at the lower right-hand corner, where it has to be blanked out again and for a longer period of time for it to return to the upper left-hand corner. A picture with proper vertical definition needs a total of 482 to 495 lines. The time required for 525 lines is allowed, giving between 30 and 43 lines of picture so the receiver can be properly synchronized, to assure that

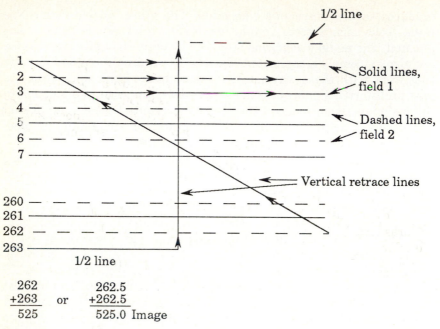

Figure 2.64   Principle of interlaced scanning.

the scanning process starts at the top of the screen. The transmitted picture has what is called a *4:3 aspect ratio*, which means that the picture shape is 4 units wide by 3 units high. Therefore, all picture tube screens are rectangular with a 3:4 ratio. Refer to Fig. 2.64 for a diagram of the screen.

For a square picture, i.e., an aspect ratio of 1:1, 485 dark and light units per line are needed. Therefore, a positive and a negative current of 485 ÷ 2 or 242.5 Hz produces these light and dark units. Recall that with an aspect ratio of 4:3 the width of the screen is one-third longer. This is equivalent to one-third more dark and light units and can be calculated as follows: 242.5 + (242.5 ÷ 3) = 242.5 + 80.8 = 323 Hz. For 525 lines at 323 Hz per line, approximately 180,000 Hz would be sufficient for one scan, i.e., from the upper left-hand corner of the screen to the lower right-hand corner. If this is done at 30 times per second, then a frequency of 30 × 180,000 or 5.4 MHz results. In actual practice a bandwidth of 4.2 MHz will give adequate definition. For a flicker-free picture that does not have to scan the scene more than 30 times a second an interlace technique is employed. To accomplish this, the scene is scanned by using only the even-numbered lines, consisting of 262.5 lines, which is known as a *field*. Immediately afterward the scene is again scanned, using the odd-numbered lines; this results in field number 2. Both fields are completed in ⅟₃₀ s. As no more lines are added, the 4.2-MHz bandwidth is sufficient. It should be evident that the field rate is 60/s and the frame rate of 2 fields is 30/s. Synchronizing pulses to time the camera and the receiver accurately is necessary so the picture will start with the half-line at the top and produce vertical sides to reproduce the image properly on the screen. Since the field rate is 60/s, which is the same rate as that of public utility power, this power source is chosen as a source of synchronization. Now if the frame is scanned horizontally at 30 frames/s and at 525 lines/frame, the waveform of the horizontal current must have a frequency of 525 × 30, or 15,750 Hz.

The source of the timing information necessary to synchronize the television cameras and the receiver has to be at the studio location, i.e., the source of the television system. The timing information is generated by a master synchronizing pulse generator, which locks the cameras together and becomes part of the video signal transmitted by the television station. Fig. 2.65 illustrates these synchronizing pulses just before the position at the lower right-hand corner of the field and the upper left-hand corner at the start of a new field.

1. H = time from start of one line to start of next line.

2. V = time from start of one field to start of next field.

3. Start of field 1 is defined by a whole line between first equalizing pulse and preceding H sync pulse.

4. Start of field 2 is defined by a half-line between the first equalizing pulse and the preceding H sync pulses.

5. Field 1 line numbers start with the first equalizing pulse in field 1.

6. Field 2 line numbers start with the second equalizing pulse in field 2.

The horizontal blanking pulses instruct the receiver to shut off the beam on the television picture tube while the beam is positioned back to the left-hand side to start painting another horizontal line. The vertical sweep circuits originated at the transmitting station camera move the beam vertically down the scene to position the horizontal lines below each other. The beam in the receiver follows because of the vertical synchronizing pulses which start this sweep. The start of the horizontal sweep is controlled by the horizontal synchronizing pulses, which are narrower and sit on top of the horizontal blanking pulses. The beam current in the receiver is varied in intensity (brightness) by the video information contained between the horizontal blanking pulses. During the time it takes, 830 to 1330 µs for the field to end at the lower right-hand corner and a new one to begin at the upper left-hand corner, some means must be taken to keep the horizontal circuits in synchronization. This is accomplished by *equalizing pulses*,

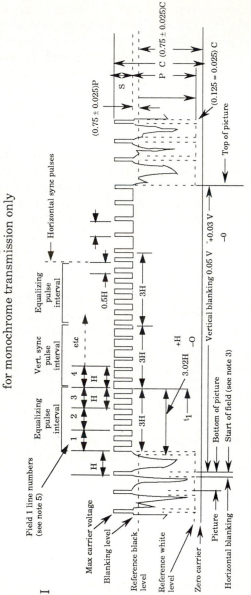

Figure 2.65 FCC model of composite video signal.

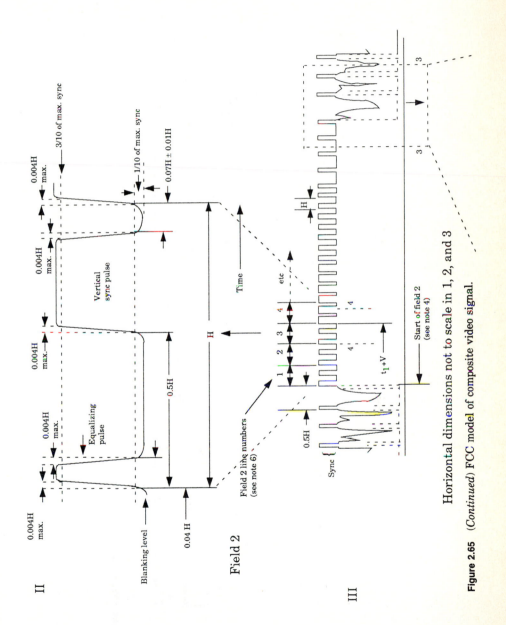

II

0.004H
max.

3/10 of max. sync

0.004H
max.

0.004H
max.

0.004H
max.

1/10 of max. sync

0.07H ± 0.01H

Vertical
sync pulse

Time

H

0.5H

0.004H
max.

Equalizing
pulse

Blanking level

0.04 H

Field 2

Field 2 line numbers
(see note 6)

1 2 3 4 etc

III

H

Sync {

0.5H

4

4

$t_1 + V$

Start of field 2
(see note 4)

3

3

Horizontal dimensions not to scale in 1, 2, and 3

Figure 2.65 (*Continued*) FCC model of composite video signal.

83

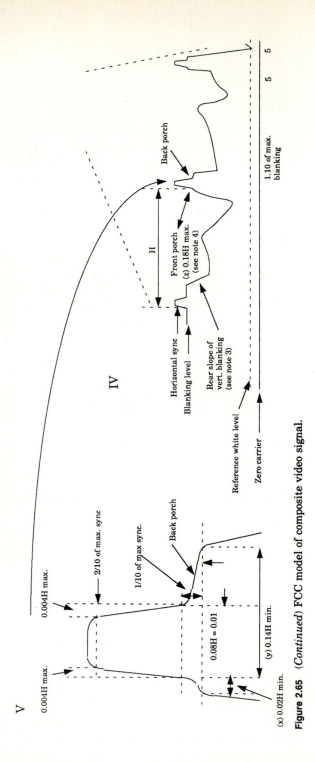

Figure 2.65 (*Continued*) FCC model of composite video signal.

which appear as pulses on a portion of the vertical blanking pulse and as serrations on the vertical sync pulse. The frequency of these pulses is double the 15,750 Hz, or 31,500 Hz, at the transmitter. Every other pulse is dropped out before transmission, giving an overall horizontal pulse rate of 15,750 Hz. During the vertical blanking interval the beam is turned off while it is being positioned to the upper left-hand corner to begin a new field. To prevent hum bars in the picture, the synchronizing pulse generator at the television studio is often locked to the 60-Hz public electric source. For remote field equipment the synchronizing pulse generator has its own built-in precision 60-Hz crystal oscillator, which can lock the synchronizing pulse generator at the television studio.

The video signal containing the line luminance and synchronizing pulses has to modulate a television transmitter by the amplitude modulation technique. A diagram of the various levels of modulation for the various pulses and picture illumination is given in Fig. 2.66. Note that the percentage modulations for the vertical and horizontal pulses are at the same level, i.e., 100 percent. Two complete consecutive lines are shown in the diagram.

The television transmitter system consists of two transmitters, one for the video signal, which is amplitude-modulated (AM), and the other for the sound or aural signal, which is frequency-modulated (FM). Both of these transmitters feed a common antenna array. The

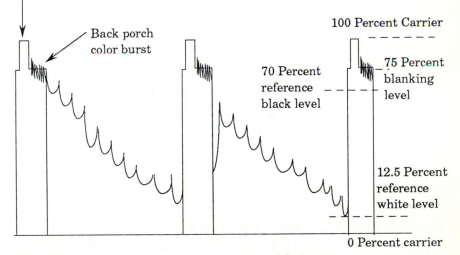

Figure 2.66 Two video lines showing percentage modulation.

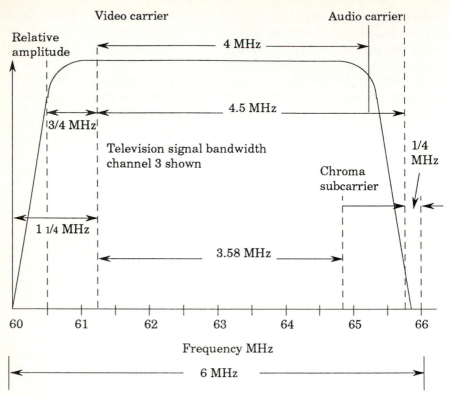

Figure 2.67

aural or sound transmitter has a frequency that is 4.5 MHz higher than that of the video transmitter. The whole bandwidth required for one television transmitting system must be contained in a 6-MHz space. A diagram of this space indicating the positions of the video and aural frequencies is depicted in Fig. 2.67.

It should be noted that the video transmitter is required to pass 0.75 MHz below the video carrier and 4.2 MHz above the video carrier, giving a bandpass of 4.95 MHz. The filter used to attenuate the video sidebands below the video carrier is termed the *vestigial sideband filter*. This filter is necessary to prevent video sidebands from interfering with the lower adjacent aural carrier.

2.10.2   Chroma or color television transmission

The color television image is reproduced on a color television screen by illuminating color dots or stripes of blue, green, and red. The mixture of the lights produced by these phosphors when activated by the electron beam in the picture tube creates the illusion of yellow, orange, cyan (light blue), magenta (red-purple), etc. The color television signal originates from the television camera at the studio, where blue, red, and green components of the image are separated into red, green, and blue signals. The signals are then corrected (gamma correction) for the proper brightness corresponding to the brightness recognition of the human eye. These signals are then combined via a matrix corresponding to 59 percent green, 30 percent red, and 11 percent blue. These percentages better discriminate against noise in the transmitter receiver link and produce a good white, gray, and black signal when viewed on a black-and-white receiver. This matrix output signal is called the Y or luminance signal, which amplitude modulates the video transmitter. Thus the color signal is compatible with black-and-white (monochrome) television sets. Another matrix in the transmitting system combines 60 percent red and 32 percent blue, which is known as the B-Y, or blue chrominance, or the I (in-phase) signal. A third matrix combines the R-Y, or red chrominance, or Q (quadrature) signal, which is made of 52 percent green, 21 percent red, and 31 percent blue.

To summarize, the Y signal controls the brightness, and the amplitude of the I and Q signals controls the saturation (purity) of the colors. The phase difference between the I and Q signals determines the *hue*, or actual colors, of the reproduced image.

To understand how the color signal is transmitted without increasing the video bandwidth the sidebands must be examined. Consider Fig. 2.68.

As can be seen, the sidebands are clustered around the line frequency of 15,750 Hz and its harmonics, leaving spaces in between. If a

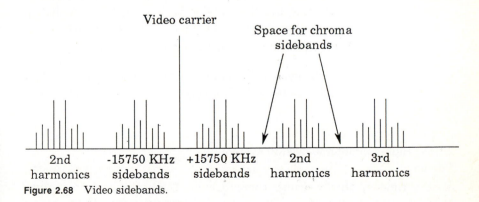

Figure 2.68 Video sidebands.

color subcarrier of 3,579,545 Hz is chosen above the video carrier, it will produce sidebands that will appear in these spaces between the Y sidebands. It was found that to obtain the desired subcarrier frequency a slight change in both the line and field frequencies was necessary. The color line frequency was set at 15,734.26 Hz and the field frequency at 59.94 Hz. This change was slight enough to allow television receivers to hold synchronism.

The color subcarrier frequency is amplitude modulated by the R-Y or Q signal, which generates sidebands in the spaces provided by the Y or luminance signal. Therefore both the luminance and chrominance signals can be transmitted in the 4.2-MHz bandwidth.

This section and the previous section complete the discussion of the baseband audio and video signals used in today's radio and television transmission systems. The problem now before us is to determine how to transmit these signals, through the process of modulation of transmitted carriers. Since the 3.58- (3.579545-) MHz chroma subcarrier is actually not transmitted as a fixed-amplitude carrier, some means to transmit this signal to color television receivers is needed.

A 3.58-MHz carrier is not transmitted because severe video interference would result. To make color television receivers lock on the 3.58-MHz chroma subcarrier eight or nine cycles of the subcarrier are inserted on the so-called back porch of the horizontal synchronizing pulse for the video baseband signal. Circuits in the color television receivers retrieve this signal, phase lock an internal oscillator to this signal, and hence reconstruct the chroma subcarrier signal from this short burst of eight cycles of the chroma subcarrier signal. Fig. 2.69 illustrates the process.

Since this method of reconstituting the chroma subcarrier by the receiver may seem far-fetched, some calculations may clarify the method. From Fig. 2.69 the width of the back porch of the horizontal sync pulse is determined as $0.06H$, where H distance is calculated as about 1/15,734 s, or is 63.5 μs. Now 0.06 × 63.5 μs = 3.8 μs wide.

The approximate time for one cycle of the 3.58-MHz subcarrier is

$$1/(3.58 \times 10^6) = 0.28 \times 10^{-6}\,\text{s} = 0.28\ \mu\text{s/cycle}$$

For a minimum of eight cycles the time needed is 8 × 0.28 μs = 2.24 μs. Therefore, the eight cycles will fit nicely in the 3.8-μs space of the back porch.

2.10.3 High-definition television system signals

At the time of this writing, there are no U.S. standards for high-definition television systems. The Federal Communications Commission

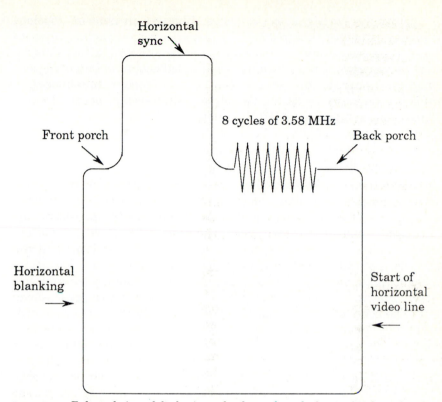

Figure 2.69 Enlarged view of the horizontal pulse to show the burst of eight cycles of chroma subcarrier.

(FCC) has indicated in some reports that compatibility with the present NTSC system will be a requirement. The change in picture aspect ratio from the present 4:3 to 16:9 means that cameras and all production facilities will have to be changed. If this new aspect ratio is accepted, ideally a wide, bright, clear screen will be developed by new technologies. Present projection systems change brightness and focus as the view angle is increased, as many of us know. One of the systems developed in Japan uses a 1125-line/60 Hz noninterlaced system and the 16:9 aspect ratio. This signal format requires more signal bandwidth than the NTSC 6-MHz system. Another system proposed in the United States calls for only 6 MHz of bandwidth using a novel modulation scheme.

The main course of action the cable television industry should take is to improve the signal-carrying capabilities of the cable system by increasing the system bandwidth and improving the system noise and distortion performances. This is the main function of a cable system. Improved cable amplifiers, cable, and equipment, as well as use of fiber-optic systems to control amplifier cascade lengths, are some of the means to improve system signal performance. Changes in network topology should be examined first to determine whether some amplifier cascades can be decreased simply by rearranging some system routing. Controlling system signal leakage will have the added benefit of improving system signal linearity, i.e., peak-to-valley ratio. A proper maintenance program is mandatory to assure the high-quality system needed by any high-definition television (HDTV) system.

2.10.4 Modulation

Modulation is actually the interaction of two sources of ac power, one low-frequency source and one high-frequency source. As we have learned, high-frequency power tends to escape electronic/electric circuitry and propagate through space. The practical application of ac energy propagating through space is radio wave transmission. In the early days of radio, modulation signals were rudimentary. Therefore, to transmit a message the ac source of high-frequency energy was merely interrupted, i.e., turned on and off, according to a prearranged code. The code, of course, was Morse code, nearly the same code railroad telegraphers used. The source of ac high-frequency energy was a pulsed spark gap exciting a parallel resonant LC circuit, which as we have learned acts as an electric flywheel. The tank circuit, as it was often called, was coupled to a long wire antenna. The Morse code pulses that turned the carrier frequency on and off were referred to as *modulation*. Essentially the carrier was on at full power or off at no power.

2.10.4.1 Amplitude modulation.

When a vacuum tube was used to pulse the parallel resonant LC circuit (tank circuit) and microphones were invented, the next step was to vary the carrier amplitude with the audio power source. Thus amplitude modulation resulted. The audio signal source generated by varying air currents on the diaphragm of the microphone produced complicated electrical wave shapes. To clarify the process of amplitude modulation, let us consider that the carrier and source of modulation are both sine waves. Fig. 2.70 shows the carrier voltage V_c and the modulating voltage V_m as a high-frequency sine wave and a low-frequency sine wave, respectively.

At point 1 V_m is positive maximum (+) and V_c is also at a maximum, i.e., V_m + carrier is on full. When V_m is negative (−) then V_c is at a minimum, i.e., V_m − carrier is nearly or fully off. Therefore, the modulated carrier wave should look as it does in Fig. 2.71.

A measure of the effective amount of modulation is the *index of modulation M*. The index of modulation equals 1 when the modulating

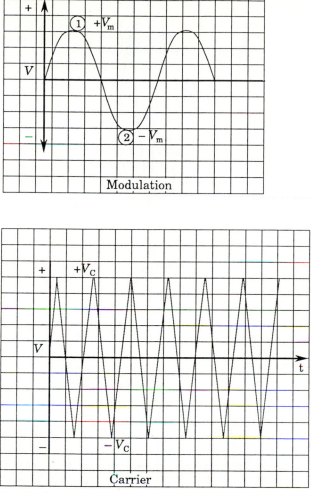

Figure 2.70

signal turns the carrier to the exact off position, when it turns a negative maximum. Mathematically $M = (V_{m_{p-k}}/V_{c_{p-k}})$. M is the ratio of the peak value of the modulating voltage to the peak value of the carrier voltage. Therefore, if we can arrange a plot of the modulated carrier as seen on the screen of a high-frequency oscilloscope, the value of M can be determined as indicated in Fig. 2.72.

The percentage of modulation is simply $M \times 100$ percent. In terms of V_{max} and V_m percent modulation = $(V_{max} - V_{min})/(V_{max} + V_{min}) \times 100$ percent.

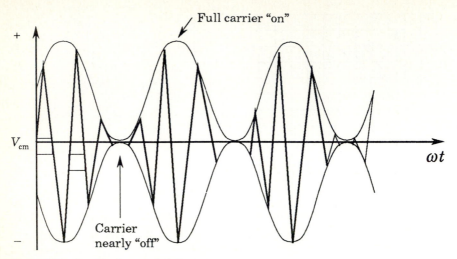

Figure 2.71   Amplitude-modulated carrier.

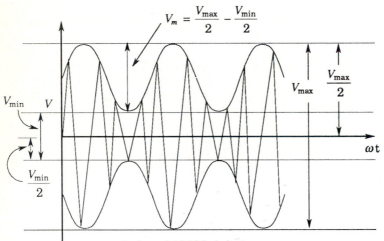

Index of AM Modulation

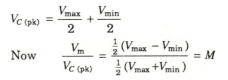

$$V_{C\,(pk)} = \frac{V_{max}}{2} + \frac{V_{min}}{2}$$

Now $\dfrac{V_m}{V_{C\,(pk)}} = \dfrac{\frac{1}{2}(V_{max} - V_{min})}{\frac{1}{2}(V_{max} + V_{min})} = M$

Figure 2.72

Example Suppose we had a peak reading voltmeter and the modulating peak voltage feeding the final carrier amplitude was found to be 750 V_{ac} peak and the peak radio frequency carrier voltage were measured at 1500 V peak. To calculate M simply take the ratio

$$M = \frac{750 \text{ V}}{1500 \text{ V}} = \frac{1}{2}$$

$$\% \text{ modulation} = \frac{1}{2} \times 100\% = 50\%$$

An oscilloscope screen is illustrated in Fig. 2.73. The oscilloscope can be used to measure M and percentage modulation. Refer to Chapter 6 for more information about oscilloscope measurements. In brief, the oscilloscope screen presents plots of amplitude versus time. Therefore, the screen can present the signal as a time function or wave shape such as sine waves, square waves, pulses, triangular waves, or sawtooth waves.

Another useful plot is the amplitude versus frequency plot, called a *spectrum*. The instrument capable of making this type of measurement is the *spectrum analyzer* (also discussed in Chapter 6). A spectrum plot of the various voltages present in amplitude modulation is shown in Fig. 2.74.

The only energy that is propagated, of course, is the higher frequencies: V_{LSB}, V_c, and V_{USB}. The signals designated V_{LSB} and V_{USB} are called the *lower sideband signal* and the *upper sideband signal*, respectively. When no modulation is present, they do not occur. Since the modulating signal in what is known as AM radio is usually speech and music, which certainly is not a simple sine wave, both upper and

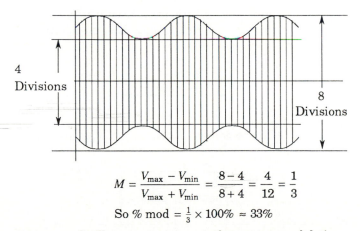

$$M = \frac{V_{max} - V_{min}}{V_{max} + V_{min}} = \frac{8-4}{8+4} = \frac{4}{12} = \frac{1}{3}$$

So % mod = $\frac{1}{3} \times 100\% \approx 33\%$

Figure 2.73 Oscilloscope measurement of percentage modulation.

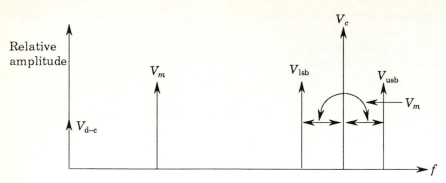

Figure 2.74   Spectrum of an amplitude-modulated (AM) signal.

lower sidebands change position and amplitude with the modulating signal. Recall that at zero modulating signal the carrier has some output and during loud crashes of music will then go from maximum carrier to minimum. The upper and lower sidebands are sort of images of themselves, and it soon became evident that since the message (modulating signal) occurred both above and below the carrier only one was needed. Hence single-sideband voice transmission resulted. It should also be evident that the message is in the sideband information and not in the carrier. Essentially the carrier signal is just a reference to determine the sideband location. Therefore, the carrier power was decreased to a value just low enough to determine the sidebands. Thus, what is known as *single-sideband suppressed carrier radio* came into use. This is a really efficient form of AM radio. A simple plot of the required bandwidth is shown in Fig. 2.75.

It should be seen that just passing the carrier and one sideband can result in a substantial saving of spectrum space. This is the method used by commercial television-broadcast stations. Essentially the carrier provides a reference needed to substantiate the positions and

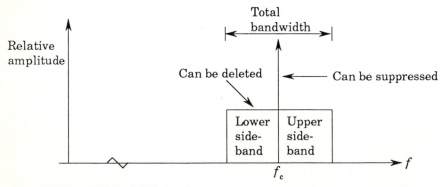

Figure 2.75   Bandwidth of AM signal.

hence the information contained in the sidebands. Therefore, in radio single-sideband (SSB) the carrier is suppressed and not actually transmitted. At the receiver a tunable oscillator is adjusted in frequency to act as an inserted carrier needed to detect the sidebands. If the oscillator is not tuned precisely correctly, the detected signal becomes distorted. This oscillator must also be stable in frequency and amplitude because any drift causes distortion. The SSB radio systems in use today are efficient in that all the transmitted power is spent on the sidebands. Long-distance radio communications use this form of transmission.

2.10.4.2 Frequency modulation.   So far the topic of varying the amplitude of a radio frequency (RF) carrier with a baseband signal power source such as audio and video signals has been discussed. In *frequency modulation* frequency, rather than amplitude, is varied at the signaling rate. It must be remembered that with amplitude modulation (AM) the larger the modulating signal amplitude becomes the larger the sideband amplitude becomes. For example, where the modulating signal is audio the larger the audio signal becomes, the larger the sideband signals become, and hence the receiver produces maximum loudness. Of course, 100 percent amplitude modulation produces maximum detected loudness. In the case of FM the amplitude of the audio signal moves the carrier above and below the carrier center frequency. Therefore, this swing in carrier frequency corresponds to maximum loudness. To avoid using too much bandwidth for the frequency modulation case, some limits have to be imposed. The FCC standards for commercial FM broadcast radio limit the carrier swing to 7±5 kHz. This is sufficient to allow good audio frequency response. The ±75-kHz carrier deviation corresponds to 100 percent modulation for commercial FM radio. The rate of moving the carrier is at the rate or frequency of the audio signal.

Any change in carrier frequency is actually not a pure sine wave because during the course of change the positive half-cycle is in the process of frequency change at a different point than the bottom half. This so-called shift or change in wave shape causes sideband pairs above and below the center carrier frequency. The greater the frequency swing the greater the number of sideband pairs.

For the standard FM radio broadcast system the highest audio frequency used is 15 kHz and the maximum frequency shift is ±75 kHz. Therefore, 75 kHz ÷ 15 kHz is what is known as *deviation ratio* of the FM carrier, which is 5. Since 15 kHz is the upper limit, the ratio of the maximum allowed deviation of 75 kHz to any other modulating frequency, i.e., 5 kHz, etc., is called the *index of modulation*. It should be evident that when the deviation ratio equals the modulation index,

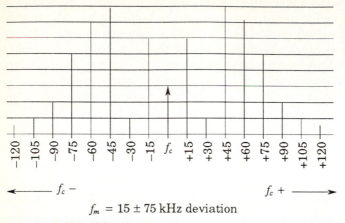

$f_m = 15 \pm 75$ kHz deviation

Figure 2.76   FM sideband spectrum.

the modulating frequency has to be 15 kHz. Theoretically FM carrier deviation should produce an infinite number of sidebands. However, all of them are not big enough to be significant. It has been found in practice and by mathematical calculation that a 15-kHz modulating sine wave signal and a deviation ratio of 5 will produce eight sidebands of a large enough amplitude to be significant. The spectrum plot of this example is shown in Fig. 2.76.

Notice the symmetry of the sideband spectrum about the carrier frequency. Note that the upper limit of 15 kHz is the spacing between adjacent sidebands. In addition, the maximum spectrum extends to 120 kHz for a 15 kHz shift to ± 75 kHz, thus requiring a bandwidth of ± 120 kHz or 240 kHz wide. The sidebands ± 120 kHz are quite small and since in the usual music of today the amplitude of the audio signal is quite small for say 5 to 15 kHz, these sidebands under actual conditions are below significant amplitude.

The commercial FM radio band consists of the overall frequency band of 88 to 108 MHz, which gives a 20-MHz total bandwidth. In this 20-MHz bandwidth space 100 FM radio channels are spaced every 200 kHz, beginning at 88.1 MHz for the first carrier center frequency. A chart of the 100 channel center frequencies is shown in Table 2.8.

Since a variety of electronic devices such as varactor diodes which act as voltage controlled capacitors have been discussed, a word about FM modulation techniques is in order. When this device is placed in an *LC* resonant circuit controlling the frequency of an oscillator and the voltage across the capacitor is the modulating signal, then the oscillator frequency will be deviated at the modulating rate, thus producing frequency modulation. For example, if the varactor has a fixed dc bias voltage that sets the center oscillator frequency and the audio

TABLE 2.8 FM Broadcast Frequencies

Channel	Frequency	Channel	Frequency	Channel	Frequency
1	88.1	35	94.9	69	101.7
2	88.3	36	95.1	70	101.9
3	88.5	37	95.3	71	102.1
4	88.7	38	95.5	72	102.3
5	88.9	39	95.7	73	102.5
6	89.1	40	95.9	74	102.7
7	89.3	41	96.1	75	102.9
8	89.5	42	96.3	76	103.1
9	89.7	43	96.5	77	103.3
10	89.9	44	96.7	78	103.5
11	90.1	45	96.9	79	103.7
12	90.3	46	97.1	80	103.9
13	90.5	47	97.3	81	104.1
14	90.7	48	97.5	82	104.3
15	90.9	49	97.7	83	104.5
16	91.1	50	97.9	84	104.7
17	91.3	51	98.1	85	104.9
18	91.5	52	98.3	86	105.1
19	91.7	53	98.5	87	105.3
20	91.9	54	98.7	88	105.5
21	92.1	55	98.9	89	105.7
22	92.3	56	99.1	90	105.9
23	92.5	57	99.3	91	106.1
24	92.7	58	99.5	92	106.3
25	92.9	59	99.7	93	106.5
26	93.1	60	99.9	94	106.7
27	93.3	61	100.1	95	106.9
28	93.5	62	100.3	96	107.1
29	93.7	63	100.5	97	107.3
30	93.9	64	100.7	98	107.5
31	94.1	65	100.9	99	107.7
32	94.3	66	101.1	100	107.9
33	94.5	67	101.3		
34	94.7	68	101.5		

alternating voltage is superimposed on the dc bias, the oscillator will be deviated plus or minus about the center frequency in step with the audio voltage. When this audio voltage happens to be at 0 V, then the oscillator will be at center frequency. When it swings positive, then the frequency will be increased above center frequency, and when negative, below center frequency. This type of device has simplified FM techniques over some of the previous methods.

2.10.5 Receiving systems

Since our discussion of FM is from a cable television or transmission point of view, receiver circuitry will not be taken up in any detail. Essentially for AM a simple diode circuit may be used to recover the

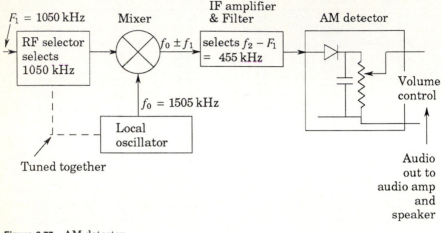

Figure 2.77   AM detector.

modulation from an RF carrier. A simple so-called envelope detector is shown in Fig. 2.77.

The input signal to the detector is at the intermediate fixed frequency to which the carrier signal is converted by the tuned converter circuits. Standard AM radio broadcast systems usually use the intermediate frequency of 455 kHz. The tuning dial selects the input signal and adjusts the local oscillator to a frequency of 455 kHz higher than the selected carrier frequency. Both signals are fed to the converter circuit, which selects the difference frequency. This method is known

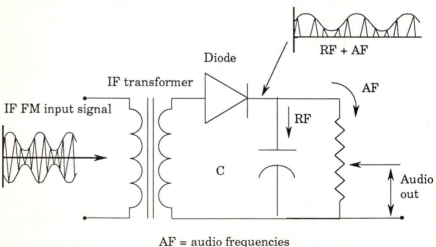

AF = audio frequencies
RF = radio (carrier) frequencies

Figure 2.78   Superheterodyne principle.

as *superheterodyning*. The *heterodyne* combines two signals to obtain sum and difference frequencies. A filter is then used to select either the sum or difference frequency. The *superheterodyne* is the frequency that is higher than, or super to, the selected frequency (the local oscillator frequency). An example of this method is shown in Fig. 2.78.

The FM receiver is very nearly the same, with the addition of one circuit called a *limiter* and an FM detector substituted for the AM detector. The FM receiver version of the circuit is shown in Fig. 2.79.

The limiter essentially amplifies the intermediate frequency (IF) signal to nearly the saturation point, thus increasing the immunity of noise entering the detector. This circuit is needed between the IF amplifier and an earlier type of detector called a *discriminator*. The discriminator detects the frequency shifts, thus producing a voltage proportional to the frequency shifts, but it also responds to any changes in amplitude. The ratio detector also responded to the modulated-frequency shifts but exhibits a self-limiting action, thus deleting the need for a separate limiter circuit. The quadrature detector is a common type of FM detection circuit used today. This type of circuit lends itself to integrated circuit techniques. This circuit is shown in Fig. 2.80.

The amplifier in the circuit in Fig. 2.80 multiplies the signal at input 1 by the signal at input 2. The voltage V_{in} is fed directly to input 1. The signal appearing at amplifier input 2 is actually V_{in} shifted 90°, which is said to be in quadrature to V_{in}. In other words, a demodulator detects modulation on a carrier discriminator and ratio detectors are demodulators. Hence this is how this demodulator gets its name. Therefore, at the center frequency of the FM signal, L, C, R, and C_2 shift the input frequency 90° with the signal at input terminal 1. When this condition is met, the output V_{out} is 0. When the input frequency is varied plus or minus about the center frequency, then the

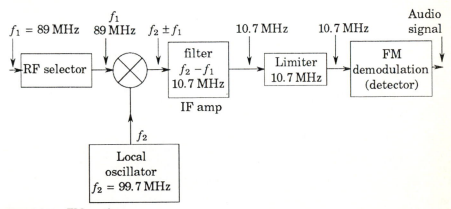

Figure 2.79   FM receiver.

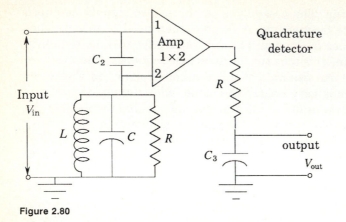

Figure 2.80

phase shift of the signals between terminals 1 and 2 is varied about 90°, causing the product of these signals V_{out} to respond to the frequency changes. This results in a signal voltage proportional to the frequency changes. It should be recalled that when the product of two signals of the same waveform and frequency is multiplied by the cosine 90°, the phase angle between them is zero ($\cos 90° = 0$). Therefore, the output signal V_{out} is proportional to the cosine of the angle between the signals and hence the frequency changes. A useful by-product of a frequency modulation detector is automatic frequency

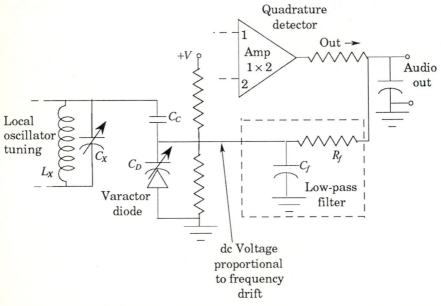

Figure 2.81   Automatic frequency control.

control of the local tuning oscillator of the receiver. Automatic frequency control (AFC) is used to keep the local oscillator accurately tuned to the desired input center frequency so the correct IF center frequency can result. Such drift can make the selected FM station drift too far off frequency, thus causing a distorted demodulated signal. If a sample of the demodulated FM signal is passed through a low-pass filter, then only the very-low-frequency changes will get through the filter. These slow changes are proportional to any frequency drift of the local oscillator. This voltage is adjusted to the level necessary to act as the controlled bias voltage for a varactor diode placed in the local oscillator tuning circuit.

Such a circuit is shown in Fig. 2.81. D is a varactor diode; $+V$ and dc drift voltage cause C_D to vary. The effects of C_D are coupled through C_C to correct the tuning of $L_X C_X$ for the frequency drift due to temperature, etc. Another type of circuit that can demodulate FM signals is the phase-locked-loop (PLL) circuit. This type of circuit also appears as an integrated circuit using only a few external components. A PLL circuit configured as an FM detector is shown in Fig. 2.82; f_c is set by the dc bias on the voltage-controlled oscillator (VCO). Frequency f_c is set to the intermediate frequency center frequency, 10.7 MHz for commercial FM radio. The input FM changing in frequency is fed to the phase detector, which is also fed the center frequency f_c. The output of the phase detector is a voltage consisting of FM $\pm f_c$, which is connected to the low-pass filter, which in essence selects the difference frequency FM $-f_c$. The amplifier adjusts the magnitude of this signal and drives the voltage-controlled oscillator, which varies the frequency f_c to follow the input frequency shifts. The voltage V_{out} is actually the demodulated FM voltage, which is developed by the changes in input frequency. An example of the output voltage as a function of input frequency for this type of circuit is illustrated in Fig. 2.83.

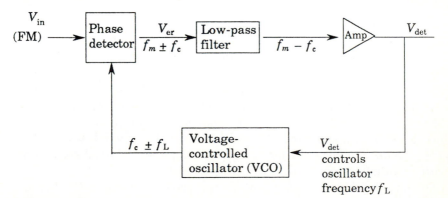

Figure 2.82 Phase-locked-loop (PLL) FM detector.

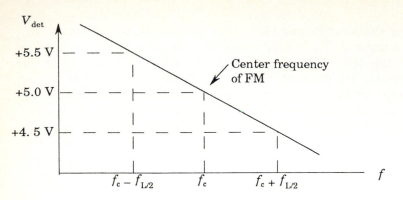

Maximum VCO signal swing called the frequency lock range.

Figure 2.83   Phase-locked-loop operational curves.

2.10.6   Preemphasis and deemphasis

It should be remembered that the amplitude of the modulating signal deviates the frequency of the carrier signal, producing a large number of sideband pairs about the carrier. In the course of changing the carrier frequency a certain amount of phase modulation results. When the modulating signal is at high frequencies during high-pitched music passages and at low volume the sideband pairs will be low-amplitude and spread far out from the carrier, thus becoming hard to demodulate. In order to preserve the higher-pitched passages in music, the amplitude is accentuated in the modulating procedure by a calibrated amount. Therefore, to have the music accurately preserved in the receiver, the higher-pitched sounds have to be deemphasized after the demodulating process. This procedure is diagrammed in Fig. 2.84. The RC time constant of both the preemphasis and deemphasis networks is specified at 75 μs. It should be clear that the two networks compensate each other and produce, when coupled together, a flat amplitude curve.

2.10.7   Stereo frequency modulation

Experimentation about how human beings hear sounds with two ears indicated that more realistic methods of sound reproduction might be realized. Binaural sound, in which two microphones are spaced apart and amplified separately to drive a set of earphones, produced astounding results. However, the inconvenience of wearing earphones led to more experiments using two loudspeakers placed apart to produce a left and a right sound reproducing system, which was called *stereophonic*. Stereo, short for stereophonic, uses a left and right separate channel system as shown in Fig. 2.85.

The only mixing of signals is at the source (band) and at the listen-

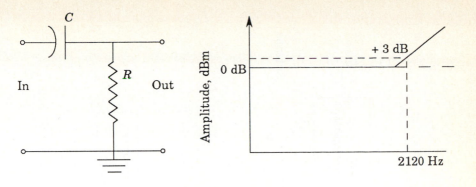

Preemphasis

$$f = \frac{1}{2\pi RC} = \frac{1}{2\pi \times 75 \times 10^{-6}}$$

$$= \frac{10^6}{150\pi} = 2120 \text{ Hz}$$

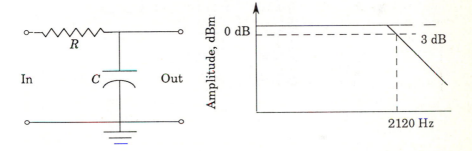

Deemphasis

Figure 2.84

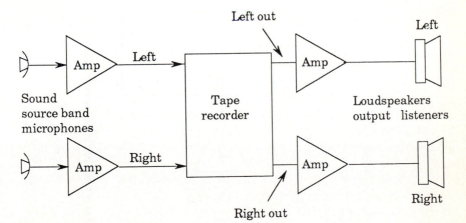

Figure 2.85 Basic stereo system.

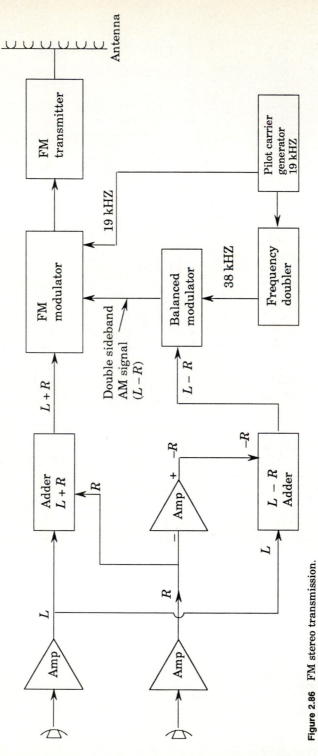

Figure 2.86 FM stereo transmission.

er's location speaker. The placement of the microphones and the speakers is important for good stereo separation (difference between left and right channels). Also the listeners should not be too far left or right in the listening area.

The trick now is to be able to broadcast both the left and right channels using our commercial FM radio system and maintain high-quality sound plus left and right channel separation. Naturally this is not an easy task, and there is significantly more circuitry for the transmitter and receiver systems. Essentially the monophonic single-channel system, as discussed, can be thought of as the output of the left and right channels combined, $L + R$. Now if the $L - R$ can be transmitted as well, then two simple adding circuits and one inverter circuit should be able to separate the left and right signals at the receiver.

$$\begin{array}{ccc}
\begin{array}{r} L + R \\ + \; L - R \\ \hline +2L \end{array} &
\begin{array}{r} L + R \\ - (L - R) \\ \hline \end{array} = &
\begin{array}{r} L + R \\ + (-L) + R \\ \hline +2R \end{array}
\end{array}$$

For the transmitter to transmit two channels separately, channel 1, $L + R$, channel 2, $L - R$, consider the transmission in Fig. 2.86.

The balanced modulator (bal mod) produces an AM double-sideband signal with the carrier frequency nearly nulled out (low-amplitude). This signal covers the bandwidth of 23 to 53 kHz (38 kHz ± 15 kHz). Therefore, the frequency bandwidth needed to pass high-pitched music passages at 15 kHz is ±15 kHz = 30 kHz wide.

The receiver is also significantly more complicated. A simple monophonic FM receiving system is shown in Fig. 2.87.

The stereo receiver is shown in Fig. 2.88, which illustrates the stereo separation technique.

Notice that the balanced demodulating circuit produces two outputs $L - R$ and $-(L - R)$ and requires three inputs: (1) the 38-kHz signal needed to demodulate the (2) 23- to 53-kHz signal, as well as (3) the $(L + R)$ monophonic signal. The baseband pass structure is shown in Fig. 2.89.

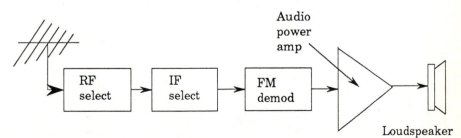

Figure 2.87 Simple monophonic FM receiver.

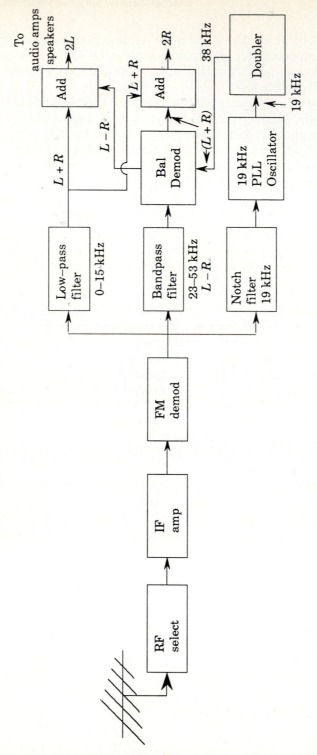

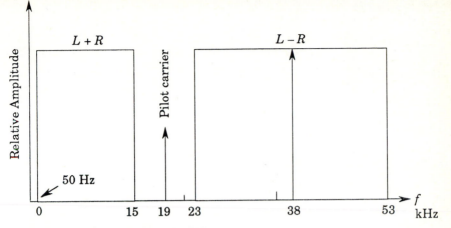

Figure 2.89 Bandpass structure for FM stereo.

Selected Bibliography

Barker, Forrest. *Communications Electronics, Systems, Circuits, and Devices*. Prentice Hall, Inc., Englewood Cliffs, N.J., 1987.

Benson, K. Blair, editor in chief. *Television Engineering Handbook*. McGraw-Hill Book Co., New York, 1986.

Freeman, Roger L. *Telecommunication Transmission Handbook*, ed. 2. John Wiley & Sons, New York, 1981.

Hoyt, William H. Jr., Kemmerly, Jack E. *Engineering Circuit Analysis*, ed. 3. McGraw-Hill Book Co., New York, 1978.

John, Walter C. *Transmission Lines and Networks*. McGraw-Hill Book Co., New York, 1950.

Jordan, Edward C., editor in chief. *Reference Data for Engineers: Radio, Electronics, Computers and Communications*. ed. 7. Indianapolis, Howard W. Sams, 1985.

Orr, William I. *Radio Handbook*. Indianapolis, Howard W. Sams, 1988.

Schrader, Robert L. *Electronic Communication*. ed. 2, McGraw-Hill Book Co., New York, 1967.

Williams, Arthur B., Taylor, Fred J. *Electronic Filter Design Handbook*. ed. 2, McGraw-Hill Book Co., New York, 1988.

3

Coaxial Cables and Cable Systems

3.1 Development of Coaxial Cables

With the development of radar and very-high-frequency (VHF) radio systems during World War II, the need for better wires and cable became apparent. The usual war priorities produced the copper braided over rigid polyethylene dielectric, with a standard twisted copper center conductor. Because of corrosion and abrasion of the outer copper braid, a plastic, rubber, or polyethylene jacket was used. Various connectors were tried, and the PL-SO 259 connector and the BNC connector were developed and became standard.

After the war the U.S. government sold surplus equipment to industry, schools, and colleges at low prices in an effort to help the country to get back on its feet. The supply of coaxial cable was large enough, and for many users not needed so this early type of coaxial cable was used by the fledgling community-antenna television (CATV) industry. This type of cable did not fare well when used exposed and under tension in aerial plants. The jacket would split and open up, allowing moisture to invade the copper braid outer shielding. Also, when it was used in buried plants, moisture from the ground that infiltrated the small cracks or breaks in the outer plastic jacket made during handling and installation quickly ruined the cable. The cable industry's need for better cable and equipment became apparent to manufacturers who were looking at new products to make. Thus the seamless aluminum tubing cable was developed. This cable, without the normal holes of the woven-braided type, offered lower signal attenuation, constant characteristic impedance throughout the length of the cable, and better standing wave ratio (SWR) characteristic. Often in areas not near the ocean salt atmosphere or industrial corrosive air, the cable was used bare with no covering or outer jacket. The dielectric insulating material was the newly developed polyethylene foam, which supported the center conductor in place. This cable was improved as time went by and is basically what is used in present-day coaxial cable systems. These improvements included the addition of an outer protective polyethylene jacket with a sealant sticking it to the aluminum shield. For underground use a flooding compound was placed between the aluminum shield and the outer jacket. A steel flexible tubing was placed over the outer plastic jacket and a second jacket placed over the steel, thus giving the industry what is known as *armored cable*. This cable was used in underground plants buried in rocky or mountainous terrain. Sealant was often formed over the center conductor and on the underside of the aluminum tubing to keep the dielectric foam in place. The center conductor was either a solid copper piece of wire or a copper clad conductor. The copper metal was drawn over the aluminum core by a special process to form the copper clad conductor. This lowered the cost by reducing the amount of expensive copper, thus making the most expensive part of the cable the more plentiful, cheaper, and lighter aluminum.

Other improvements were in the composition of the insulating dielectric. Styrofoam was tried, and although lower-loss cable resulted, other problems surfaced, e.g., water migration through the insulation caused by a defective connector, which allowed water to enter. When certain gases were injected into the polyethylene foam insulation during manufacturing a superior cable resulted which is now often used in new builds and rebuilds. Addition of an integral steel messenger cable molded to the outer plastic jacket allowed the cable to be self-supporting on the poles, thus eliminating the need for the galvanized steel messenger strand and the subsequent lashing. The cable can also be ordered on reels already placed in a flexible plastic duct, thus allowing direct burial in rocky areas without the expensive armor.

Manufacturers of coaxial cable provide electrical specifications and mechanical specifications to aid the user in selecting the proper type of cable. The electrical specifications include such parameters as nominal characteristic impedance, return loss, attenuation versus frequency, loop resistance (usually in $\Omega/1000$ ft), capacitance (usually in picofarads/foot), and velocity of propagation (percentage). Mechanical specifications often include minimum bending radius (in inches) and maximum pulling tension (in pounds). More and more lengths and weights are also specified in the metric systems. Conversion charts and calculators aid in converting from one system to the other. When conversion to metric units takes place, it will be found that many calculations can be made in one's head.

3.2 Coaxial Cable as a Transmission Line

The word *coaxial* means common axis or common center, often referred to as *concentric*. A cross section of coaxial cable is shown in Fig. 3.1.

Essentially coaxial cable is a type of transmission line used to transport high-frequency electric power. In free air we know that high-frequency electric power can be transmitted. The radio and television industries make use of this fact, placing their information on this transmitted power, which is termed *modulation*. Therefore, this transmitted power carries the television picture and sound information.

In free air or a vacuum alternating power is transmitted (propagated) through space at the speed of light, which is about 3×10^8 m/s. Consider an ac generator with an output that varies sinusoidally. A sinusoidal alternating current alternates from plus to minus smoothly as in Fig. 3.2.

The distance this current will propagate through the air in the time of one cycle is called the *wavelength*. Wavelength, denoted mathematically by λ, the Greek letter lambda, equals velocity times period where the velocity of transmission is 3×10^8 m/s. The *period* is the time the sinusoid passes from one [+] excursion through one [−] excursion, equal to (1/frequency). *Frequency* is the speed at which the sine wave varies in 1 s. Therefore,

$$\lambda = V \times P = V \frac{1}{f} \tag{1}$$

Example. A sinusoidal electric generator produces alternating power at a frequency of 60 cycles/s. Find the corresponding wavelength λ in miles.

$$\text{Recall } V = 3 \times 10^8 \text{ m/s,}$$

$$1 \text{ ft} = 30.48 \text{ cm} = 0.3048 \text{ m}$$

$$V \text{ (ft/s)} = \frac{3 \times 10^8 \text{ m/s}}{0.3048 \text{ m/ft}} = 9.843 \times 10^8 \text{ ft/s}$$

$$V \text{ (mi/s)} = \frac{9.943 \times 10^8 \text{ ft/s}}{5280 \text{ ft/mi}} = 1.864 \times 10^5 \text{ mi/s}$$

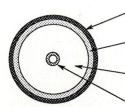

Polyethylene jacket

Aluminum sheath

Polyethylene foam (gas injected)

Copper clad aluminum center conductor

Figure 3.1 Cross section of coaxial cable.

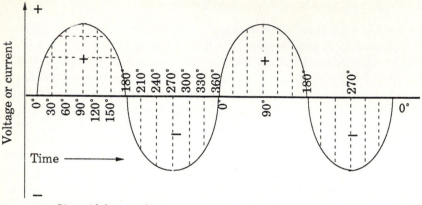

Figure 3.2 Sinusoidal wave shape.

Therefore, the speed of electromagnetic radiation in English system units is 186,400 mi/s (speed of light). Now

$$\lambda = \frac{186,400 \text{ mi/s}}{60 \text{ cycles/s}} = 3106 \text{ mi} = \text{distance of one wavelength}$$

The frequency in cycles/second is given the name hertz (abbreviated Hz) in honor of Rudolph Heinrich Hertz, the German physicist who discovered radio propagation.

The previous example illustrates that a commercial power transmission line is segmented by transformers to overcome line loss and prevent the transmission line from becoming an antenna one wavelength long. After all, 3106 mi is nearly the distance across the United States. Essentially a transmission line is supposed to distribute electric energy with minimum possible loss. In cable television coaxial cable is the transmission line of choice. Of course, coaxial cable does indeed have losses, which are specified in decibels/100 ft at certain frequency points. Table 3.1 is a loss chart for PIII-type cable for four sizes: 0.500-in, 0.750-in, 0.875-in, and 1.000-in-diameter cable.

3.2.1 Coaxial cable losses

Notice that as the frequency of alternating current signals increases so does loss. Also notice that television channel 2 has a picture carrier of 55.250 megahertz (MHz); hence for 0.500-in cable loss is nearly 0.52 dB/100 ft and at channel 7 (175.250 MHz) nearly 1.00 dB/100 ft. For more accurate results, a curve can be plotted to make finding the loss between the values given in the table easier. When designing a cable system the losses at channel 2 (the lowest-frequency signal) and the highest-frequency signal loss are used in signal level calculations.

TABLE 3.1   Cable Attenuation* versus Frequency

Frequency, MHz	Cable type			
	0.500	0.750	0.875	1.000
5	0.16	0.11	0.09	0.09
30	0.40	0.26	0.24	0.23
50	0.52	0.35	0.32	0.30
110	0.75	0.52	0.47	0.44
150	0.90	0.62	0.55	0.52
180	1.00	0.68	0.60	0.59
250	1.20	0.81	0.72	0.68
300	1.31	0.90	0.79	0.75
400	1.53	1.05	0.91	0.87
450	1.63	1.12	0.98	0.92
500	1.70	1.17	1.04	0.97
550	1.80	1.23	1.09	1.02
600	1.87	1.29	1.15	1.06

*Maximum in dB/100 ft.

Now that the loss is known for a 300-MHz signal frequency and 0.750-in-diameter cable as 0.90 dB/100 ft, the loss will be 9 dB/1000 ft.

For the unity gain building block using an amplifier of 20 dB gain the cable should have a maximum loss of −20 dB, which will occur at the highest frequency specified for use. For 0.75-in cable at an upper limit of 400 MHz the loss is 1.05 dB/100 ft. Therefore, 20 dB of cable will be calculated as follows:

$$\frac{20 \text{ dB}}{1.05 \text{ dB}} \times 100 \text{ ft} = 19.05 \times 100 \text{ ft} = 1905 \text{ ft}$$

This means that if the amplifier is connected to a piece of cable 1905 ft long with nothing else causing any loss the unity gain section of cable with associated sample signal levels will look like the case illustrated in Fig. 3.3a.

For unity gain the amplifier has to have as much gain as the cable has loss at the upper-frequency limit where the cable loss is greatest. Since the loss of the cable is less at the low-frequency limit, clearly the sections are not unity gain. In practice the amplifiers are reasonably flat, or constant gain, over the design amplifier bandwidth. Now that the cable loss versus frequency characteristics are known, one can calculate what the amplifier will have as a signal input at the lower-frequency bounds. Consider the preceding cable section, in which the loss, gain, and signal levels are shown at the low- and high-frequency limits for 0.750 cable at 20 dB loss at 400 in Fig. 3.3b. At 400 MHz cable loss is 1.05 dB/100 ft = 10.5 dB/1000 ft. Therefore, 20-dB cable will be 1905 ft long at 400 MHz.

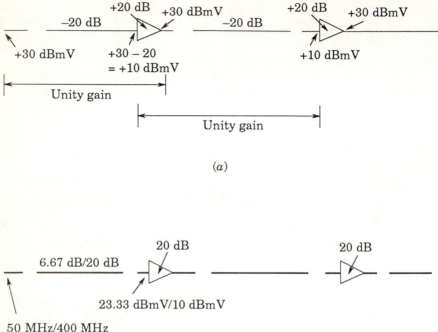

(a)

Figure 3.3 Unity gain building block.

For a cable of length 1905 ft and a cable loss of 0.35 dB/100 ft = 3.5 dB/1000 ft, the loss per foot can be calculated by

$$1905 \text{ ft (at } 0.35 \text{ dB/100 ft)} = 3.5 \text{ dB/1000 ft}$$

$$\frac{1905}{1000} \times 3.5 = 6.67 \text{ dB}$$

It has been shown that the input signal level at 50 MHz and 400 MHz is 13.33 dB difference. Since the amplifier has the same gain at 50 MHz as at 400 MHz, the output signal at 50 MHz would be 23.33 + 20 = 43.33 dBmV and at 400 MHz would be +30 dBmV as before. Clearly something must be done about the low cable loss at 50 MHz. The solution to the problem is what is known as *cable equalization*. Cable equalizers are made up of passive components consisting of capacitors, inductors, and resistors in a circuit network that provide loss characteristics that are the inverse of the cable loss. Essentially the equalizers have approximately 13.33-dB loss at 50 MHz and nearly 0-dB loss at 400 MHz. In practice large equalizing networks are difficult to make and remain stable as a function of temperature. Since this network is usually placed within the amplifier enclosure, it is not at the same temperature as the cable which it is trying to compensate. In practice the amplifier output is sloped upward, i.e., it has more gain and hence signal level at 400 MHz than at 50 MHz. Usually a 3- to 4-dB upward slope is used; i.e., the amplifier output level at 400 MHz is 3 dB higher than the level at 50 MHz. The point is illustrated in Fig. 3.4.

Since the cable used in a cable television system is essentially a transmission line and is of the coaxial type, a study of other types of transmission lines and transmission line general characteristics is in order.

3.2.2 Flat transmission line

Probably one of the most common types of transmission line is the two-wire line commonly known as *flat lead* to television antenna people. Ham radio operators fashioned two-wire transmission lines made from two parallel copper wires separated by insulators spaced to keep the wires parallel and equidistant from each other. A cross section of two-wire lines, showing the electric and magnetic field lines, appears in Fig. 3.5.

Notice that (1) the electric and magnetic field lines although curved are at right angles to each other; (2) both the electric and the magnetic field lines surround the parallel wires and hence can be influenced by outside electrical forces.

The two-wire line usually had very low losses over the television band, but it was subjected to outside electrical interference and could be influenced by metal objects coming in close proximity to the line.

The flat two-wire television transmission line was prone to ignition interference by passing cars, as well as other power-company-related noise that entered the electric-magnetic fields carrying the television signals. Metal objects such as drain pipes, water pipes, and electric wires near this transmission line interfered with the signal-carrying electric-magnetic fields of the two-wire lines.

Coaxial cable does not have some of the problems of noise pick-up and nearby metal objects, because all of the signal-carrying electric-

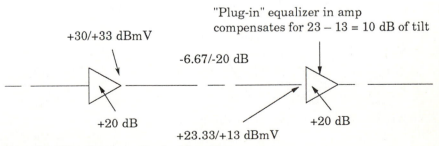

Figure 3.4 Sloped and equalized cable section.

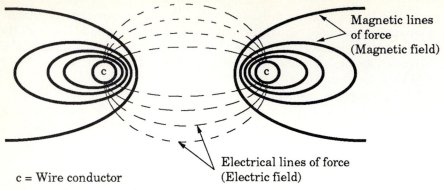

c = Wire conductor

Figure 3.5   Cross section of a two-wire transmission line.

magnetic fields are inside the cable. Since coaxial cable is more difficult to construct, it is more expensive. It is also more lossy than the open wire line. A cross section of coaxial cable is shown in Fig. 3.6. For cable television purposes as well as a great many industrial uses coaxial cable is the favored type of transmission line. The schematic diagram of a transmission line system is shown in Fig. 3.7.

3.2.3   Transmission line parameters and impedance matching

The transmission line is said to be terminated at the receiving end by terminating impedance Z_r, which is usually a 75-Ω resistor for cable television work. At the sending end an electrical generator supplies power to the transmission line, through the sending and impedance Z_g, which again is usually a resistance, 75 Ω for cable television. For telephone work transmission line distance and length are usually cal-

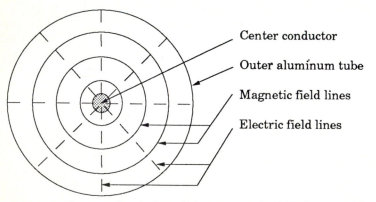

Figure 3.6   Cross sectional view of electromagnetic fields in a coaxial cable.

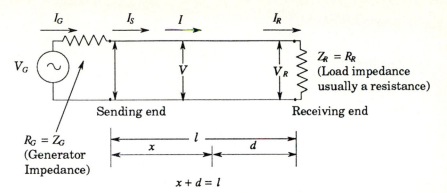

Figure 3.7 Schematic diagram of a transmission line.

culated in miles and in radio work in meters. The electrical constants of the line are usually given as a function of length. For example,

L: inductance/unit of length

C: capacitance/unit of length

R: resistance/unit of length

In cable television work the foot or 100 ft is the unit of length. Recall that the loss table was in decibels/100-ft units. A table of line constants for various transmission lines will give the capacitance, usually in picofarads/foot, the inductance in microhenrys/foot, and the loop resistance in either ohms/foot or ohms/1000 ft.

To simplify the behavior of electrical signals on a transmission line, consider the following example of a transmission line connected to a simple battery with a switch shown in Fig. 3.8.

What is taking place is that a wave V volts high is traveling down the transmission line at a velocity v m/s. At any point in time T the wave front point can be found as a function of distance down the line from the sending end. The speed of propagation for transmission lines is a function of the inductance and capacitance parameters of the line. These parameters depend on the material in the space surrounding the center conductor, called the *dielectric* or *insulating plastic foam*, as well as the conducting material of the center conductor and the surrounding metallic tubing. The parameters along with the geometry of the cable determine what is known as the *characteristic* or *surge impedance* of the transmission line. The flat television antenna wire had an impedance of 300 Ω; all cable television coaxial cable has a characteristic impedance of 75 Ω. Radio transmission work as well as industrial applications use various types of transmission lines with a characteristic impedance of 50 Ω. When a transmission line is driven

Figure 3.8   Transmission line charged by a battery.

by a generator with an impedance equal to the characteristic impedance and also equal to the load or terminating impedance, the transmission line acts like a line of infinite length.

3.2.4   Transmission line reflections

If the terminating impedance is different from the characteristic impedance, when the wave arrives at the receiving end, a portion of the
wave is reflected toward the sending end. This reflected portion is a
function of the difference between the characteristic impedance and
the receiving end impedance. The value of this reflected wave is determined by the reflection coefficient, which is a function of the impedance mismatch.

$$K_r = \frac{Z_r - Z_o}{Z_r + Z_o} \text{ at the receiving end}$$

where Z_o = characteristic impedance (Ω)
Z_r = receiving end impedance (Ω)

Consider the following example, diagrammed in Fig. 3.9.
As soon as the switch is closed the voltage V appears across the
transmission line and the wave at voltage V travels down the line at
velocity v m/s. If the cable is coaxial and a perfect insulating material
fills the area around the center conductor and outer aluminum sheath,
or the transmission line is the open wire type operating in a perfect
vacuum, the velocity of this wave will be at the speed of light, which

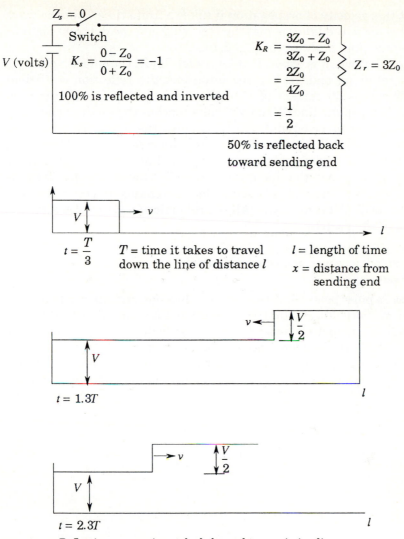

The diagram shows:

$Z_s = 0$

Switch

V (volts)

$K_s = \dfrac{0 - Z_0}{0 + Z_0} = -1$

100% is reflected and inverted

$K_R = \dfrac{3Z_0 - Z_0}{3Z_0 + Z_0}$

$= \dfrac{2Z_0}{4Z_0}$

$= \dfrac{1}{2}$

$Z_r = 3Z_0$

50% is reflected back
toward sending end

$t = \dfrac{T}{3}$ $T =$ time it takes to travel
down the line of distance l

$l =$ length of time
$x =$ distance from
sending end

$t = 1.3T$

$t = 2.3T$

Figure 3.9 Reflections on a mismatched charged transmission line.

we have calculated at nearly 186,400 mi/s. By now it must be evident that since nothing is perfect, including transmission line cable, the velocity of propagation must be less than the speed of light. For modern coaxial cable in which the insulating material, often called the dielectric, is gas-injected polyethylene foam, the velocity of propagation is 87 percent. This means that a wave of energy will travel down the line at 87 percent of the speed of light.

If at this velocity of propagation it takes a finite time T for the wave to travel down the line, then at the receiving end the diagram for voltage versus time is shown in Fig. 3.10.

When the switch is closed V volts travels down the line at velocity v to the receiving end. At the receiving end the reflection coefficient of $+1/2$ causes 50 percent of V to be added to V and $V/2$ volts then travels back up the line. When $V/2$ volts reaches the sending end, then 100 percent is inverted and sent back down the line to the receiving end ($K_s = -1$). When $V/2$ volts reaches the receiving end then $+1/2$ or 50 percent is added and sent back up the line. Each time the wave travels the length of the line it takes time T. That is why for the voltage at the receiving end it takes $2T$ for any change to occur after the initial time T. After nearly $9T$ all the reflections have ceased after the closing of the switch.

3.2.5 Standing waves and voltage standing wave ratio

By now it must be evident that when television signals are transmitted on a poorly terminated transmission line, signals are reflected back and forth up and down the line. These reflected signals add to or subtract from the incident signal, causing areas of voltage peaks and voltage minimums. In short, the root mean square (rms) voltage has points of maximums and minimums along the line. This will happen

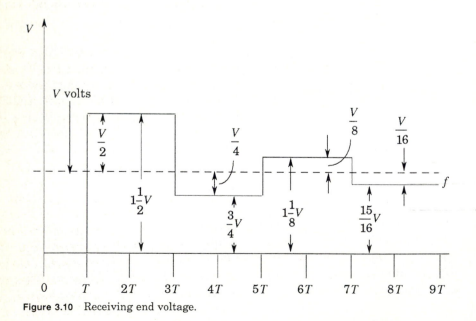

Figure 3.10 Receiving end voltage.

for each radio frequency carrier carried on the cable. The ratio of the maximum value to the minimum value of the alternating rms voltage on the line is defined as the *standing wave ratio* or *voltage standing wave ratio* (VSWR).

Mathematically VSWR = (V_{max}/V_{min}), where V_{max} is the maximum voltage and V_{min} is the minimum voltage.

V^+ is the incident wave and V^- is the reflected wave.

$$|V_{max}| = |V^+| + |V^-|$$

$$|V_{min}| = |V^+| - |V^-|$$

So
$$\text{VSWR} = \frac{|V_{max}|}{|V_{min}|} = \frac{|V^+| + |V^-|}{|V^+| - |V^-|} \times \frac{|V^+|}{|V^+|}$$

$$= \frac{1 + |V^-|/|V^+|}{1 - |V^-|/|V^+|}$$

The reflective coefficient is defined as $K = |V^-|/|V^+|$.

Then

$$\text{VSWR} = \frac{1 + K}{1 - K} = \begin{cases} \infty \ (\text{infinity}) & \text{when } K = 1 \\ 1{:}1 & \text{when } K = 0 \end{cases}$$

When $K_s = 0 = K_r$, we also have VSWR = 1:1, and hence we have a perfect sending and receiving end impedance match. Then

$$\frac{Z_s - Z_o}{Z_s + Z_o} = 0 \quad \text{and} \quad Z_s = Z_o$$

In like manner $Z_r = Z_o$. No reflections will result.

3.2.6 Return loss

Return loss is a measure of the reflection coefficient in decibels. Essentially the return loss is expressed as the ratio of the $|P^+|$, the incident power to $|P^-|$, the reflected power in decibels.

$$RL = 10 \log \frac{|P^+|}{|P^-|}$$

However,

$$|P^+| = \frac{|V^+|^2}{Z_o} \quad \text{and} \quad |P^-| = \frac{|V^-|^2}{Z_o}$$

so

$$\frac{|P^+|}{|P^-|} = \frac{\dfrac{|V^+|^2}{Z_o}}{\dfrac{|V^-|^2}{Z_o}} = \frac{|V^+|^2}{|V^-|^2}$$

$$RL = 10 \log \frac{|V^+|^2}{|V^-|^2} = -20 \log \frac{1}{K}$$

Now for 0.500 in good-quality coaxial cable the nominal impedance is $75 \pm 2\Omega$. Therefore, as an example the return loss for this cable will be calculated.

Assume $Z_o = 77\ \Omega$ $\qquad$ (75 + 2)

$$K_r = K_s = \frac{Z_s - Z_o}{Z_s + Z_o} \qquad Z_s = Z_r = 75$$

So $K = \dfrac{75 - 77}{75 + 77} = \dfrac{-2}{152} = -0.013$

$$RL = -20 \log \frac{1}{0.013} = -20 \log 76.9$$

$$= -20 \times 1.886 = 37.7\ dB$$

Assume $Z_0 = 73\ \Omega$ $\qquad$ (75 − 2)

$$K = \frac{75 - 73}{75 + 73} = \frac{2}{148}$$

So $RL = -20 \log \dfrac{1/2}{148}$

$$= -20 \log \frac{148}{2} = -20 \log 74$$

$$= -20 \times 1.87 = 37.4\ dB$$

Essentially the return loss is 37 dB, which is very good. In practice various manufacturing runs of cable vary from batch to batch. This variation can range from a return loss of 26 dB to nearly 37 dB. Cable with a return loss of less than 26dB really should not be used. Most cable has a return loss between 28 and 32 dB.

Since cable can be manufactured with an impedance tolerance of ± 2 Ω, the reduction in the return loss can be accounted for by the packaging of the cable on the reels, shipping and handling, plus the installation process. Poor installation practices can and unfortunately often do cause considerable damage to the cable, thus causing a poor return loss. When cable is used with a poor return loss (RL), signal voltage reflections are caused along the line and in turn cause signal voltage variations along the length of the line. These signal voltage variations can be observed at taps, or amplifier input/output test points, and the maximums and minimums of signal voltage are a function of the carrier frequencies. Often signal voltage variations are referred to as *peak-to-valley frequency response* or sometimes as *signal ripple*. The worse the return loss the deeper the minimums and higher the maximums. This peak/valley ratio of signal level is due in part to cable with poor return loss as well as taps, splitters, and directional couplers with poor return loss and poor amplifier frequency response.

By now it should be evident that good cable and proper handling and installation procedures are necessary to construct a good cable television system. Often the cable is tested for return loss before installation to ensure that the cable delivered is of the same specification as ordered and that there is no shipping damage. This procedure is quite easy and fast to perform on the reels of cable. This procedure is shown in Fig. 6.10.

Cable installation can use many techniques, and the expertise of the installation crews should be properly researched before any contractual agreements are reached. Once a type of cable has been selected, it is a good idea to consult the cable manufacturer as to their recommendations on handling and installation. Many manufacturers of cable have installation manuals available to their customers. Connector manufacturers also often have installation procedures, tooling, and techniques either in a manual or in a reference bulletin.

In general cable is usually reeled off the top of the reel and a reel brake is used to prevent the reel from freewheeling on the support bar of the reel trailer or reel stand. If the reel is allowed to coast on its axle more than one-half turn, the cable can be spilled out on the ground and kinked.

If the cable is damaged by either crushing or flattening, then the return loss decreases beyond acceptable limits. This is called *structural return loss*, because its cause is a change in the cable structure.

Up to now, cable loss and return loss have been discussed for television frequencies. The cable is also activated by electric energy at 60-V_{rms}, 60-Hz alternating current, which is used to power the line distribution amplifiers. This electric power is inserted onto the trunk cable through a power inserter. In most cases a ferroresonant voltage regulating transformer power supply supplies power to the trunk cable through the power inserter. There are fuses in the power inserter for protection. Usually the power inserter can feed the 60-V, 60-Hz power to two different legs through its two output ports. Since the current-carrying capacity of the cable is limited, the power supplies are limited as well. The standard current output from this type of power supply ranges from 12 to 15 A for a single-cable system and from 18 to 24 A for a dual-cable plant.

3.3 Cable Powering and Loop Resistance

For cable to carry alternating current in these ranges the low-frequency power-carrying capabilities for the cable must be understood. What is known as the *loop resistance* of the cable is the limiting factor. This is another published electrical quantity found on the cable electrical specification. The sample chart in Table 3.2 shows the variation of low-frequency, dc resistance parameters of the cable.

An example of amplifier power distribution is shown in Fig. 3.11.

It is desired to select a cable size to power several amplifiers some distance away. The amplifier needs to obtain a minimum of 45 V_{ac} to operate correctly. Find the cable size.

For 2500 ft. of 0.412 copper clad cable the loop loss will be calculated as

$$\frac{2.5 \times 1000 \times 1.93}{1000} = 4.825 \; \Omega \text{ for 2500 ft}$$

The current drawn through this resistance will drop the voltage by $V = IR$, $4.825\Omega \times 7 \text{ A} = 33.78 \text{ V}$.

Therefore, if 33.78 V of 60 is lost, then only approximately 26.23 V will be left to power the amplifier. Clearly this is less than the minimum required voltage of 45.

Next try solid copper 500 cable, which may be a good compromise. The loop resistance is

$$1.23 \times 2.5 = 3.075 \; \Omega$$

$$\text{Now } 3.075 \; \Omega \times 7 \text{ A} = 21.53 \text{ V}$$

So 60 − 21.53 = 38.48 V still is not enough.

Since copper clad 0.750 cable is less expensive than solid copper 0.750 cable, it will be tried next.

$$\text{Loop resistance } 0.76 \times 2.5 = 1.9 \; \Omega$$

$$\text{Now } 7 \text{ A} \times 1.9 \; \Omega = 13.3 \text{ V}$$

TABLE 3.2   Loop Resistance in Ω/1000 ft at 68°F

	Solid copper center conductor	Copper clad aluminum center conductor
0.412	1.75	1.93
0.500	1.23	1.68
0.750	0.56	0.76

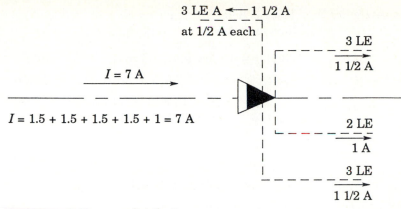

Figure 3.11 Cable power distribution.

$$\text{So } 60\text{V} - 13.3 \text{ V} = 46.7 \text{ V}$$

which will do the job.

If in the preceding situation 0.750 solid copper did not do the job, 0.875 or 1-in cable should do it. However, the cost of a separate power supply, including installation costs and possibly the cost of having the electric utility company install the secondary power lines to the supply may be prohibitive. Hence using 0.750 solid copper center conductor may be the best choice.

3.3.1 Cost factors

Possibly one of the most important cable considerations is the cost. The price of cable and components has a direct impact on the feasibility of building a cable system or extensions to a cable system. The question of using so-called fourth- or fifth-generation cable as opposed to using the tried and true may be resolved by the factor of cost alone.

Suppose now that a cable system of 100 mi with a feeder-to-trunk ratio of 2.5:1 is to be built. An investigation of the cable order quantity and pricing follows. Best-estimate costs will be used.

0.750 gas-injected polyethylene jacketed cable copper clad aluminum center conductor $610/1000 ft

0.500 same type cable $300/1000 ft

0.860 super-type cable $640/1000 ft

0.500 super-type cable $270/1000 ft

From the preceding information one might decide to use the older type of trunk cable and the newer type of distribution cable.

Cost saving for this choice is $30/1000 ft for trunk cable and $30/1000 ft for the feeder cable. Now for a feeder-to-trunk ratio of 2.5:1:

Trunk distance = (100/2.5) = 40 mi, and 100 mi of feeder cable

Cost savings for trunk: 40 × 5.28 × 1000 ft × 30/1000 ft = $6336

Cost savings for feeder: 100 × 5.28 × 1000 ft × 30/1000 = ($15,840). On a per-mile basis 22,176/100 = 221.76/mi. This still may be the best choice.

However, let us proceed further.

40 mi of trunk and amplifier spacing at 22 dB

0.750 loss at 450 MHz 1.10 dB/100 ft

0.860 loss at 450 MHz 0.95 dB/100 ft

For 22 dB of cable the length of each type of cable is computed. This will be the amplifier spacing in feet for 22 dB of cable.

$$0.750: \frac{22 \text{ dB}}{1.1 \text{ dB}} \times 100 = 2000 \text{ ft/A}$$

$$0.860: \frac{22 \text{ dB}}{0.93 \text{ dB}} = 2316 \text{ ft/A}$$

$$40 \text{ mi} \times 5.28 \times 1000 \text{ ft} = 211{,}200 \text{ ft}$$

The number of amps needed for each type is as follows:

$$0.750: \frac{211{,}200}{2000} \times 105.6 \text{ A} = 106$$

$$0.860: \frac{211{,}200}{2316} \times 91.2 \text{ A} = 92$$

Savings in amplifiers (14). If amplifiers are priced at $1500 each, then 14 × 1500 = $21,000.

It must be remembered that the cost savings of using the 0.750 cable was $6336 and the cost savings resulting from needing less amplifiers by using more expensive 0.860 super cable is $21,000 for a difference of $14,664. Also it must be remembered that the reduced number of amplifiers in the trunk, means that fewer connectors are used, maintenance costs are lower, and fewer spare amplifiers are needed. It pays to play with the numbers and do your homework. Do not jump to conclusions.

3.3.2 Mechanical considerations

Another consideration certainly worth discussing is the difference in installation techniques and resulting plant characteristics. The 0.500-

in super-type cable has a weight of approximately 110 lb/1000 ft as compared to 135 lb/1000 ft. for jacketed gas-injected polyethylene-type cable. This amounts to nearly a 23 percent increase in weight for the gas-injected polyethylene cable. Of course, the reason for the difference in weight is mainly the use of a thin wall aluminum outer conductor. This thin wall also results in a smaller bending radius, 5 in for the super cable and 8 in typically for the gas-injected polyethylene cable. Care must be exercised when performing the coring operation during splicing of the thin wall type cable. The coring tool must be of the type recommended or sold by the cable or connector manufacturer and must be sharp and properly adjusted. As in any construction operation a little practice usually pays off. Care must also be used in performing the bending operation when forming the expansion loops. Usually the thin-wall cable can be formed by using the plastic-type bending forms and making the loop by hand, rather than using a Jackson-type bending tool for forming the loops for gas-injected polyethylene type cable. The lighter-weight thin-wall cable in practice seems to withstand the rigors of expansion and contraction quite well and the connectors seem to remain tight, thus preventing signal leakage.

There has been much discussion of the amount of cable used in forming the more widely accepted flat bottom drip loops. A short consideration of the geometry of forming the loops should clarify this question. Refer to Fig. 3.12.

The perimeter of each circle is $\pi \times$ diameter, where $\pi = 3.142$; therefore, $3.142 \times 6 = 18.85$ in. Arc a (in Fig. 3.12) is therefore one-fourth of the perimeter or $\frac{1}{4} \times 18.85$ in $= 4.71$ in. The distance through the curves from x to y is 2×4.71 in $= 9.42$ in. The same calculation for the other side gives another 9.42 in. The amount of cable used from point x to z is $(9.42 \times 2) + 12$ in $= 30.84$ in. The linear distance from point x to z is $(6\frac{1}{2} \times 2) + 12 = 25$ in. So $30.84 - 25 = 5.84$ in more cable is used in forming the usual smooth 6-in-deep, 12-in-long flat bottom expansion loop. This amounts to nearly a 25 percent increase in cable length per number of loops. Therefore, if a loop for

1/2-in cable

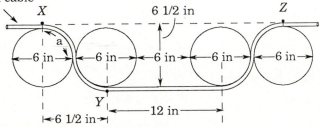

Figure 3.12   Geometry of the flat bottom cable loop.

this size of cable is going to be used at every pole, the cable order should be adjusted beyond the usual 10 percent spoilage allowance for normal construction practices.

Uses of cable as impedance matching devices and circuit elements will be discussed in the section on antennas (in Chapter 4).

3.4   Temperature Factors

So far the discussion of coaxial cable parameters has been limited to the ambient temperature of 68°F (approximately 20°C). Since most temperature measurements are still reported in degrees Fahrenheit instead of degrees Celsius a conversion chart is given in Fig. 3.13.

3.4.1   Loss variation with temperature

Cable parameters do change as a function of temperature. Cable loss increases as temperature increases. Essentially a 1 percent change in loss occurs for every 10°F change.

Percentage change in attenuation in decibels = 1 percent/10°F. Therefore, for a span of cable, change in cable loss in decibels = loss at 68°F × percentage change in cable loss.

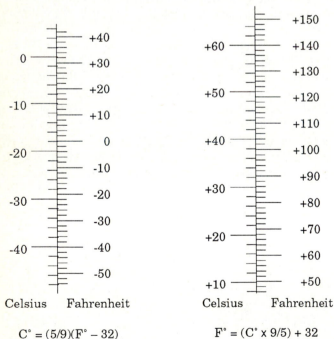

Celsius   Fahrenheit   Celsius   Fahrenheit

$$C° = (5/9)(F° − 32)$$ $$F° = (C° \times 9/5) + 32$$

Figure 3.13

Example Suppose the normal spacing of amplifiers in a cable system trunk system is 22 dB of cable loss at 68°F. When the temperature decreases to +40°F at night what will the span attenuation become?

The temperature change is 68°F − 40°F = 28°F decrease. Use −28°F.

$$\% \text{ Change in attenuation } \frac{28°F}{10°F} = -2.8 \text{ percent}$$

$$\text{Change in 22-dB span } = -0.028 \times 22 = -0.616 \text{ dB}$$

Now the 22 dB cable span becomes

$$22 - 0.62 = 21.36 \text{ dB} \qquad 21.4 \text{ dB}$$

If the amplifier gain is fairly constant, the signal on the input and output will be 0.62 dB high. In most modern cable systems built in the last 15 years thermal compensation will adjust the amplifier gain to essentially eliminate any change in signal level. However, if the temperature change is very drastic, then the better control systems known as automatic gain control (AGC) and automatic slope control (ASC) should be used.

It is now apparent that if the temperature increases 28°F the change will be +0.62 dB and the span loss will increase to 22.62 dB. To counteract this loss increase, the gain will have to be increased by the same amount. For such a swing of +28°F and −28°F, which spans a total of 56°F, the loss will be twice 0.62, or 1.24 dB total change. During winter and summer in certain parts of the country the seasonal temperature can change as much as 70°F in a 24-h period.

3.4.2 Loop resistance variations

Another cable parameter in which one would expect some variation with temperature is the loop resistance. Again, there is an increase in resistance with increase in temperature. For a new temperature t (°F), loop resistance in ohms/1000 ft at t = loop resistance in ohms/1000 ft at 68°F × (1 + 0.0022 × [−68°F]).

Example Recall that for 0.750 gas-injected polyethylene cable with a copper clad aluminum center conductor the loop resistance was 0.76 Ω/1000 ft at 68°F. If the temperature increases to 98°F, the change in temperature from −68°F is 98° −68° = 30°F. New loop resistance at 98°F = 0.76 (1 + 0.0022 × [30]) = 0.76 (1 + 0.066) = 0.76(1.066) = 0.81 Ω/1000 ft.

Recall in the previous example for a 2500-ft span of 0.750 cable with the copper clad aluminum center conductor that the loop loss was

$0.76 \times 2.5 = 1.9 \, \Omega$ at 68°F, and if this span would have to carry 7 A of current then the voltage drop $7 \times 1.9 = 13.3$ V will supply $60 - 13.3 = 46.7$ V to the amplifiers. Also recall that a minimum of 45 V was needed for the internal amplifier voltage regulators to remain within the normal regulation limits.

Now at this new temperature of 98°F the loop resistance should be recalculated to determine whether the amplifiers will be starved for power.

Now the loop resistance will be increased to $0.81 \times 2.5 = 2.025 \, \Omega$, and for 7 A the drop will be $7 \times 2.025 = 14.175$, so $60 - 14.825$ or 45.825 or 45.83 V.

Most likely the amplifiers would keep operating at this voltage, which is near the limit. However, I think this is too close for comfort. So the more prudent choice would be either to use the solid copper center conductor 0.750 cable or relocate the power supply or add another separate unit. Again prudent designers should make all calculations accurately so all alternatives can be explored and the best choices made.

3.4.3   Cable expansion and contraction

Last but not least, the cable physically expands with increase in temperature. Most metals have what is known as a *positive temperature coefficient of linear expansion*: That is, the hotter they get the longer they get. Since most of the metal used in modern coaxial cable is aluminum, the coefficient of linear expansion for aluminum will be the factor governing the cable expansion.

The coefficient of linear expansion for aluminum is 24×10^{-6}/°C which means that the change in expansion is 24 parts per million per ° Celsius change in temperature.

Since the Celsius degree is larger than the smaller Fahrenheit degree, the coefficient of linear expansion must be a smaller number, hence 5/9 of the 24 parts/million/°C. Mathematically $24 \times 10^{-6} \times (5/9) = 13.33 \times 10^{-6}$/°F.

Example Suppose a span of 2000 ft of aluminum coaxial cable goes through a temperature change of 0°F to +40°F during the winter months. Find the amount of cable expansion that will occur.

$$\Delta l = \alpha l \Delta t$$

where Δl = amount of expansion
l = length of span before temperature change
Δt = change in temperature
α = coefficient of linear expansion

Substituting:

$$\Delta l = 13.33 \times 10^{-6}/°F \times 2000 \text{ ft} \times 40°F$$

$$= 13.33 \times 10^{-6} \times 2 \times 10^3 \times 40$$

$$= 26.66 \times 10^{-3} \times 40 = 1066.4 \times 10^{-3} + 1.0664 \text{ ft}$$

$$= 1.0664 \text{ ft} \times 12 \text{ in/ft} = 12.8 \text{ in}$$

Thus 12.8 in is the amount of elongation that will occur in a 2000-ft span undergoing a 40°F change in temperature.

Including seasonal changes of 90°F, the overall change in length due to expansion and contraction can be calculated.

$$\text{Total change} = 13.33 \times 2000 \times 90°F = 26.66 \times 10^{-3} \times 90$$
$$= 2399.4 \times 10 - 3 = 2.3994 \text{ ft}$$

If the cable was installed at 68°F and if a +40°F change occurs during the summer, 68 + 40 = 108°F, then the cable will elongate 12.8 in. When the temperature decreases 40°F from 68°F, 68 − 40 = 28°F, the cable will shrink 12.8 in. The usual flat bottom expansion loops properly placed at each device on the strand will take care of these amounts of change in length.

The coefficient of linear expansion for the steel messenger strand is less than that for aluminum (about half as much). So if the change in length of the strand accounts for one-half of the change in aluminum, then sag in span will occur. Now the added change in length of the aluminum cable will be a force on the expansion loop. If in the 2000-ft span there are approximately 65 poles and 25 expansion loops, then each one will only have to account for approximately ½ in. Since the lashing process usually allows for some cable movement, all of the forces should be distributed so the cable will not cause excessive forces on the cable connectors. Proper connector design should be able to withstand the pushing due to expansion and pulling due to contraction, with the forces adjusting the shape of the expansion loops.

When selecting cable for either a new build or rebuild effort should be made to study the manufacturer's specification information. If the loss charts provided by the manufacturer do not give enough frequency values a curve can be plotted to make interpolation between values easy or the formula which makes a fair approximation may be used.

$$\frac{L_1 \text{ at } f_1}{L_2 \text{ at } f_2} = \sqrt{\frac{f_1}{f_2}}$$

where L_1 at f_1 = loss at frequency f_1 (dB)
$\quad\quad L_2$ at f_2 = loss at frequency f_2 (dB)

Example   Suppose the loss of cable at 300 MHz is 1.31 dB/100 ft for 0.500 in gas-injected polyethylene and the loss at 400 MHz is desired. Using the formula

$$\frac{L_1}{L_2} = \sqrt{\frac{f_1}{f_2}}$$

Let 300 MHz = f_2 and $L_2 f_2$ = 1.31 dB/100 ft.

$$\frac{L_1}{1.31} = \sqrt{\frac{400}{300}} = \sqrt{\frac{4}{3}} = 1.155 \qquad L_1 = 1.155 \times 1.31 = 1.5127 \text{ dB}$$

Rounding off, the value is 1.513 dB. From the chart given previously, the value at 400 MHz is 1.53, which is quite close. The difference is 1.530 − 1.513 = 0.017;   percentage   error = (0.017/1.53) × 100 = 0.011 = 1.1 percent error.

Everyone should now be satisfied that this error is indeed small, so the approximate formula is accurate for all practical purposes.

3.4.3.1 Summary.   The coaxial cable in any broadband cable communications system is clearly one of the most important components in the system. It makes up one-half of the cable-amplifier unity gain building blocks. The cable should be chosen by taking into account its electrical qualities, its physical or mechanical qualities, and its costs. The cable connector marriage, so to speak, is equally important. So far almost no manufacturers of cable also make connectors. It definitely is a must to solicit samples of cable and connectors and use properly specified cable connector preparation tools to practice installing the sample connectors on the cable.

It is necessary to find and identify as many present and future problems as possible before construction begins. The proper connectors, tooling, and cable are probably the most important decisions to make before design and construction begin. The frequency response, signal peak-to-valley response, ingress (noise in system)/egress (leakage from cable), carrier-to-noise, and system longevity and maintenance costs all depend greatly on this decision. Cost alone should never be the deciding criterion.

3.5   Cable Measurements

3.5.1   Attenuation tests

To test cable for attenuation a simple sweep system can be used. This sweep system can be of the same type used on the cable system, i.e., a simultaneous sweep system or a sweep recovery system. A normal bench-type sweep system used for amplifier alignment or a generator and signal level meter can be used. This latter method takes more time but produces accurate results. The equipment setup for this measurement is shown in Chapter 6, Fig. 6.10, where the device under test is a roll of cable.

3.5.2 Return loss

The return loss test requires the use of a bridge device. One manufacturer provides a wideband noise generator as a signal source to drive the bridge and a bridge circuit all in one instrument. The equipment connections for this measurement are shown in Fig. 3.14.

The procedure is as follows: First, connect signal level meter (SLM) to port 1; on port 2 set the signal level meter to full scale by setting the attenuators on SLM. Second, disconnect SLM from port 1 and connect it to port 2. Rotate the SLM dial over the desired bandwidth for the return loss measurement, reading the return loss directly from the meter scale. If pads have to be removed to keep an on-scale reading, add the pad values to the meter scale reading.

When using the noise generator with a return loss bridge, the return loss is taken from the impedance mismatch of the device under test from the 75-Ω standard. Refer to Chapter 6 for the mathematical basis of this test. One manufacturer of test equipment produces a noise generator and bridge combined in a small portable system which is very handy for field work.

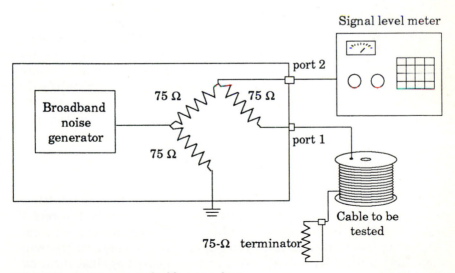

Figure 3.14 Measurement of cable return loss.

3.5.3   Time domain reflectometer

3.5.3.1   Distance fault tests.   The time domain reflectometer (TDR) appears on the commercial market in both analog and digital forms. Many engineers and technicians feel that in today's technical world the digital methods are the best. Not necessarily so. The basic principle of operation is similar to the radar principle, in which a pulse of energy is transmitted toward a target and an echo is received in return. Since the returned energy has to make a round-trip, i.e., twice the distance, and the speed of transmission is the same both ways, the elapsed time is divided by 2 and multiplied by the speed and the distance is then calculated. In the time domain reflectometer a pulse of known energy and shape is sent down a transmission line, and if an open- or short-circuited cable is found the energy is reflected toward the instrument. When the echo arrives at the sending end, the instrument has counted clock pulses and totaled the sum from the transmitted pulse to the returned pulse. The count of these pulses is a function of distance from the instrument to the fault (open or short circuit). This count is modified by a scale factor to take into account the propagation constant of the cable.

In its digital form, the propagation constant is entered in the instrument usually by thumb-wheel decimal digits in percentage velocity. Then the instrument is connected to the cable to be measured via clip leads or coaxial connector and turned on. The distance to the fault is read on a light-emitting diode (LED) or liquid crystal display (LCD) numerical display, usually in feet, hundreds of feet, or meters. A LED indicator denotes open or short circuit.

The analog instrument displays the returned pulse on an oscilloscope display. Usually the positive return pulse indicates an open circuit and the negative one a short circuit. Starting at the transmitted pulse on the display the distance dial is rotated, increasing the distance down the transmission line as shown on the display. When the echo pulse seen on the display is placed on the same scope line as the transmitted pulse, the distance to the fault can be read on the dial. This dial reading has to be multiplied by the appropriate scale factor to arrive at the correct distance. The display will also show any small abnormalities in the cable that will usually be ignored on the digital-type instrument.

Some TDR manufacturers of digital instruments have connections for an oscilloscope, thus providing the best of both worlds. If a reel of cable is measured with either a short circuit or the opposite end unconnected, a TDR will then give the distance of cable on the reel. The length printed on the cable reel can be checked against the measured value. If the velocity of propagation is not known, it can be measured by laying out a precisely known length of cable and measuring this length with the TDR. The velocity of propagation dial is set so the measured lengths agree. This value should be recorded for future use. The connection for TDR use is shown in Fig. 3.15.

3.6   Cable Handling and Transportation

Cable is delivered to a user on wooden reels, often with a cardboard covering. One end of the cable protrudes from an access hole near the center of the reel and a plywood patch usually covers the area. To test the cable on the reel the patch is removed and a connector installed

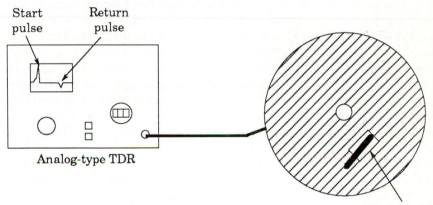

Test for length of cable on a reel

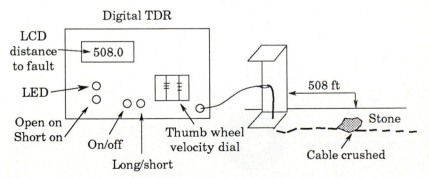

Figure 3.15   Measurement of distance to cable fault for buried plant.

with a terminator in place. The cardboard covering can then be removed and another connector installed on the cable end so the test equipment can be connected. Admittance testing for attenuation (sweeping) and return loss (bridge) can assure the end user that the cable has survived the trip from the manufacturer. Reels of cable should never be dropped from the tailgate of a truck or a loading dock because a fall can affect the return loss adversely. Proper unloading should involve a forklift truck, a ramp, or a hoisting technique.

To transport the cable reels to the job site a vehicle-hauled reel trailer is used. Such a trailer is shown in Fig. 3.16. A reel of cable is loaded onto the reel trailer by first placing the steel pipe axle through the center hole of the reel. Then the trailer is tipped up by lifting the hitch end so the "U" brackets connect on the pipe. Placing the locking pins in U bracket and lowering the trailer tongue will fasten the reel in place as the trailer is brought down to its hauling position. The sketch in Fig. 3.17 demonstrates the procedure. The cable is taken from the top of the reel, not the bottom, to prevent kinking the cable if the reel does not stop coasting immediately. Spring-loaded reel brakes keep enough friction on the reel flanges to prevent coasting of the reel. Good construction crews well versed in pole line construction techniques are necessary to assure a good and safe job. Often it may be desired either to TDR test the cable or to sweep test it before splicing commences, particularly if the cable mounting and lashing procedure has been difficult. This will indicate whether the cable has sustained any damage.

3.6.1 Aerial construction

Once the utility poles have been cleared by the electrical utility company and the telephone company, i.e., the pole owners, and released to

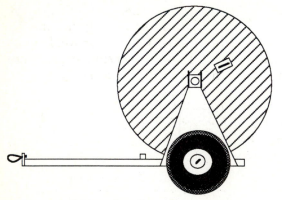

Figure 3.16   Cable reel trailer.

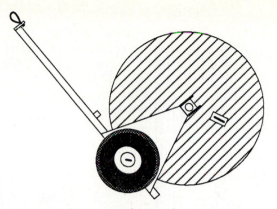

Figure 3.17   Tip up to pick up reel.

the cable television company, construction can begin. The cable television strand crews usually climb the poles and measure with the preferred wooden or plastic folding ruler where the location of the three-
bolt clamp will be. Then a hole is bored and a proper-length standard
⅝ inches galvanized machine bolt with a square washer is placed
through the hole. The three-bolt clamp is then fastened to the pole
with a ⅝-in machine nut. The two clamping bolts and the ⅝-in front
nuts are left loose so the strand can then be passed through. Usually a
good section of a street will have the poles framed with the three-bolt
clamps. Three-bolt strand clamps are either straight or curved as the
need may be. A straight three-bolt clamp is shown in Fig. 3.18, with
the pole arrangement showing straight-line pole-to-pole construction.

After the galvanized steel messenger strand used to support the cable has been placed in the clamps, one end is fastened and a winching
system, often called a "come along" or coffing hoist, is used to tension
the strand. This winching system is a hand lever operated device, with
which the expertise of the operator determines the strand tension. A
slight sag in the tensioned strand should be seen by a visual inspection of the tensioned strand. A more sophisticated method would be to
use one of the more common dynamometer-type tension gages. Once
the tension is set, the other end of the strand is fastened to a pole eye
bolt by means of a dead-end strand splice. This is a preformed piece of
strand formed in an "eye" configuration; the ends are flattened and
formed to fit the end of the strand. Sand is cemented to these flattened
preformed ends so when the strand end is wrapped with the preformed
ends a good tight grip will be formed. These eye-type dead-end strand
connections, called *preforms*, have a variety of configurations and
strand sizes. The usual configurations are dead-end eyes and strand
splices. The usual strand size used in the cable television industry is

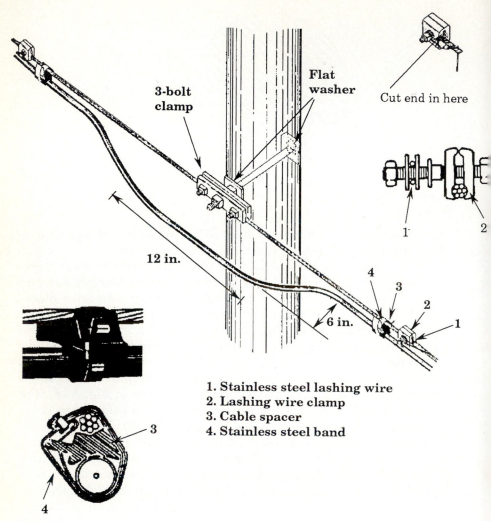

3-bolt clamp

Flat washer

Cut end in here

12 in.

6 in.

1. Stainless steel lashing wire
2. Lashing wire clamp
3. Cable spacer
4. Stainless steel band

Figure 3.18 Pole hardware for straight pole-pole construction.

¼ in or ⁵⁄₁₆ in; the more common size is the ¼-in. The steel content determines the maximum tensile strength of the strand; the usual classifications are standard high strength and extra high strength. The minimum tensile strength for the extra high strength strand is 6650 lb for the ¼-in strand and 11,200 lb for the ⁵⁄₁₆-in type. The usual length of strand on a reel is about 5000 ft, which is nearly 1 mi; weight for ¼-in strand extra high strength is about 650 lb. Care in lifting and handling should be observed so injuries to personnel can be prevented. A strand and cable ending on a guyed pole (dead end) is shown in Fig. 3.19, with the hardware types indicated. Various assorted pole arrangements with the appropriate hardware are shown in Figs. 3.20 through 3.24.

The cable is lashed to the steel messenger strand by means of a *lasher*, which spins stainless steel lashing wire around the cables and the strand. The ends of the lashing wire are fastened to the strand with lashing wire clamps. To separate the cable and prevent it from rubbing against the strand at the drip loops (expansion loops), a plastic spacer is used. It is fastened to the cable and strand with a stainless steel strap. These are known as *bands* and *spacers*.

To place the cable on the poles, rollers are mounted and locked on the strand so they will not fall off. The cable is pulled from the cable reel mounted on the reel trailer and hung on the rollers. One method of performing this task is shown in Fig. 3.25. Usually several pole spans are prepared in this manner before lashing begins.

The lashing procedure is shown in Fig. 3.26. Of course, there may be several variations on these procedures, and some are easier than others, depending on availability of personnel and equipment. The main point to remember is that construction can be and is often hazardous and safety precautions must be taken at all times.

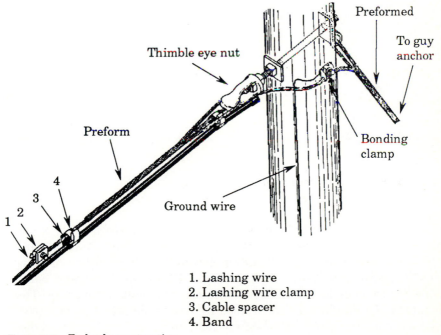

1. Lashing wire
2. Lashing wire clamp
3. Cable spacer
4. Band

Figure 3.19   End pole construction.

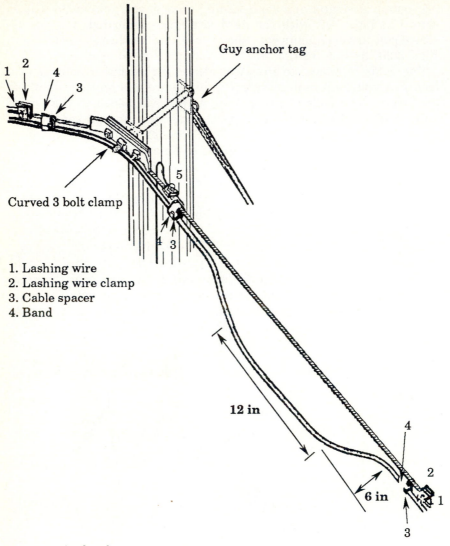

Figure 3.20 Angle pole.

The grounding and bonding of the steel strand are usually made with soft drawn No. 6 copper single-strand wire clamped to the strand and the Telco ground with bonding clamps often referred to as "bug nuts" because they resemble bugs. Fig. 3.19 shows the strand being grounded to a vertical ground rod because this is an ending pole. The usual rule for bonding as understood in many pole attachment agreements is bond to Telco plant the first, last, and every tenth pole. Sometimes at the end pole, if no Telco plant is available the cable company

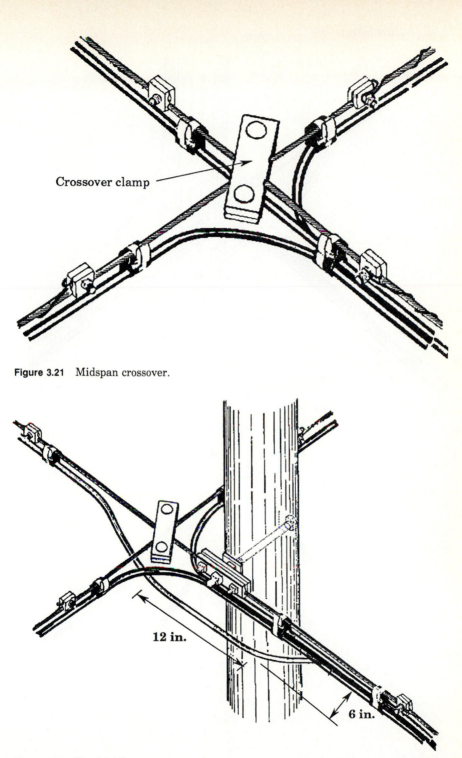

Crossover clamp

Figure 3.21 Midspan crossover.

12 in.

6 in.

Figure 3.22 Straight line crossover.

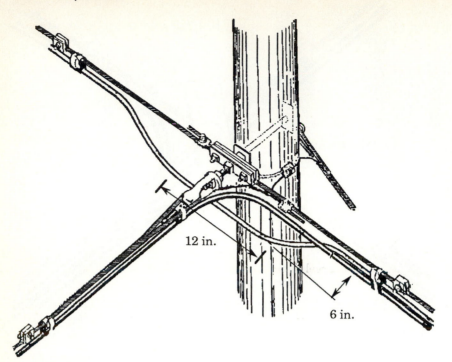

Figure 3.23   Dead-end combination.

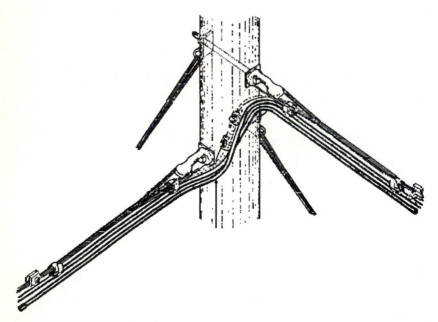

Figure 3.24   Double dead end.

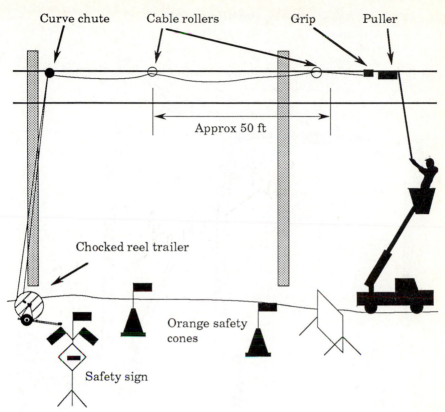

Curve chute   Cable rollers   Grip   Puller

Approx 50 ft

Chocked reel trailer

Orange safety
cones

Safety sign

Figure 3.25   Cable pulling procedure.

drives a ground rod called a *vertical ground*. Proper grounding and bonding procedures should be followed according to the pole attachment agreement so the electrical, telephone, and cable plants can exist together with damage due to lightning and power surges minimized.

3.6.2 Underground cable plant

Two features account for the increased cost of buried cable plant. One is that the cost of opening and closing a trench along with any restoration work, such as loaming, seeding, and raking, is quite high compared to that of aerial plant. The second factor is the road crossings, which either require boring under the road and placing conduit or cutting and patching the road, which is very expensive. Another factor adding to the cost is that more cable is used to cover both sides of the road, and in this case more taps, passives (couplers and splitters), and amplifiers are also often needed. Before any underground construc-

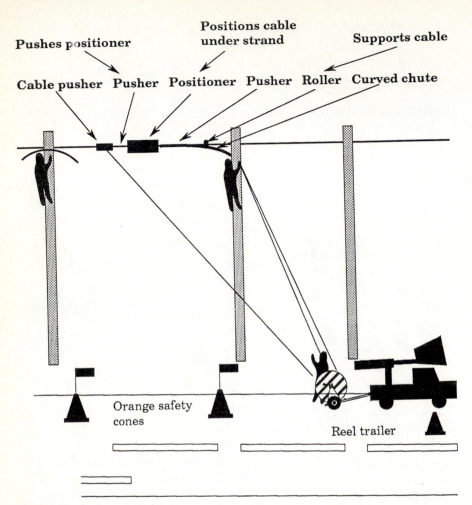

Figure 3.26 Lashing procedure.

tion, the telephone company, electrical company, gas company, and community water and sewer departments have to be notified of the proposed work and time schedule. Once notified, these utility companies and the municipal water and sewer departments have the responsibility of locating and marking on the ground the location of their plant by appropriately color-coded spray paints. A cooperative agency providing service to underground utility companies is called "Dig Safe," which when contacted by anyone desiring to install underground plant will contact all the other utilities, thus acting as a clearinghouse for the necessary notification process.

3.6.3 Trenching standards and methods

If a cable television company is given permission to place a trench containing only their own plant, usually a 4- to 6-in-wide trench is made with one of the fairly common trenching machines to a depth of 18 to 24 in. When a suitable length of trench has been opened, the bottom should be cleaned of any stones. Then a pad of sand should be placed in the bottom of the trench. The cable can be pulled out from the reel mounted on the reel trailer and then placed on the ground near the trench. When the trench bottom has been prepared, the cable is carefully placed in the trench, followed by about 4 in of stone-free sand or gravel. If the trenching machine is fitted with a backfill blade, then the earth removed during the trenching operation can be used to backfill the trench. The ends of the cable are brought up from the trench in a smooth curved arc into a pedestal enclosure. Neoprene caps should be fitted over the cable ends to keep out moisture. Fig. 3.27 illustrates the placement of cable in a trench with the appropriate pedestal hous-

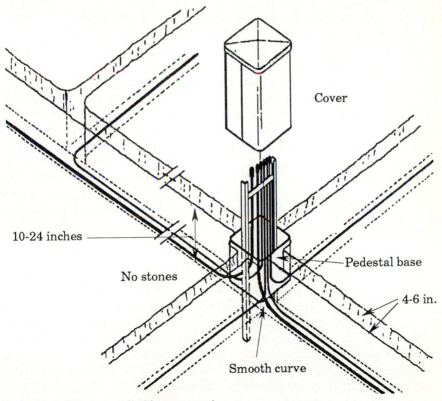

Figure 3.27 Placement of cable in a trench.

ing. Some trenching machines have a boring attachment that enables the machine to do double duty. In some instances where the soil conditions are sandy, the cable can be directly buried by plowing it into the ground. Most cable plowing machines are the vibratory type, in which a vibrating blade is placed in a starting hole; when the machine is moved forward, this blade loosens the earth, allowing a stationary blade with the cable chute to place the cable underground. If conduit is desired, cable already placed in flexible conduit can be purchased and then can be vibratory plowed. The plowing procedure is generally lower in cost because no backfill is needed, nor in many cases restoration. Generally the plowing process works well in rural areas. In cities buried cable plant costs can skyrocket. Pavement sawing, conduit, concrete slurry backfill, and road surface restoration are some of the reasons. In some instances the telephone company may have conduit space available. The cable company and/or their contractors have to work in manhole and vault areas using expensive techniques and equipment. Conduit rental costs can be very expensive. Buried plant often has minimum make-ready costs and no pole rental. However, the up-front cost can be and usually is large. Grounding and or bonding still has to be used for underground plant, particularly in lightning-prone areas. Often the cable company has to use more ground rods than it does for aerial plant.

In some instances when housing areas are developed it is advantageous to use a trench shared with the other utilities. Often water and sewer lines will be on one side of the street, where sewer lines may be located at a 6-ft depth and water at a 4-ft depth. Where a utility trench is shared by the cable company and the power and telephone company the trench configuration may look like the one shown in Fig. 3.28. Shared use of a utility trench is often beneficial to all the companies concerned mainly because of the cost saving. The work is usually done jointly, with each company's methods and rules observed and learned.

3.6.4 Underground subscriber drops

The feeder cable is brought up from the cable trench into a pedestal containing a tap. Such a pedestal containing a tap and a line extender amplifier is shown in Fig. 3.29. The subscriber drop cable should be the flooded-direct burial type. This cable can be plowed by using a smaller variety of vibratory plow available on the market today or can be placed in a small 2-in-wide by 6- to 8-in-deep trench made by one of the small light-duty trenchers in common use today. Some cable companies allow subscribers to dig their own trench or cut a 6-in-deep slice using a lawn edger or flat spade. Cable can be forced into the slit

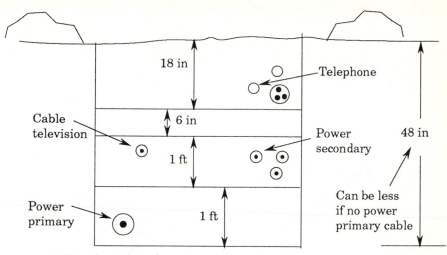

Figure 3.28   Shared trench configuration.

with a thin piece of wood. Once the drop cable is placed under the subscriber's lawn, it can be threaded into the pedestal and connected to the tap by using a watertight "F" fitting. The cable enters the subscriber's house by the appropriate ground block in the usual manner.

3.7   Amplifier Cascade Theory

The word *amplifier* usually indicates that some input signal to a device emerges at the output point larger than at the input: amplification occurred. Unfortunately the input signal may contain noise mixed in with the intelligence which will also be amplified. When amplifiers are cascaded together through a cable system the noise also builds up. To investigate this phenomenon one should start with a simple system, as shown in Fig. 3.30.

The noise figure of the device is simply given as

$$\text{Pure number } F = \dfrac{\dfrac{S_i}{N_i}}{\dfrac{S_o}{N_o}} \quad \begin{matrix} \leftarrow \text{Input signal-to-noise ratio} \\[1em] \leftarrow \text{Output signal-to-noise ratio} \end{matrix} \tag{3.1}$$

Rewriting Eq. (3.1):

$$F = \frac{S_i}{N_i} \times \frac{N_o}{S_o} = \left(\frac{N_o}{N_i}\right)\left(\frac{S_i}{S_o}\right)$$

The gain of the device is defined as (S_o/S_i). Rewriting Eq. (3.1) again

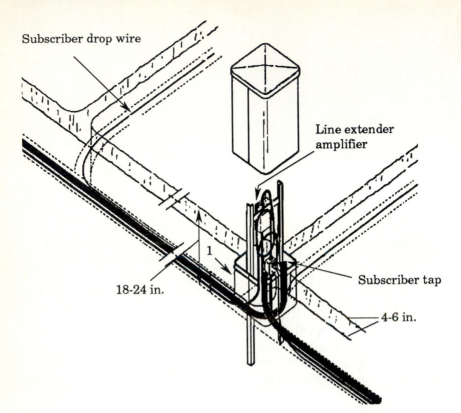

Figure 3.29 Underground distribution system.

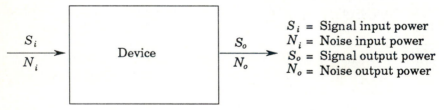

S_i = Signal input power
N_i = Noise input power
S_o = Signal output power
N_o = Noise output power

Figure 3.30 Signal and noise power for a device.

$$F = \frac{N_o}{N_i} \times \frac{1}{G}$$

If the device were perfect, the output noise would simply be the input noise multiplied by the device gain G. However, in the real world the device will always contribute some noise N_n. Therefore, we may write

$$N_o = N_i(G) + N_n \tag{3.2}$$

Combining Eq. (3.1) and Eq. (3.2) and calculating F the noise figure

$$F = \frac{N_i(G) + N_n}{N_i(G)} = 1 + \frac{N_n}{N_i(G)} \tag{3.3}$$

Since cable people always deal in decibels the noise figure can be written simply as

$$F_{dB} = 10 \log F$$

Suppose several devices are cascaded together as in Fig. 3.31, which shows three devices. The overall noise figure of the system is

$$F = \frac{N_o}{N_i(G)}$$

The output noise power N_o can be calculated as

$$N_o = N_i(G_1)(G_2)(G_3) + N_{n_1}(G_2)(G_3) + N_{n_2}(G_3) + N_{n_3}$$

Overall $G = (G_1)(G_2)(G_3)$

We may write

$$F = \frac{N_i(G_1)(G_2)(G_3) + N_{n_1}(G_2)(G_3) + N_{n_2}(G_3) + N_{n_3}}{N_i(G_1)(G_2)(G_3)}$$

Writing a different way by dividing the terms in the numerator by the denominator:

$$F = 1 + \frac{N_{n_1}}{N_i G_1} + \frac{N_{n_2}}{N_i(G_1)(G_2)} + \frac{N_{n_3}}{N_i(G_1)(G_2)(G_3)} \tag{3.4}$$

Recall Eq. (3.3). This is the noise figure for the first stage. Therefore,

$$F_1 = 1 + \frac{N_{n_1}}{N_i G_1}$$

In like manner we may write

$$F_2 = 1 + \frac{N_{n_2}}{N_i G_2}$$

and

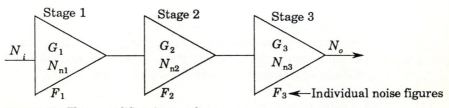

Stage 1   Stage 2   Stage 3

N_i | G_1 | G_2 | G_3 | N_o
N_{n1} | N_{n2} | N_{n3}
F_1 | F_2 | $F_3 \leftarrow$ Individual noise figures

Figure 3.31   Three amplifiers in cascade.

$$F_3 = 1 + \frac{N_{n_3}}{N_i G_3}$$

Rearranging $\qquad F_1 - 1 = \dfrac{N_{n_1}}{N_i G_1} \qquad$ and $\qquad F_2 - 1 = \dfrac{N_{n_2}}{N_i G_2}$

and

$$F_3 - 1 = \frac{N_{n_2}}{N_i G_3}$$

Substituting these values in Eq. (3.4) we then have a simpler cascade formula for three stages.

$$F = F_1 + \frac{F_2 - 1}{G_1} + \frac{F_3 - 1}{G_1 G_2}$$

so for n number of stages the more general formula is

$$F = F_1 + \frac{F_2 - 1}{G_1} + \frac{F_3 - 1}{G_1 G_2} + \cdots + \frac{F_n - 1}{G_1 G_2 \cdots G_{n-1}} \qquad (3.5)$$

An example should clarify the relationship. For a two-stage amplifier the values are stage 1, $G_1 = 20$ dB, $F_1 = 3$ dB, and stage 2, $G_1 = 15$ dB, $F_2 = 4$ dB. In Eq. (3.5), the noise and gain are pure numbers. Therefore, one has to convert the decibel values to pure numbers.

Stage 1		Stage 2	
$F_{dB} = 3$ dB	$G_{1\,dB} = 20$ dB	F_2 dB $= 4$ dB	G_2 dB $= 15$ dB
$3 = 10 \log F_1$	$20 = 10 \log G_1$	$4 = 10 \log F_2$	$15 = 10 \log G_2$
$\log F_1 = 0.3$	$\log G_1 = 2$	$\log F_2 = 0.4$	$\log G_2 = 1.5$
$F_1 = $ antilog 0.3	$G_1 = 100$	$F_2 = 2.5$	$G_2 = 31.6$
$F = 2$			

Now these values are substituted in Eq. (3.1).

$$F = 2 + \frac{2.5 - 1}{100} = 2 + \frac{1.5}{100} = 2.015$$

$$F_{dB} = 10 \log 2.015 = 10 \times 0.304$$

Notice that the gain for stage 2 was not used in this calculation.

If these amplifiers are connected by a piece of cable, then this would represent a negative gain in the calculation. It should also be noted that the first stage contributes most to the overall noise figure.

3.7.1   Noise in cable television systems

To study noise problems more completely, consider the following simple diagram of a resistor at 68°F (about 20°C) in Fig. 3.32. In general,

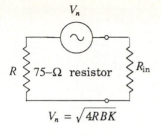

$$V_n = \sqrt{4RBK}$$

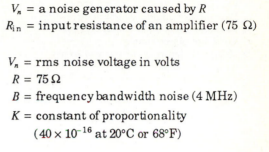

$R = 75\text{-}\Omega$ noiseless resistor

$V_n =$ a noise generator caused by R

$R_{in} =$ input resistance of an amplifier ($75\ \Omega$)

$V_n =$ rms noise voltage in volts

$R = 75\ \Omega$

$B =$ frequency bandwidth noise (4 MHz)

$K =$ constant of proportionality

$\quad$ (40×10^{-16} at $20°C$ or $68°F$)

Substituting values

$$V_n = \sqrt{4 \times 75 \times 4 \times 40 \times 10^{-16}}$$

$$= \sqrt{4.87 \times 10^{-12}}$$

$$= 2.2 \times 10^{-6}\ V$$

$$= 2.2\ \mu V$$

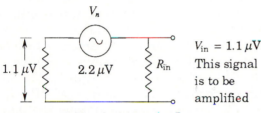

$V_{in} = 1.1\ \mu V$

This signal
is to be
amplified

Figure 3.32 Cable television noise floor.

hot resistors generate more noise than cold ones. This noise is due to molecular movement of the substance of which the resistor is made, which increases as temperature increases.

It should be evident that if $2.2\ \mu V$ is generated, then half of this value appears across each resistor. Since in cable systems signal levels are represented by decibel-millivolts (dBmV) one can refer to a conversion chart or calculate the value by the formula

$$\text{dBmV} = -20 \log \frac{1\ mV}{0.0011\ mV} = -20 \log 909.1$$

This is a negative value because the measure value is less than the 1-mV reference. The 1-mV reference is now placed in the numerator. $\text{dBmV} = -20 \times 2.959 = -59.17$ dBmV ≈ -59 dBmV.

It should be evident that if the voltage were $1\ \mu V$, the level would be -60 dBmV. Also, $+60$ dBmV $= 1\ V = 1{,}000{,}000\ \mu V$. This value of -59 dBmV is the so-called noise floor for $75\text{-}\Omega$ cable systems.

The Television Allocations Study Organization (TASO) set up by the FCC, studied noise by a viewing test using identical television receivers and many viewers. For the approximate television video signal bandwidth of 4 MHz, the following signal-to-noise (S/N) ratios were established.

TASO rating	S/N ratio
1. Excellent picture quality (no snow)	45 dB
2. Fine picture quality (slight snow)	35 dB
3. Passable (acceptable) picture quality (snow not objectionable)	29 dB
4. Marginal picture quality (snow somewhat objectionable)	25 dB

The term *signal-to-noise ratio* indicates the signal power divided by the noise power; in the units of decibels the power ratio becomes carrier to noise

$$(C/N) \text{ dB} = 10 \log \frac{P_{signal}}{P_{noise}}$$

Since sources of noise accumulated by the cascade of amplifiers build up, the minimum S/N ratio at the end of the system can cause unacceptable picture quality. Because slight snow or picture degradation can be seen at 35 dB, the FCC minimum specification is set at 36 dB. This is the overall carrier to noise from CATV antenna to subscriber's television receiver, so the CATV cascade usually is limited to a 42-dB carrier-to-noise ratio. This allows a 6-dB margin for the antenna preamplifier modulator or head-end noise sources (42 dB − 36 dB = 6 dB). For HDTV systems with the wide-screen format the snow effect caused by noise at a 42-dB value will be unacceptable. HDTV cable systems are going to have to hold the carrier-to-noise ratio to approximately 47 dB at minimum.

The noise accumulated in a system cascade (series string of repeater amplifiers) is essentially caused by the amplifiers themselves. The amount of noise contributed by the amplifiers is defined as the *amplifier noise figure.*

Consider the elementary circuit for a properly terminated amplifier in Fig. 3.33. The noise input at best is −59 dBmV. If the amplifier

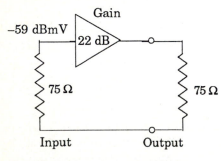

Noise output level = −30 dBmV

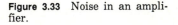

Figure 3.33 Noise in an amplifier.

were perfect and did not generate any noise then the noise output would be $-59 + 22 = -37$ dBmV. However, the measured noise level in our example is -30 dBmV. Therefore, 7 dB of noise is contributed by the amplifier so the amplifier noise figure is said to be 7 dB. Mathematically, $-30 - (-37) = -30 + 37 = 7$ dB. This was referred to as F_{dB} previously.

Now we can define the *noise figure* as the difference between the measured noise output level and the thermal noise (-59 dBmV) plus the gain of the amplifier.

To see how the accumulated noise increases, consider the example of two identical amplifiers connected by a piece of coaxial cable shown in Fig. 3.34.

Recall that the noise figure of an amplifier is the amount of noise contributed by that amplifier and that the noise input to the terminated amplifier is -59 dBmV. Therefore, the noise output of the first amplifier is -59 dBmV (noise) + G_{dB} (gain) + NF_{dB} (noise figure). At the input to the second amplifier the cable attenuates the noise by G_{dB}. The second amplifier, assuming it and the cable are at the same temperature, will again have its own noise generator NF.

To see how power adds in a system, consider the identical power generators feeding a common load in Fig. 3.35. Essentially, if power is doubled, the increase is 3 dB. The input noise to the second amplifier is $-59 + (NF + 3$ dB) and at the output it is $-59 + NF + 3$ dB $+ G$. Therefore, the noise figure of two identical amplifiers in cascade is 3 dB higher than the noise figure of one amplifier since we have doubled the number of amplifiers in cascade. This can be proved another way by using $F = F_1 + (F_2 - 1)/G_1 \ldots$, for example, as shown in Fig. 3.36.

What has been said so far can be indicated mathematically as follows.

$$NF_n = NF_1 + 10 \log n$$

where NF_1 = noise figure of each identical amplifier

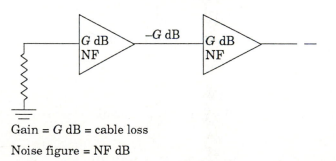

Gain = G dB = cable loss

Noise figure = NF dB

Figure 3.34 Two amplifiers in cascade connected by a cable section.

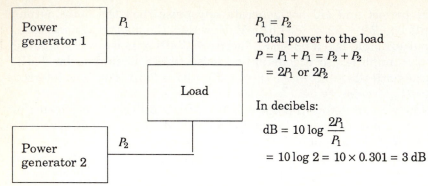

Figure 3.35   Double power is 3 dB greater.

NF$_n$ = noise figure of cascade of identical amplifiers
n = number of identical amplifiers in cascade

This, of course, assumes that all of the n amplifiers have the same gain and noise figure. In practice with today's modern manufacturing processes most commercial CATV-type amplifiers are pretty much alike. Any departure from published specifications is usually small, disregarding obviously failed units. A few examples should illustrate the point.

Example 1

$$n = 1 \quad \text{then} \quad F_n = F_1 + 10 \log n$$

where $10 \log n$, which is the cascade factor in the equation, is $10 \log 1 = 0$. Therefore, $F_n = F_1$ where $n = 1$.

Example 2

$$n = 2 \qquad 10 \log 2 = 10 \times 0.301 = 3 \text{ dB}$$

$$F_2 = F_1 + 3 \text{ dB}$$

Example 3

$$n = 3 \qquad F_3 = F_1 + 10 \log 3 = F_1 + 10 \times 0.477$$

$$F_3 = F_1 + 4.77 \text{ dB}$$

Example 4

$$n = 4 \qquad F_4 = F_1 + 10 \log 4 = F_1 + 6.02 \text{ dB}$$

Example 5

$$n = 8 \qquad F_8 = F_1 + 10 \log 8 = F_1 + 9.13 \text{ dB}$$

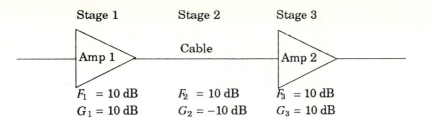

Stage 1   Stage 2   Stage 3

Cable

Amp 1   Amp 2

$F_1 = 10$ dB $F_2 = 10$ dB $F_3 = 10$ dB
$G_1 = 10$ dB $G_2 = -10$ dB $G_3 = 10$ dB

Amplifier 1 is identical to amplifier 2 and is connected by a
piece of cable with a loss equal to the amplifier gain

$$F = F_1 + \frac{F_2 - 1}{G_1} + \frac{F_3 - 1}{G_1 G_2}$$

Amp 1   Cable   Amp 2
$10 = 10 \log F_1$ $F_2 = 10$ $F_2 = 10$
$\log F_1 = 1$ $-10 = 10 \log G_2$ $G_2 = 10$
$F_1 = 10$ $\log G_2 = -1$ Same as
Also $G_1 = 10$ $G_2 = 0.1$ amp 1

Substituting $F = 10 + \dfrac{10 - 1}{10} + \dfrac{10 - 1}{(10)(0.1)} = 10 + 0.9 + 9 = 19.9 \approx 20$

$F_{dB} = 10 \log 20 = 10(1.3) = 13$ dB   3 dB more than 10 dB F_1, F_2

Figure 3.36   Two identical amplifiers connected by a piece of cable with loss
equal to amplifier gain.

These examples state that the cascade factor $10 \log n$ is added to the
noise figure given for one of the identical amplifiers. This method
gives the system noise figure.

The system carrier-to-noise ratio can now be computed by simply
subtracting the decibel value of the signal level at the end amplifier
and the system noise figure as calculated by the formula. To test for
carrier-to-noise ratio, a signal level meter with the appropriate 4-MHz
bandwidth test should be used. First, the normal signal level is measured at a tap or high signal level test point. Often a tap plate can be
temporarily changed to provide a signal level of a minimum of +20
dBmV. After recording this signal level the meter should be tuned to
a space at least one channel wide near the test carrier frequency.
Meter attenuation pads should be removed until a midscale meter
reading for the noise is indicated. The switch usually labeled S/N

should be activated to widen the meter bandwidth to the prescribed 4 MHz and record the value. The algebraic differences is the carrier-to-noise (C/N) ratio value at that location.

To summarize the change in carrier-to-noise ratio along a cascade of amplifiers, consider the example shown in Fig. 3.37. From amplifier 1 through 32 the C/N ratio has been degraded by 15 dB. The first four amplifiers have degraded the C/N by 6 dB, which is about 40 percent of the overall C/N ratio. This fact should be remembered because the first four amplifiers are extremely important.

If during troubleshooting it was found that amplifier 4 was noisy and causing picture degradation toward the end of the system, this would tell us that in the first part of the system the C/N must be at least above 36 dB and at the end of the system less than 36 dB. If the last amplifier that is known to be good is swapped with amplifier 4, then better picture will be seen by more subscribers. The noisier amplifier at the end will not cause the buildup of noise through the cascade or only the few subscribers at the end will be affected. Of course, if a good spare amplifier is available—and it should be—the defective amplifier 4 should be replaced.

Often during construction amplifiers should be pretested before installation, and testing for noise is definitely a good idea. A diagram for such a test is given in Fig. 3.38. A signal level meter with noise test feature has to be used or a meter correction factor is needed.

When coaxial switch SW_1 is in position 1, the generator will provide a test signal of the proper level so as not to overdrive the amplifier. With SW_2 closed the signal is measured at the output of the amplifier. The output level (dBmV) minus the input level (dBmV) from the generator will be the gain G (dB). When SW_1 is placed in position 2, thus terminating the amplifier input, and SW_2 is opened, inserting the postamplifier of known gain, NF, the noise level, can be measured by placing the S/N switch into the noise position and reading the noise level from the instrument. Now the noise figure can be calculated. Using the values given in the previous examples after measuring the amplifier gain under test to be 22 dB gives

-59 dBmV + NF + 22 (measured gain) + 20 (postamplifier gain)

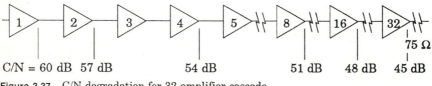

C/N = 60 dB 57 dB   54 dB   51 dB   48 dB   45 dB

Figure 3.37   C/N degradation for 32-amplifier cascade.

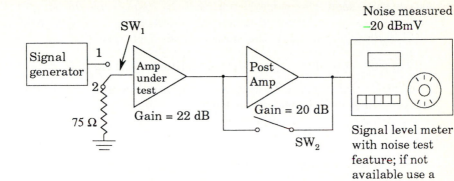

Figure 3.38 Amplifier noise test.

$$+7 \text{ (postamplifier NF)} = -20 \text{ dBmV}$$

$$-10 + NF = -20 \text{ dBmV}$$

$$NF = 10 \text{ dB}$$

During the system design phase of an amplifier cascade the C/N ratio at several points in the frequency band will be calculated. It would definitely be valid to measure the C/N ratio along several points in the cascade and record these values in a maintenance log book. The amplifiers that double would certainly be advisable as reference points. These would be at the outputs of amplifiers 1, 2, 4, 8, 16, etc. The difference between amplifiers 1 and 2, 2 and 4, 4 and 8, 8 and 16, etc., should, of course, be 3 dB. If problems occur, this information can be useful in finding the cause.

Combining C/N ratios where line extenders and bridging amplifiers are connected to the trunk system, we have a case in which the amplifiers are nonidentical and we must find a way to combine C/N ratios in such a system.

First the C/N ratio of a single amplifier is given by the mathematical expression

$$C/N = 59 - NF_1 + \text{input level (dBmV)}$$

What is seen here is that the C/N ratio at the amplifier's output should be the same as the input modified by the amplifier's noise figure. Typical noise figures of trunk amplifiers range from 7 to 9 dB; typical input level of signal is +10 dBmV. Therefore, the single-amplifier C/N ratio is, with a noise figure of 7 dB, $59 - 7 + 10 = 62$ dB.

For a typical line extender amplifier the typical noise figure is 9 dB and the input signal level is +15 dBmV. So the C/N ratio for a single line extender is $59 - 9 + 15 = 65$ dB.

As we have seen, the noise builds up in a cascade as the single-amplifier noise figure is added to the cascade factor of 10 log N. Now the C/N ratio is degraded by the same cascade factor, so

$$\text{C/N (cascade)} = \text{C/N (single-amplifier)} - 10 \log N$$

To combine C/N ratio for a system consisting of unlike amplifiers, we should remember the theory for logarithms; we must actually combine the numerical power ratios and then calculate the resulting C/N ratio.

$$\text{Recall dB} = 10 \log \frac{P_1}{P_n} \quad \text{and} \quad \log \frac{P_1}{P_2} = \frac{\text{dB}}{10}$$

and the logarithm of a number is the power, or *exponent*, to which the base 10 is raised. Therefore,

$$\frac{P_1}{P_2} = 10^{\text{dB}/10}$$

Where P_2 is greater than P_1 then

$$\frac{P_2}{P_1} = 10^{\text{dB}/10}$$

So this is the expression that has to be combined. To calculate the system C/N ratio by combining the individual C/N ratios, consider the following example.

Example A cascade of 20 identical amplifiers in a trunk system has a single-amplifier C/N ratio of 62 dB and is bridged through one bridging amplifier with a C/N ratio of 63 dB and two line extender amplifiers with a C/N ratio of 65 dB. First, calculate the 20 amplifiers of trunk.

$$\text{C/N cascade} = 62 - 10 \log 20 = 62 - 10 \, (1.3)$$
$$= 62 - 13 = 49$$

At this time we can calculate the C/N ratio of the two line extender amplifiers

$$\text{C/N (cascade)} = 65 - 10 \log 2 = 65 - 10 \, (0.3) = 65 - 3 = 62 \text{ dB}$$

Now combine the three C/N ratios by converting to the power ratios

$$\begin{aligned}
&49 \text{ dB (system trunk cascade)} & 10^{-49/10} &= 10^{-4.9} \\
&63 \text{ dB (bridger)} & 10^{-63/10} &= 10^{-6.3} \\
&62 \text{ dB (two line extenders)} & 10^{-62/10} &= 10^{-6.2}
\end{aligned}$$

Using a scientific pocket calculator, one can compute the following power ratios and add them.

To calculate $10^{-4.9}$ press 1 and 0, then press the y^x key, followed by 4 and point and 9 for 4.9. Then press the $+ / -$ key and then the $=$ key to perform the calculation. The number 0.0000126 should appear on the display.

So
$$10^{-4.9} = 0.0000126$$
$$10^{-6.3} = 0.0000005$$
$$10^{-6.2} = \underline{0.0000006}$$
$$0.0000137 = \text{total}$$

Now determine the system C/N ratio $-10 \log 0.0000137$ using the calculator, add the numbers; the sum is 0.0000137. Press log key: -4.8632794 appears in the display. Then press the × key followed by 1 and 0 and = . Then -48.632794 will be the combined C/N ratio. This number should be rounded to -48.63 dB.

This C/N ratio should be expected because the trunk C/N ratio, which is 49, should only be degraded slightly because there are only one bridger and a short cascade of two line extender amplifiers. Knowing what to expect and what the results should be provides a way to check the method for deriving the answer.

This process can, of course, be carried one step further by considering the converter. If the converter manufacturer provides a C/N ratio for the converter, it can be used, or if the input signal and noise figure are known, the C/N ratio can be calculated.

Example   A converter has a noise figure of 15 dB and a typical input signal of $+5$ dBmV; then the C/N ratio (converter) $= 59 - 15 + 5 = 49$ dB.

Now, combining the previous C/N ratio of 48.63 with the 49 of the converter, the same method can be used.

$$\frac{-48.63}{10^{10}} = 0.0000137 \qquad \text{as before}$$

$$\frac{-49}{10^{10}} = \underline{0.0000126}$$
$$0.0000263$$

$10 \log 0.0000263 = -45.8$ dB, which is still an excellent picture.

To summarize the effect of noise on CATV systems, we have learned that the picture impairment due to noise is dependent on the C/N ratio, which also governs the baseband S/N ratio. Unfortunately, C/N and S/N ratios in many cases are used interchangeably. This is actually not correct. C/N, or carrier-to-noise, ratio is the carrier level to the noise level over a 4-MHz bandwidth for the power ratio. In decibels the C/N ratio is the carrier level in decibel-millivolts minus (because we are dividing carrier power by noise power) the noise level in decibel-millivolts. If the signal level meter does not have a C/N ratio test option, one can correct the noise reading if the bandwidth of the signal level meter is known.

Recall En = $\sqrt{4KRB}$. En at different bandwidths can be calculated by

$$\frac{En_1}{En_2} = \frac{\sqrt{4KRB_1}}{\sqrt{4KRB_2}}$$

where En = the noise voltage in volts
K = Boltzman's constant
R = resistance in ohms
B_1 = bandwidth of 4 MHz
B_2 = meter bandwidth of 200 KHz = 0.200 MHz

So
$$\frac{En_1}{En_2} = \frac{\sqrt{4KR} \times \sqrt{B_1}}{\sqrt{4KR} \times \sqrt{B_2}} \quad \text{and} \quad \frac{En_1^2}{En_2^2} = \frac{B_1}{B_2}$$

in dB
$$20 \log En_1 - 20 \log En_2 = 10 \log \frac{B_1}{B_2}$$

so
$$10 \log \frac{4}{0.2} = 10 \log 20 = 10 \times 1.3 = 13\text{-dB meter correction}$$

3.7.2 Noise combining

In previous sections noise and signal-to-noise ratio have been discussed. In cable system cascades of amplifiers, carrier-level-to-noise-level ratio is an important parameter. Usually in radio frequency systems such as broadcast television, microwave links, and earth stations carrier-to-noise ratio is the parameter the cable operator measures and maintains. In the video measurement section of Chapter 6 baseband signal-to-noise measurements and control are discussed. Generally a poor carrier-to-noise ratio causes a degraded signal-to-noise ratio.

To examine how noise combines, let us experiment with two noise sources feeding a simple CATV splitter connected as a combiner as shown in Fig. 3.39. Assuming the splitter loss is 3 dB per port, the noise entering the two ports at 0 dBmV will be attenuated at −3 dB but will power combine at +3 dB, causing the signal level meter to read 0 dBmV.

To test the theory, if one noise source is disconnected and the port terminated, the signal level meter will read −3 dBmV.

Noise generators are commercially available from one manufacturer of CATV instruments; they consist of a flat noise generator coupled with a return loss bridge. This type of instrument is low in cost and extremely useful in making noise tests, return loss tests, and frequency response tests.

Suppose now that we are lucky enough to have two signal sources

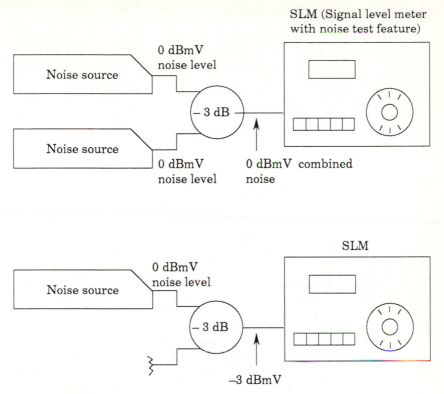

Figure 3.39 Noise combining experiment.

feeding into the two ports of the splitter as a combiner and each generator is operating on a different television frequency, as shown in Fig. 3.40. Now suppose we tune one generator to the exact frequency of the other generator, for example, tune the channel 3 generator to channel 7. We will find that the meter indicates a +3 dBmV reading.

What has happened is that the noise has combined on a power basis, overcoming, so to speak, the −3-dB splitter (combiner) loss. The signals combine on a voltage basis when on the same frequency, thus gaining 6 dB and losing 3 dB for a net gain of 3 dB. This causes our signal level meter to read +3 dBmV when the generator for channel 3 is tuned to channel 7. For the most part we never see this condition because we use splitters as combiners at different frequencies so only the port loss is seen. The phenomenon of the net gain of 3 dB is expected by a power combiner, and this is why the power doubling cable amplifier works. Power doubling involves parallel amplifiers which give an improvement in signal level of 3 dB without adding any noise level. Therefore, there is an improvement of 3 dB in output carrier to noise over the conventional push-pull type of CATV amplifier.

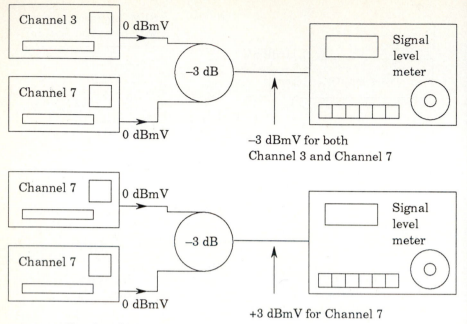

Figure 3.40 Signal combining experiment.

To investigate noise combining, often referred to as funneling, let us consider the common two port splitter used as a combiner and the directional coupler as a combiner as shown in Fig. 3.41a and b.

Now we can see that the through leg (-1 dB) is only degraded in C/N ratio at 0.5 dB over the splitter, but the tap leg (-10 dB) is dropped to a C/N ratio of 35.5 dB. The lower level signal has been hurt by the high-level noise of the through port. To survive this sort of combining, the signal level entering the tap port should be increased by at least the amount of the loss.

Now that we have examined noise combining for the splitter and directional coupler, it should be evident that when combining signals using either device, broadband noise into both ports will combine to cause deterioration in the overall carrier-to-noise ratio. The signal sources to be combined should have good bandpass filtering to allow the signal bandwidth to be combined.

Since many head ends do not use splitters but directional couplers arranged as a combiner, such a system should be examined for the noise problem. This type of combiner is made from 10-dB directional couplers arranged according to the diagram shown in Fig. 3.42. Each through port is assumed to have 0-dB loss. N_1, N_2 combined with N_3 gives

$$N_{1,2} = -32 \text{ dBmV} = 10 \log x \qquad \log x = -3.2 \qquad x = 0.000631$$
$$N_3 = -35 \text{ dBmV} = 10 \log y \qquad \log y = -3.5 \qquad \underline{y = 0.000316}$$
$$\overline{0.000947}$$

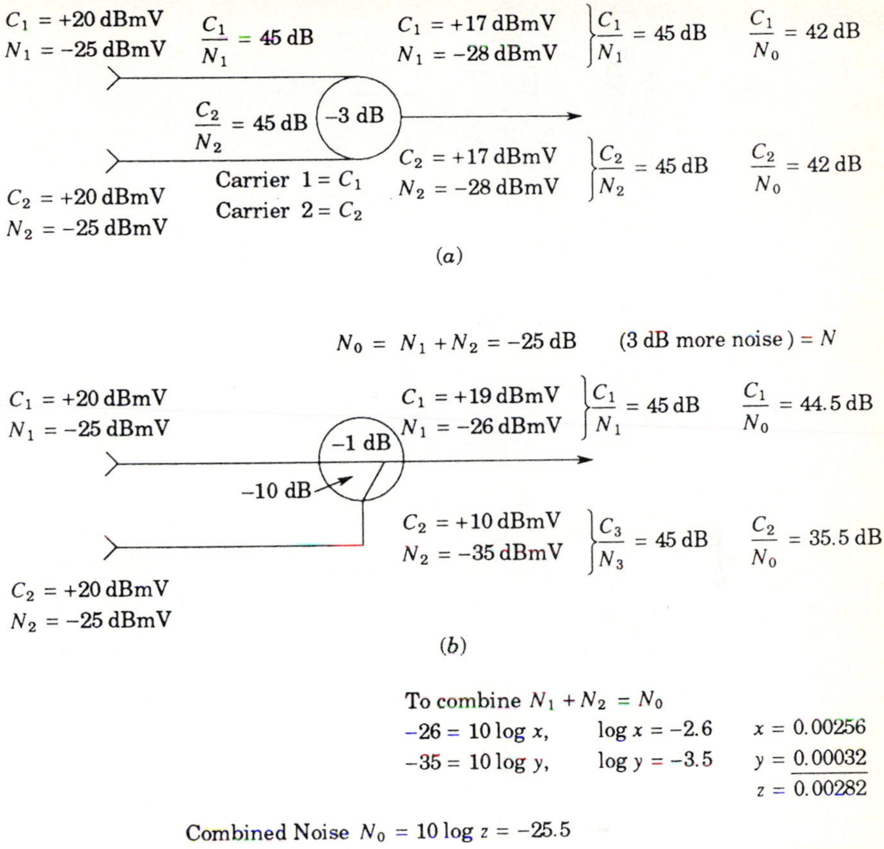

$C_1 = +20\,\text{dBmV}$ $\dfrac{C_1}{N_1} = 45\,\text{dB}$ $C_1 = +17\,\text{dBmV}$ $\left.\begin{array}{l}\end{array}\right\}\dfrac{C_1}{N_1} = 45\,\text{dB}$ $\dfrac{C_1}{N_0} = 42\,\text{dB}$
$N_1 = -25\,\text{dBmV}$ $N_1 = -28\,\text{dBmV}$

$\dfrac{C_2}{N_2} = 45\,\text{dB}$ $\boxed{-3\,\text{dB}}$

$C_2 = +17\,\text{dBmV}$ $\left.\begin{array}{l}\end{array}\right\}\dfrac{C_2}{N_2} = 45\,\text{dB}$ $\dfrac{C_2}{N_0} = 42\,\text{dB}$

Carrier $1 = C_1$ $N_2 = -28\,\text{dBmV}$

$C_2 = +20\,\text{dBmV}$ Carrier $2 = C_2$
$N_2 = -25\,\text{dBmV}$

(a)

$$N_0 = N_1 + N_2 = -25\,\text{dB} \qquad (3\,\text{dB more noise}) = N$$

$C_1 = +20\,\text{dBmV}$ $C_1 = +19\,\text{dBmV}$ $\left.\begin{array}{l}\end{array}\right\}\dfrac{C_1}{N_1} = 45\,\text{dB}$ $\dfrac{C_1}{N_0} = 44.5\,\text{dB}$
$N_1 = -25\,\text{dBmV}$ $N_1 = -26\,\text{dBmV}$

$\boxed{-1\,\text{dB}}$

$-10\,\text{dB}$

$C_2 = +10\,\text{dBmV}$ $\left.\begin{array}{l}\end{array}\right\}\dfrac{C_3}{N_3} = 45\,\text{dB}$ $\dfrac{C_2}{N_0} = 35.5\,\text{dB}$
$N_2 = -35\,\text{dBmV}$

$C_2 = +20\,\text{dBmV}$
$N_2 = -25\,\text{dBmV}$

(b)

To combine $N_1 + N_2 = N_0$
$-26 = 10 \log x, \qquad \log x = -2.6 \qquad x = 0.00256$
$-35 = 10 \log y, \qquad \log y = -3.5 \qquad y = 0.00032$
$$z = 0.00282$$

Combined Noise $N_0 = 10 \log z = -25.5$

Figure 3.41 Noise combining.

$$N_{1,2,3} = 10 \log 0.000947 = -30.2\,\text{dBmV}$$

N_4 combined with $N_{1,2,3}$ gives N_0.

$$\begin{array}{lll} & & 0.000947 \\ 10 \log N_4 = -35\,\text{dBmV} & \log N_4 = -3.5 & N_4 = 0.000316 \\ & & \overline{0.001263} \end{array}$$

$$N_0 = 10 \log 0.001263 \qquad N_0 = -29\,\text{dBmV}$$

Now Out: $\dfrac{C_1}{N_0} = \dfrac{C_2}{N_0} = \dfrac{C_3}{N_0} = \dfrac{C_4}{N_0} = 39\,\text{dB}$

Before: $\dfrac{C_1}{N_1} = \dfrac{C_2}{N_2} = \dfrac{C_3}{N_3} = \dfrac{C_4}{N_4} = 45\,\text{dB}$

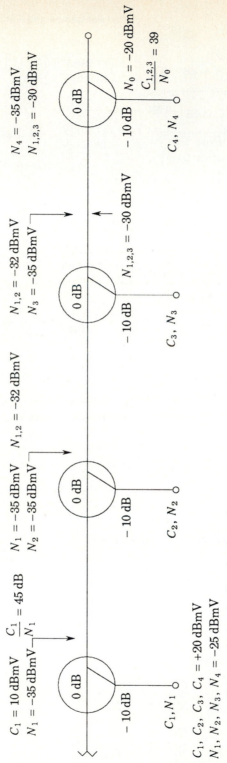

Figure 3.42 Noise combining for directional couplers.

$C_1, C_2, C_3, C_4 = +20\,\text{dBmV}$
$N_1, N_2, N_3, N_4 = -25\,\text{dBmV}$

$C_1 = 10\,\text{dBmV}$ $\dfrac{C_1}{N_1} = 45\,\text{dB}$ $N_1 = -35\,\text{dBmV}$ $N_{1,2} = -32\,\text{dBmV}$
$N_1 = -35\,\text{dBmV}$ $N_2 = -35\,\text{dBmV}$

$N_{1,2} = -32\,\text{dBmV}$ $N_{1,2,3} = -30\,\text{dBmV}$
$N_3 = -35\,\text{dBmV}$

$N_4 = -35\,\text{dBmV}$
$N_{1,2,3} = -30\,\text{dBmV}$

$N_0 = -20\,\text{dBmV}$
$\dfrac{C_{1,2,3}}{N_0} = 39$

0 dB

−10 dB

C_1, N_1

C_2, N_2

C_3, N_3

C_4, N_4

The result is the same as that for using splitters with inputs, it is the same for carrier level, noise level, and, of course, C/N ratio.

Using directional couplers also has some problems in a head-end combining network because the through loss though fairly small can accumulate. Fig. 3.43 illustrates the point.

The through loss is 1 dB and the port loss is 10 dB. Now for a flat output we must taper the input signals. Notice that port 10 is hottest because it is closest to the output. Input port 1 has to have 9 dB more signal than input port 10. Now if the frequency of the carriers has to be set according to, e.g., a 6-dB upward tilt, then the channels that have to be of higher amplitude should be connected to the lower loss input ports. Again all connections in the head end should be made correctly and tightened with a wrench. Care should also be taken to check out the frequency response of the devices. This can be done by using either a sweep generator setup or the white (all-band) noise source and a spectrum analyzer or signal level meter.

3.8   Types of Amplifiers

3.8.1   Push-pull

Most amplifiers used today are of the push-pull type, and, as we have learned, second-order distortion has been decreased significantly, thus causing third-order distortion to be the limiting factor for the cascade number. When third-order distortion increases to the point where the carrier-to-distortion ratio approaches 50 dB, the effects of distortion are visually perceptible in the television pictures. The buildup of third-order distortion increases rapidly with the cascade number, that is, $20 \log N$. When the value of 50 dB carrier-to-distortion ratio (C/D ratio) is reached, the cascade length should be stopped. If the cable plant has to be extended beyond that point, then the procedure is as follows: starting at the head end, the amplifiers should be changed, proceeding down that branch of the system, changing the amplifiers to either the parallel hybrid type or the feed-forward type. All of the amplifiers in the branch do not have to be replaced, just the number needed to keep the end of the trunk and line cascade within the 43-dB C/N ratio and the 51-dB C/D_{3rd} (3rd order distortion).

The combining of the unlike C/N and C/D_{3rd} ratios has to be performed as before by combining the power ratios and voltage ratios to calculate the new C/N and C/D_{3rd} ratios.

3.8.2   Power hybrid

To understand the improvement of using the parallel hybrid amplifier over the push-pull type, consider the example in Fig. 3.44.

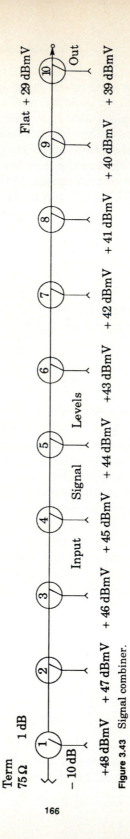

Figure 3.43 Signal combiner.

Term
75 Ω   1 dB

− 10 dB

Input   Signal   Levels

Flat + 29 dBmV   Out

+48 dBmV   + 47 dBmV   + 46 dBmV   + 45 dBmV   + 44 dBmV   +43 dBmV   + 42 dBmV   + 41 dBmV   + 40 dBmV   + 39 dBmV

166

For the push-pull hybrid amplifier:

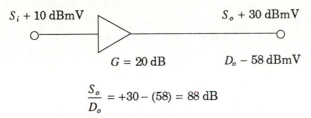

S_i + 10 dBmV

S_o + 30 dBmV

G = 20 dB

D_o − 58 dBmV

$$\frac{S_o}{D_o} = +30 - (58) = 88 \text{ dB}$$

Figure 3.44 Push-pull-type amplifier.

Now consider what happens when two splitters are connected back to back, i.e., splitter to combiner, using short, equal lengths of cable as shown in Fig. 3.45a and b. Short equal lengths of cable have negligible loss. Assume that each splitter has a loss of 3 dB. Now an amplifier is connected in each line between the splitters. This gives the parallel hybrid type. For the output signal O_s, the combination

$$O_s = S_o + 3 \text{ dB} = 30 + 3 = 33 \text{ dBmV}$$

and the output distortion level (O_d) for the combination

$$O_d = D_o + 3 \text{ dB} = -58 + 3 \text{ dB} = -55 \text{ dBmV}$$

So now the C/D ratio can be calculated

$$C/D = O_s - O_d = 33 - (-55) = 33 + 55 = 88 \text{ dB (as before)}$$

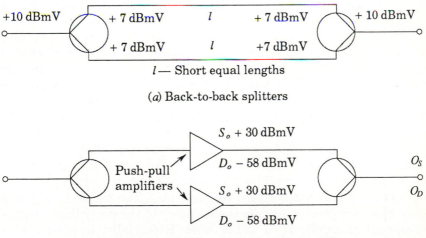

+10 dBmV   + 7 dBmV l + 7 dBmV   + 10 dBmV

+ 7 dBmV l +7 dBmV

l— Short equal lengths

(a) Back-to-back splitters

S_o + 30 dBmV

Push-pull amplifiers

D_o − 58 dBmV

S_o + 30 dBmV

D_o − 58 dBmV

O_S

O_D

(b) Development of the parallel hybrid amplifier

Therefore, more output level has been obtained without hurting the overall C/D ratio.

To investigate the output noise figure (O_n) for this type of amplifier again consider the single push-pull amplifier in Fig. 3.44. Now if two are connected in parallel by a splitter and combiner as before and shown in Fig. 3.45b and noise is randomized, the splitter has no through loss or gain, hence

$$N_o = -30 \text{ dBmV}$$

Recall that the output signal, calculated previously, was +33 dBmV so C/N = 33 − (−30) = 63 dB. This C/N ratio for a parallel hybrid module is 3 dB better than the single push-pull amplifier because 3 dB more signal was obtained without the buildup of noise. It must be remembered that to compare apples to apples the input signal to the first splitter has to be 3 dB higher than the single push-pull amplifier to overcome the input splitter loss or must be at a +13 dBmV level. Then it can be said the module gain will be 33 dBmV − 13 dBmV = 20 dBmV, which is the same as the single push-pull hybrid gain.

3.8.3 Feed-forward amplifier

Another type of amplifier often used in present-day CATV systems is the feed-forward amplifier. This type of amplifier was actually invented many years ago, but its principles were not very well understood. However, feedback amplifier types found use as operational amplifiers, in analog computer circuitry, in filters, etc. Feed-forward theory was applied to the wideband radio frequency amplifiers used in CATV systems which greatly reduced the limitations caused by third-order distortion. Consider the diagram for the feed-forward amplifier configuration where G_m and G_e are push-pull ordinary gain blocks, as shown in Fig. 3.46. The signal is applied to the input of the module and is applied to the main amplifier G_m; it is also applied to delay line 1 (DL_1), where it is delayed 180°, causing $-D$ (distortion level). Amplifier G_m amplifies the signal S and also causes distortion D, hence $S + D$. This signal is fed to G_E through DC_2 and DC_3 and adjusted in level by pad P. Amplifier G_E has only the distortion signal D appearing at its input terminals because the signal S is essentially cancelled. The error amplifier G_E amplifies the distortion and adds this signal to the output signal through DC_4. The signal level paths have to be very precise, as does the accuracy of the delay lines, to cancel and not cause distortion and/or noise. Essentially the noise generated by G_m is cancelled just as the distortion is. However, G_E generates noise and since the distortion signal is small, the added distortion

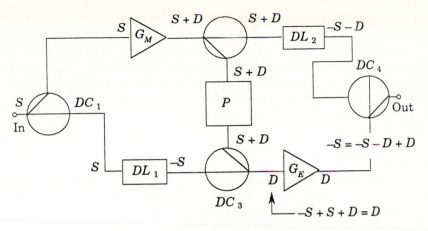

Feed-forward amplifier connection

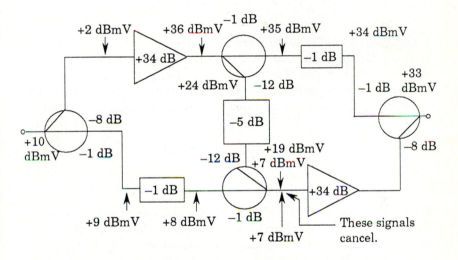

Feed-forward amplifier operational example

Figure 3.46 Feed-forward amplifier operational example.

of amplifier G_E is negligible. The output noise is essentially controlled by G_E. This noise figure can be calculated by adding the input losses to the noise figure of the error amplifier G_E. Mathematically

$NF_{FF} = NF_{GE}$ (6 dB) − input directional coupler loss (−1 dB)

− delay line 1

Loss (− 1 dB) − directional coupler loss DC_3 (−1 dB) = 6 − (−1) −

(−1) = 6 + 1 + 1 + 1 = 9 dB

Careful inspection of catalog specifications is necessary, and usually a conference with the manufacturer's representative is helpful before making a selection of the feed-forward type amplifiers. Typically, a carrier-to-composite triple beat specification is 93 dB and a carrier-to-noise ratio is 60 dB for the single amplifier unit. Instead of the usual 22-dB spacing used with push-pull-type equipment, 22-, 26-, and 29-dB spacing is available. It must be remembered that the cable size may be increased, i.e., less loss and 26-dB spacing used, instead of 29-dB, which could be more cost-effective and possibly allow a longer cascade as well.

Now if the single amplifier carrier-to-noise ratio is 60 dB, then to find out what the maximum cascade number will be for a 43-dB cascade carrier-to-noise ratio we may calculate as follows

$$43 \text{ dB} = 60 \text{ dB} - 10 \log N \quad \text{and} \quad 10 \log N = 60 - 43 = 17$$

$$\log N = \frac{17}{10} = 1.7 \quad N = 50.12$$

So cascade number = 50 amplifiers.

The scientific calculator can find the inverse log by entering 1.7 followed by the inv key and then pressing the log key.

Now for the composite triple beat calculation, recall that the carrier-to-composite triple beat ratio for a single amplifier is typically 93 dB and the minimum allowable for the cascade is 53 dB. Therefore,

$$53 \text{ dB} = 93 \text{ dB} - 20 \log N$$

So

$$40 = 20 \log N \quad \log N = 2 \quad \text{and} \quad N = 100$$

Now it is evident that noise is the limiting factor, thus limiting us to 50 amplifiers in cascade for this case. Suppose these typical carrier-to-noise ratios and composite triple beat specifications for a single amplifier also gave 26 dB spacing as a specification; then the cable type would govern the physical plant length. For example, for cable with loss of 0.95 dB/100 ft at 450 MHz, the distance can be calculated for 26-dB spacing.

$$0.95 \text{ dB/100 ft} = 9.5 \text{ dB/1000 ft}$$

$$\frac{9.5 \text{ dB}}{1000 \text{ ft}} = \frac{26}{X} \quad X = \frac{26,000}{9.5}$$

$$X = 2736.8 \text{ ft}$$

Dropping the decimal number gives 2736 ft maximum distance between amplifiers. Therefore, for 50 amplifiers then $50 \times 2736 = 136,800$ total feet of longest cascade, or about 26 mi.

3.9 Amplifier Distortions

As we have seen, amplifiers do indeed produce noise which is random in nature and of nearly constant amplitude across the system bandwidth. Also this noise builds up along the amplifier cascade according to the cascade factor of $10 \log N$ where N = number of amplifiers along the cascade.

Like noise, amplifiers cause signal *distortion*, meaning that the shape of the signal, commonly called the *waveform*, is not amplified correctly. For example, if the input signal is a sine wave and the output waveform is not a sine wave but a poor replica of a sine wave, distortion of the output signal has occurred. A true amplifier with no distortion produces an output signal that looks exactly like the input signal, but is larger. If the output is different then it is said to be distorted.

3.9.1 Types of distortion

To develop a sense of how distortion occurs, consider the *transfer characteristics* of an amplifier, shown graphically in Fig. 3.47, which depicts a peak value of input signal of 1.5 mV and a corresponding output signal of 50 mV. Because the transfer curve is very linear between points x and y, the output waveform is a larger replica of the input waveform and exhibits no distortion. Beyond points x and y the transfer characteristics become nonlinear and distortion occurs. That is why setting the input level of a trunk amplifier is critical to prevent the input signal from causing the output signal to reach the curved portion of the transfer characteristic graph, causing distortion. If the transfer characteristics change shape so the linear portion is small and/or shifted, then distortion of the output signal will result. Refer to Fig. 3.48.

Third-order distortion is depicted in Fig. 3.49. Notice that the transfer characteristic plot has a curvature at the top, which, as we shall see, will cause a third-order distortion or a signal in the frequency spectrum of three times the input frequency f_1.

Unfortunately the foregoing illustration is an oversimplification. In the real world many carriers appear in the system bandwidth, for example, 54 to 450 MHz. Second-order distortion is usually accompanied by fourth-order, sixth-order, etc. Third-order distortion also has higher-order distortion frequencies. However, the higher the order distortion the lower the amplitude of the distorting frequency signals. Also, because of the distorting transfer characteristic plots, a certain amount of signal mixing occurs among the various carriers carried on the system; this also causes sum and difference frequencies which may fall in our 54- to 450-MHz bandwidth. These extraneous signals

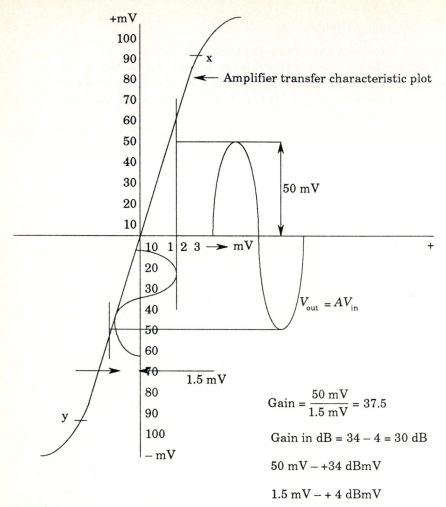

Figure 3.47

caused by amplifier distortion contain modulation which can cause picture impairment of the desired signal. Third-order distortion signals containing modulation cause picture impairment called *crossmodulation*. This term probably came from the old telephone term *crosstalk*.

There are various types of amplifiers that limit the amount of distortion produced. The push-pull amplifier essentially cancels the predominant second-order distortion. This type of amplifier basically has a separate acting amplifier for the positive and negative halves of the sine wave and thus a reasonably linear and long transfer characteris-

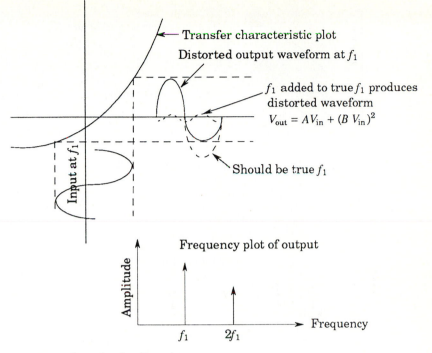

Transfer characteristic plot

Distorted output waveform at f_1

f_1 added to true f_1 produces distorted waveform

$V_{\text{out}} = AV_{\text{in}} + (B\ V_{\text{in}})^2$

Should be true f_1

Input at f_1

Frequency plot of output

Amplitude

Frequency

f_1 $2f_1$

Figure 3.48   Second-order distortion.

tic. Second-order distortion is not now the limiting factor in push-pull-type amplifiers.

Third-order distortion takes over as the culprit, as does noise, which never goes away. The feed forward type of amplifier, which is essentially push-pull, uses a third-order distortion cancelling scheme. Use of this type of amplifier is, of course, more expensive but it essentially allows noise buildup to be the limiting factor for the cascade number.

3.9.2 Distortion buildup

It is essential to discuss the manner in which distortion builds up as the cascade number increases and what is the minimum carrier-to-distortion ratio that causes picture impairment. As we have seen, the amplifiers increase the signal level voltage from input to output, and it is the output voltage waveform that is distorted. Distortion, principally third-order distortion, builds up with respect to the voltage. Therefore, the buildup of third-order distortion is given by:

$$D_{\text{3rd}} = \text{single-amplifier} + 20 \log n$$

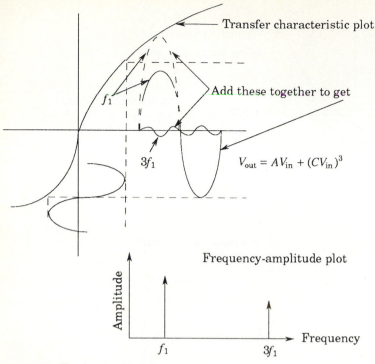

Figure 3.49   Third-order distortion.

where n = number of amplifiers in cascade.

The cascade factor for third-order distortion is $20 \log n$. Therefore, the system carrier/third-order distortion ratio = single-amplifier $C/3\text{rd} - 20 \log n$ distortion. In essence, the single-amplifier-carrier-to-third-order-distortion ratio is degraded by the cascade factor $20 \log n$ because of the buildup of third-order distortion by the same cascade factor.

In like manner, the technique of combining unlike amplifiers-carrier-to-third-order-distortion ratio is the same as the C/N ratio. Mathematically the expression is

$$C/3\text{rd} = 20 \log 10 \, \frac{-C/3\text{rd}_1}{20} + 10 \, \frac{-C/3\text{rd}_2}{20} + 10 \, \frac{-C/3\text{rd}_3}{20}$$

For example, where $\dfrac{C}{3\text{rd}_1}$ is for the trunk cascade

$\dfrac{C}{3\text{rd}_2}$ is for a bridging amplifier

$$\frac{C}{3rd_3} \text{ is for 2 line extender amplifiers}$$

Then typically $\dfrac{C}{3rd_1}$ single trunk $= 72$ dB

$$\frac{C}{3rd_2} \text{ single bridger} = 60 \text{ dB}$$

$$\frac{C}{3rd_3} \text{ line extenders} = 60 \text{ dB}$$

For a cascade of 10 trunk amplifiers at 450 MHz, 1 bridger, and two line extenders proceed as follows.

For trunk cascade $\dfrac{C}{3rd_1} = 72 - 20 \log 10 = 72 - 20 = 52 \text{dB}$

Bridger $\dfrac{C}{3rd_2} = 60$

Two line extenders $\dfrac{C}{3rd_3} = 60 - 20 \log 2 = 60 - 6 = 54 \text{ dB}$

Combining trunk $10^{-52/20} = 10^{-2.6} = 0.0025119$

Bridger $10^{-60/20} = 10^{-3} = 0.001000$

Line extenders $10^{-54/20} = 10^{-2.7} = \dfrac{0.00200}{0.00551}$

To combine

$$\text{Carrier}_{3rd \text{ order system}} = 20 \log 0.00551 = -45.18$$

The visible level of carrier to third order distortion is about 52 dB. If the head end is using an incrementally related carrier (IRC) technique, where the carrier frequencies are spaced 6 MHz apart through the band but not harmonically related, a 6-dB worse carrier-to-third-order-distortion ratio can be tolerated without picture impairment. Therefore, a 6-dB allowance would place the system at -51.18 dB, equivalent to the $C/3$rd ratio. Now if the HRC (harmonically related carrier) type of head end is used where the carrier frequencies are placed at 54 to 550 MHz and are spaced 6 MHz apart, then 9 dB will be allowed, placing the system at 54.18 dB, which now is not marginal. However, notice that the cascade is only 10 amplifiers. If longer cascades are needed, better amplifiers should be selected or the bandwidth pulled back to 450 MHz. At 450 MHz for standard, push-pull amplifiers typically run 10 dB better or carrier-to-third-order-distortion ratio is 82 dB instead of 72 dB for the trunk, 62 dB for the bridger and line extenders.

Now if the bandwidth is limited to 450 MHz and the cascade length doubled to 20 amplifiers let's see what will happen.

$$\text{Trunk } C/3\text{rd}_1 = 82 - 20 \log 20 = 82 - 20 \times 1.3 = 82 - 26 = 56$$

$$C/3\text{rd}_{2,3} = 62 - 20 \log 3 = 62 - 20 \times 0.477 = 62 - 9.5 = 52.5$$

lumping bridger and line extenders because they have the same $C/3$rd. Combining

$$\text{trunk } \quad \frac{-56}{10^{20}} = 10^{-2.8} = 0.0015849$$

$$\text{bridger and line extenders } \frac{-52.5}{10^{20}} = 10^{-2.6} = \frac{0.0035119}{0.0040968}$$

$$\text{Carrier/3rd order system} = 20 \log 0.0040968 = -47.75$$

This is nearly 48 dB.

Now using the 6-dB improvement allowance for IRC, this will yield an equivalent 54 dB, which will be good. If HRC is used, this figure will become 57 dB equivalent. It should be evident that particularly by using HRC the cascade length could be increased beyond 20 amplifiers at 450 MHz.

So far third-order distortion has been treated collectively. However, it actually causes several picture impairment problems: crossmodulation, intermodulation, and composite triple beat. These third-order distortion products differ in the manner in which they cause picture impairment. Crossmodulation shows up as a windshield wiper effect, and the carrier level to crossmodulation beat level ratio becomes partly visible between 52 and 51 dB. Modulation from other carriers falling on the desired carrier causes the problem in the picture. Third-order intermodulation is again caused by the non- linearities of the amplifier transfer characteristics. Third-order intermodulation beats are of the order $2f_1 \pm f_2$. Consider, for example, channel 3 with a video carrier frequency of 61.250 MHz and audio carrier of 65.75 MHz.

$$2f_1 \pm f_2 = 2 \times 61.25 \pm 65.75 = 188.25$$

$$2f_1 - f_2 = 2 \times 61.25 - 65.75 = 56.75$$

$$188.25 - \text{channel 9 (video carrier)} = 188.25 - 187.25 = 1 \text{ MHz}$$

$$56.75 - \text{channel 2 (video carrier)} = 56.75 - 55.25 = 1.5 \text{ MHz}$$

These resulting beat products always appear close to video carriers. Intermodulation means that the beats of 188.25 and 56.75 MHz are caused by frequencies f_1 and f_2, which belong to the same television channel. Also these beats contain modulation from f_1, which is an amplitude-modulated video signal. f_2 will contain frequency-modulated audio modulation. Typically CATV head ends are operated with the audio carrier level 15 dB lower than the video carrier level. This keeps the audio modulation out of the upper adjacent video carrier and also tends to reduce the intermodulation products.

Triple beats are third order distortion products of the $f_1 \pm f_2 \pm f_3$ type, where f_1, f_2, f_3 are video carrier frequencies containing video modulation. Again these triple beats are caused by the nonlinearities of the amplifier's transfer characteristic. All of these sum and difference frequency components increase with the number of carriers carried on the cable. The greater the bandwidth the more the channels and the greater the number of beats. When these beats increase in number, amplitude picture impairment is evident. In the discussion of testing and measurement in Chapter 6 more will be learned about amplifier distortion. Table 3.3 contains data on the buildup of beats that fall within a channel space, called *composite triple beats* for fully loaded cable systems. Data are also given for the downstream fully loaded high split systems of 450- and 550-MHz upper frequency limit.

3.10 Signal Transportation

3.10.1 Supertrunking

Supertrunking, as it is often called, is the method used by cable television systems to transport signals from HUB to HUB, where the HUB site is the center of a distribution center for a given community. Unlike conventional television system trunk lines, the supertrunk usually contains no bridging amplifiers and no signal splits; generally it just connects hub to hub points. Most supertrunk lines pass through areas where there is no required service (no houses) and sometimes limited power availability. For longer distances where no power is available and/or the roads and rights of way are limited or nonexistent, cable television signal transportation has to use means such as microwave. However, supertrunking can be and often is more economical where the distances are workable and power is available along public rights of way. Of course, the supertrunk can be either buried or aerial plant as costs and the situation dictate. If the distance is quite long and the number of

TABLE 3.3 **System Beats**

Channel No.	35 Channels	40 Channels	52 Channels	60 Channels	77 Channels	36 Channels 234–450 MHz	53 Channels 234–550 MHz
2	159	235	435	615	1104		
3	171	240	456	640	1137		
4	180	251	473	661	1167		
5	31	36	48	56	73		
6	31	36	48	56	73		
7	342	458	788	1048	1707		
8	345	464	800	1064	1731		
9	348	469	811	1079	1755		
10	349	473	821	1093	1777		
11	350	476	830	1106	1799		
12	349	478	838	1118	1819		
13	348	479	845	1129	1839		
14	274	348	644	868	1450		
15	288	384	666	894	1485		
16	299	398	686	918	1517		
17	308	409	703	939	1547		
18	316	420	720	960	1576		
19	323	429	735	979	1604		
20	329	438	750	998	1631		
21	334	445	763	1015	1657		
22	338	452	776	1032	1682		
23	345	479	851	1139	1857		
24	342	478	856	1148	1875		
25	337	476	860	1156	1891		
26	332	473	863	1163	1907	306	676
27	326	469	865	1169	1921	323	701
28	321	464	866	1174	1935	339	726
29	315	458	866	1178	1947	354	749
30	309	451	865	1181	1959	368	772
31	301	443	863	1183	1969	381	793
32	293	435	860	1184	1979	393	814
33	283	427	856	1184	1987	404	833
34	273	419	851	1183	1995	414	852
35	260	410	845	1181	2001	423	869
36	245	400	838	1178	2007	431	886
37		389	830	1174	2011	438	901
38		377	821	1169	2015	444	916
39		364	811	1163	2017	449	929
40		349	800	1156	2019	453	942
41		331	790	1148	2019	456	953
42			775	1139	2019	458	964
43			761	1129	2017	459	973
44			747	1118	2015	459	982
45			733	1106	2011	458	989
46			719	1093	2007	456	996
47			704	1079	2001	453	1001
48			688	1064	1995	449	1006
49			671	1048	1987	444	1009
50			653	1031	1979	438	1012

TABLE 3.3   System Beats (*Continued*)

Channel No.	35 Channels	40 Channels	52 Channels	60 Channels	77 Channels	36 Channels 234–450 MHz	53 Channels 234–550 MHz
51			634	1013	1969	431	1013
52			613	995	1959	423	1014
53			589	977	1947	414	1013
62				959	1935	404	1012
63				940	1921	393	1009
64				920	1907	381	1006
65				899	1891	368	1001
66				877	1875	354	996
67				854	1857	339	989

Triple beats per channel

channels limited, then supertrunking and signal transpositioning could do the job. Planning and designing a supertrunk should be done carefully with proper attention to future bandwidth requirements, minimum maintenance needs, and minimum power requirements. The choice of amplifiers, e.g., push-pull, parallel power hybrid (power doubling), or feed-forward, has to be made. Usually the push-pull types use less power. To help in making such decisions, a mathematical exercise will provide information as to the cable types and amplifier types which will accomplish the task. Then a cost analysis will provide the financial information needed to make the choice. A few practical examples should illustrate the methods and options for making such choices.

Example   It is desired to connect a HUB site to a HUB site together with a one-way feed from HUB 1 to HUB 2 a distance of 26 mi carrying 30 channels. At HUB 2 we want to add 15 more satellite channels to the 30. The 30 channels arriving at HUB 2 will remain "ON" channel (2 to 31) and the 15 just added in at the top (32 to 46). "Seat of the pants" reasoning indicates that 1-in gas-injected polyethylene foam cable will be the cable choice. This large cable has a loss at channel 36 (we are allowing 5 more channels for expansion) of 0.72 dB/100 ft. Suppose it is known that 300- or 330-MHz amplifiers are unavailable and 450 parallel hybrid amplifiers are available. Now we need to make some calculations as to the number of amplifiers required to meet the performance specifications.

The single amplifier specifications are as follows.

AGC/ASC response (flatness): 0.4 dB 50–450 MHz

Gain (minimum full gain): 26.5 dB

Gain control range: 0–8 dB

Slope control range: 2–8 dB

Required input level with equalizer inserted: 10 dBmV

Operating output level: 26/32. We will push channel 36 output to 32 dBmV

Composite triple beat: 91 dB

Crossmodulation: 91 dB

Single amplifier second-order distortion: 36 dB

Noise figure: 7.5 dB

To find out whether noise is going to be a problem we will calculate the C/N ratio for a single amplifier at the output. Refer to Section 3.7.1 for information on noise in CATV systems.

$$C/N_{\text{(single amp)}} = L_o - (N_t + N_o + G)$$

Substituting $$C/N_{\text{(single amp)}} = 32 - (-59 + 7.5 + 22)$$

$$= 61.5 \text{ dBmV at channel 36}$$

where L_o = output signal level = 32
N_t = thermal noise = -59 dBmV
N_a = noise figure of amplifier $C/N_{\text{(single amp)}}$ = 61.5 dB
G = operating gain = 22 dB

For the cascade of N amplifiers where N is not known we will proceed by estimating what our desired C/N ratio should be.

Now if at the end of the system driven by HUB 2 we need a C/N of 43 dB and the system there requires a C/N at HUB 2 of 46 to give a system C/N of 43, then our supertrunk will have to provide a 46-dB C/N ratio.

$$C/N_{\text{(cascade)}} = C/N_{\text{(single amp)}} - 10 \log N$$

$$46 \text{ dB} = 61.5 \text{ dB} - 10 \log N \qquad 10 \log N = 61.5 - 46$$

$$10 \log N = 15.5 \qquad \log N = 1.55 \qquad N = 35 \text{ amplifiers}$$

This is the maximum allowed. Now we need to calculate the number needed for our supertrunk to see if the number is less than 35.

The cable amplifier span looks like the diagram in Fig. 3.50. To find the cable span = 22 dB at channel 36 we can calculate

$$\text{Loss at channel 36} = \frac{0.72}{100 \text{ ft}} \qquad \text{so distance} = \frac{22(100)}{0.72} = 3056 \text{ ft}$$

Our marathon distance for the supertrunk is 26 mi.

$$26 \text{ mi} \times 5280 \text{ ft/mi} = 137,280 \text{ ft}$$

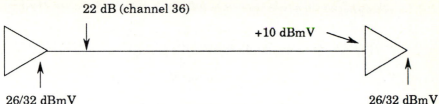

22 dB (channel 36)

+10 dBmV

26/32 dBmV   26/32 dBmV

Figure 3.50   Parallel hybrid supertrunk cable-amplifier section.

Now for each piece of cable 3056 ft long we need an amplifier. So for 137,280 ft, we will calculate the number of amplifiers for the 26-mi run.

$$\frac{137{,}280}{3056} = 45 \text{ amplifiers}$$

Therefore, we need 45 amplifiers, and our desired C/N ratio occurs at amplifier No. 35. Obviously this will not work, so it is pointless to examine the distortion factors.

The next step is to try for a higher-output-level amplifier, such as a feed-forward type.

Gain (minimum full gain): 30 dB

Output level: 32/38 dB (6-dB slope)

Input level: 11 dB

Noise figure: 8 dB

CTB (composite triple beat): 92 dB

Crossmodulation: 90 dB

Single second: 87 dB

For this situation we will run output at channel 35 to 38 dBmV for an input of 11 dBmV, giving an operational gain of 27 dBmV. Therefore, in brief we will space at 27 dB, see Fig. 3.51.

27 dB (channel 36)

+11 dBmV

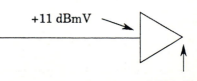

32/38 dBmV   32/38 dBmV

Figure 3.51   Feed-forward supertrunk cable amplifier section.

$$C/N_{(single\ amp)} = 38 - (-59 + 8 + 27) = 38 + 59 - 35 = 62\ dB$$

For 300-MHz spacing $\text{Span distance} = \dfrac{27(100)}{0.72} = 3750\ ft$

$$C/N_{(cascade)} = 46 = 62 - 10 \log N \qquad \log N = 1.6$$

$$N = 39 \text{ amplifiers for C/N of } 46$$

But a span is longer, so (137,280/3750) = 37 amplifiers.

Now we see that 37 feed-forward amplifiers are needed and 39 will do the job for a C/N ratio of 46 dB.

Since 37 amplifiers are needed we will calculate our expected C/N ratio for the supertrunk cascade.

$$C/N_{(cascade)} = C/N_{(single\ amp)} - 10 \log N$$

$$= 62 - 10 \log 37 = 62 - 10(1.57)$$

$$= 62 - 15.7 = 46.3\ dB$$

This is small but adequate headroom.

Now to see whether distortion will be a problem, we will calculate the CTB, X_m (crossmodulation distortion), and single second beat.

CTB for single amplifier is 92 dB for our feed-forward amplifier CTB (cascade)

$$CTB_{(cascade)} = CTB_{(single\ amp)} - 20 \log N$$

$$= 92 - 20 \log 37 = 92 - 20(1.57) = 92 - 31.4$$

$$= 60.5\ dB$$

This is great. Usually CTB at 52 dB is acceptable.

Crossmodulation for the cascade is degraded by the 20 log N factor; therefore, we can safely see that 90 dB − 31.4 dB = 58.6 dB and should not cause any picture impairment.

$$\text{Single 2nd} = \text{single 2nd/single amp} = 15 \log N$$

$$= 87 - 15 \log 37 = 87 - 15\ (1.57)$$

$$= 87 - 23.5 = 63.5\ dB$$

Great again.

Maybe it seems we are a little close to the wire on noise, but remember that if we double the amplifiers we degrade the C/N ratio by 3 dB. Since we started with 46, then 3 dB less is 43, which is acceptable for many systems. It must be remembered that it is most likely that from HUB 2 to the longest system end there will not be feed-forward amplifiers and there may be some other types. Recall we said that at the end of HUB 2 we needed a 43-dB C/N and hence we started with a 46-dB C/N at beginning of HUB 2. Also the amplifiers most likely will be sloped at 4 dB, that is, +38 dBmV output at channel 36 and 34 dB and at channel 2. For our distortion calculations the X_m, CTB, and 2nd single were given for a full 60-channel loading. We are using 450-MHz gear for 300-MHz work so when the supertrunk is tested for distortion the numbers will most likely be 2 to 3 dB better. Noise will be the problem, and the amplifiers could be operated about 1 dB hotter, potentially improving the cascade C/N ratio.

3.10.2 Supertrunk powering considerations

Another area that should be investigated is the powering of the supertrunk amplifier cascade. Usually the feed-forward type of amplifiers with just an automatic gain and slope control module and the trunk amplifier itself and the power supply module will draw from the cable 60-V power supply something on the order of 0.90 Amperes (A). Therefore, if we have 37 amplifier stations drawing 0.9 A the total current is 33.3 A. For maintenance it would seem like a neat idea to power the cascade by one 60-V power supply from each end. This would mean that each supply would have to provide 16.2 A. Most supplies on the market today provide anywhere from 12 to 15 A each. Thus our concept of a supply at each end is out. The next best would be to have one at each end and one in the middle if, of course, commercial power were available in the middle. To make any calculations the cable loop resistance has to be found either by measuring or by using the manufacturer's specifications. If the loop resistance of the 1-in copper clad cable has a typical value of 0.4 Ω/1000 ft, then recall that 26 mi corresponds to 137,280 ft. Therefore, the total resistance for the run will be on the order of 137.280 × 0.4 = 55 Ω. A schematic of our supertrunk plant is shown in Fig. 3.52. Solid copper center conductor cable will lower the loop resistance; however, in the 1-in size many manufacturers do not make it. Most likely the cost would be prohibitive.

Power supplies *A*, *B*, and *C*, providing 12 A each, can produce 36 amperes to cover the total current draw of 33.3 A. Simple circuit theory tells us that if a span of cable between amplifiers has a resistance of 1.5 Ω and if this span feeds other spans and amplifiers drawing 10 A, then the first span voltage drop will be 10 × 1.5 = 15 V. If this value were subtracted from 60 V it would leave 45 V at the first amplifier. This would amount to using up the supply at the beginning. This example tells us why the standard rule for placing power supplies is to

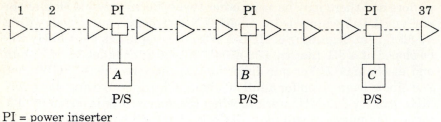

PI = power inserter
P/S = power supply

Figure 3.52   Schematic of supertrunk example.

place them in the middle of say a 10-amplifier run, feeding 5 amplifiers each side. A good method would be to place one power supply in each HUB site where some form of reliable standby power is available and then plan for the minimum number possible spaced out on the supertrunk run. Therefore, determining how far we can go from each end is the first order of business. Then the power supplies spaced along the run will be addressed.

As shown in Fig. 3.53, from each end we will proceed as follows. We have seen that we can only go five amplifiers in one direction with a current draw of 3.6 A. Thus a 12-A (720VA-volt ampere rated) type power supply will be used since it is the smallest size available. Reasoning further, since the two supplies in each HUB site will take care of 5 amplifiers each for a total of 10, then three more supplies, properly spaced, should take care of the remaining 27 amplifiers in the cascade. Each of the power supply segments should be as illustrated in Fig. 3.54a and b.

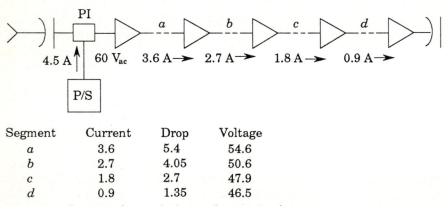

Segment	Current	Drop	Voltage
a	3.6	5.4	54.6
b	2.7	4.05	50.6
c	1.8	2.7	47.9
d	0.9	1.35	46.5

Figure 3.53   Current-voltage calculations for supertrunk system.

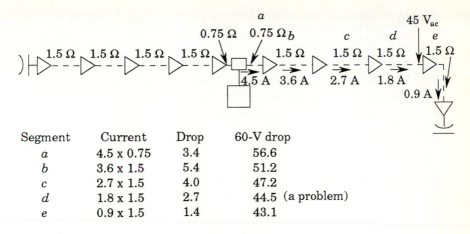

Segment	Current	Drop	60-V drop
a	4.5 x 0.75	3.4	56.6
b	3.6 x 1.5	5.4	51.2
c	2.7 x 1.5	4.0	47.2
d	1.8 x 1.5	2.7	44.5 (a problem)
e	0.9 x 1.5	1.4	43.1

(a) First try

The fix will be to drop one amplifier and feed five in the direction at the amplifier location as shown.

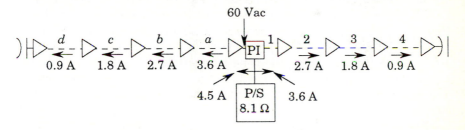

Segment	Current Drop	60-V drop
a	3.6 x 1.5 = 5.4	54.6
b	2.7 x 1.5 = 4.0	50.6
c	1.8 x 1.5 = 2.7	47.9
d	0.9 x 1.5 = 1.4	46.5
		OK

Segment	Current Drop	60-V drop
1	3.6 x 1.5 = 5.4	54.6
2	2.7 x 1.5 = 4.0	50.6
3	1.8 x 1.5 = 2.7	47.9
4	0.9 x 1.5 = 1.4	46.5
		OK

(b) Second try

Figure 3.54   Supertrunk power supply solution.

Now it should be clear that the three power supplies serving the remaining amplifiers in the cascade can handle, because of the cable resistance, 9 amplifiers each with a total current draw of 4.5 + 3.6 A, or 8.1 A. This totals 27 amplifiers; adding the 5 on each end gives us our whole 37-amplifier run. Fig. 3.55 summarizes the results.

The preceding example may seem to indicate that we lucked out, but remember that this is the minimum allowable. It could arise that the amplifiers at HUB 1 and at HUB 2 are not actually in the HUB site, since for efficient design the amplifiers should be out in the cascade. If this happens, the added cable drop will cause some amplifiers to be starved for input voltage. More prudent power design would be to place the power supply at the first and last amplifier locations. At least these two power supplies will be a cable span away from each HUB site and easily maintainable.

3.10.3 Standby power supplies

At this stage of the game, the possibility of using some form of standby power supply on the market today should be considered. The reliability of the supertrunk run is the most important factor, and any standby power supply system considered should increase that reliability, not decrease it. Standby power supplies operate on storage batteries when in the standby mode. The time allowed for the cable system to run in the standby mode (*standby time*) depends on several factors: the loading on the power supply, the number of storage batteries, the ampere-hour rating of the batteries, and the efficiency of the supply. If a 720-VA (12-A) standby power supply only has to supply 8.1 A maximum as in our example, then the power supply is only supplying eight-twelfths or two-thirds of its maximum power. Therefore, it should run about one-third longer before the batteries are exhausted and the power supply shuts down (low battery voltage cutoff). Some power supply manufacturers feature a 720-VA supply with three batteries instead of two, hence a longer standby time. If the efficiency in the standby mode is greater than another company's power supply, then it will possibly operate longer in the standby mode or operate the same amount of time with two batteries instead of three. When choosing a standby power supply, the foremost factor is reliability. Therefore, first, the features have to be examined and narrowed down; when a choice is made, a thorough investigation of the track record of the supply should be made by contacting other users of the device. Then, one should be purchased on an evaluation loan to test with a load. Since most standard cable television power supplies use a ferroresonant transformer to step down (120 to 60 V_{ac}) and regulate the voltage, this same transformer is often used in conjunction with

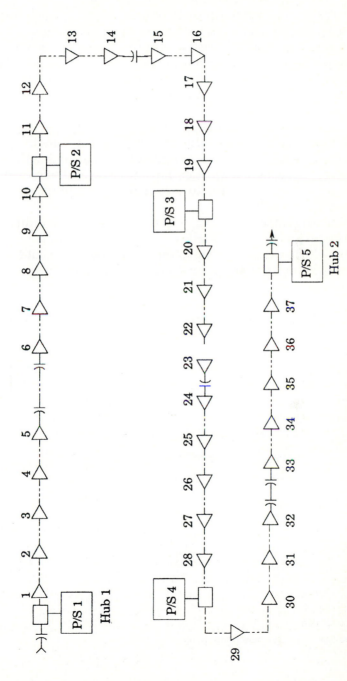

Figure 3.55   Placement of power supplies: supertrunk example summary.

an inverter to produce a standby power supply. The inverter essentially converts the battery voltage to a square waveform of alternating current, which when connected to a winding on the ferroresonant transformer is stepped up to 60 V$_{ac}$ at a wave shape termed a *quasi square wave*. The ferroresonant transformer also regulates this 60 V$_{ac}$ within the specified voltage tolerances. The only trouble is that the ferroresonant transformer has to operate with its core in the magnetic saturation region, which draws a large amount of current from the batteries. This current is taken by the supply to allow it to operate and hence is unavailable to the cable plant.

This type of standby power supply is not very efficient. Some manufacturers use another transformer type in the standby mode and operate the regulation part electronically and at high efficiency. Again features, i.e., bells and whistles, have to be carefully studied. A test system should be assembled to test a power supply; such a system is shown in Fig. 3.56.

With all switches open V_m will read 64 V$_{ac}$ on an ordinary mechanical meter and 60 V$_{ac}$ on a true root mean square (rms) meter. $A = 0$ A.

As switches are closed, ammeter A increases; V should be monitored to see whether it decreases with increasing current, i.e., regulation.

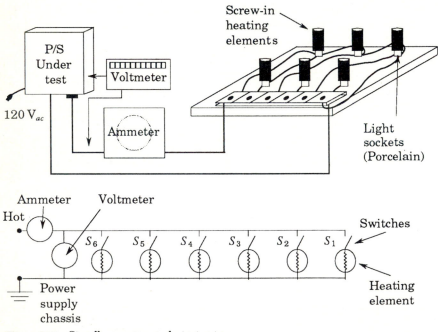

Figure 3.56 Standby power supply test setup.

$$\% \text{ regulation} = \frac{V_{NL} - V_{FL}}{V_{FL}}$$

where FL = full load
      NL = no load

Also the full load time should be measured in the standby mode, watching the voltage decrease as the batteries discharge. Often it is helpful also to monitor the dc battery voltage as the supply operates. Some power supply manufacturers supply meters on the service panel or make it available as an option. Possibly if a unit is on loan for evaluation, metering can be requested.

3.11 Microwave Line of Site

The delivery of video signals to a cable television system head end, so far, has either been by "off-air" reception of television broadcast stations using tower-mounted antennas or by satellite up-down links using parabolic receiving dish antennas. Another method often used is the microwave link.

3.11.1 Amplitude-modulated link systems

This may consist of the usual cable television head end located on a prime site and transmitting these signals to other hub sites (other CATV systems) located miles away in several directions via a microwave line of site (LOS) system of the amplitude-modulated link (AML) type. Also some cable television systems may receive one or more channels or programming via a common carrier microwave service company. Usually when a cable television operator contracts to receive programming via a common carrier microwave company, this company, as part of the contract, provides the working link, both equipment installation and maintenance, for a fixed monthly fee. All the cable operator has to provide is usually head-end building space, tower space, and electric power. FM is often used by the common carrier microwave companies. The systems cable operators have to maintain and service are their own AML-type microwave systems. The AML system takes the cable television band of carrier frequencies and converts this band to microwave frequencies and transmits the signal to the receiving site, where the microwave band of frequencies is converted back down to the standard cable television band. A listing of the AML conversion frequencies is shown in Table 3.4. There are four groups of frequencies: C, D, E, and F. Note that group F can handle

TABLE 3.4 CARS Band Frequencies

		AML Frequencies (Bandwidth = 6 MHz)							
		CARS group C		CARS group D		CARS group E		CARS group F	
CATV channel	VHF boundaries, MHz	FCC designation	Microwave boundaries, MHz	FCC designation	Microwave boundaries, MHz	FCC designation	Microwave boundaries, MHz	FCC designation	Microwave boundaries, MHz
2	54–60	C01	12700.5–12706.5	D01	12759.7–12765.7	E01	12952.5–12958.5	F01	13012.5–13018.5
3	60–66	C02	12706.5–12712.5	D02	12765.7–12771.7	E02	12958.5–12964.5	F02	13018.5–13024.5
4	66–72	C03	12712.5–12718.5	D03	12771.7–12777.7	E03	12964.5–12970.5	F03	13024.5–13030.5
PT	72–76	C04	12718.5–12722.5	D04	12777.7–12781.7	E04	12970.5–12974.5	F04	13030.5–13034.5
5	76–82	C05	12722.5–12728.5	D05	12781.7–12787.7	E05	12974.5–12980.5	F05	13034.5–13040.5
6	82–88	C06	12728.5–12734.5	D06	12787.7–12793.7	E06	12980.5–12986.5	F06	13040.5–13046.5
FM1	88–94	C07	12734.5–12740.5	D07	12793.7–12799.7	E07	12986.5–12992.5	F07	13046.5–13052.5
FM2	94–100	C08	12740.5–12746.5	D08	12799.7–12805.7	E08	12992.5–12998.5	F08	13052.5–13058.5
FM3	100–106	C09	12746.5–12752.5	D09	12805.7–12811.7	E09	12998.5–13004.5	F09	13058.5–13064.5
FM4	106–108	C10	12752.5–12754.5	D10	12811.7–12813.7	E10	13004.5–13006.5	F10	13064.5–13066.5
AUX1	108–114	C11	12754.5–12760.5	D11	12813.7–12819.7	E11	13006.5–13012.5	F11	13066.5–13072.5
AUX2	114–120	C12	12760.5–12766.5	D12	12819.7–12825.7	E12	13012.5–13018.5	F12	13072.5–13078.5
A	120–126	C13	12766.5–12772.5	D13	12825.7–12831.7	E13	13018.5–13024.5	F13	13078.5–13084.5
B	126–132	C14	12772.5–12778.5	D14	12831.7–12837.7	E14	13024.5–13030.5	F14	13084.5–13090.5
C	132–138	C15	12778.5–12784.5	D15	12837.7–12843.7	E15	13030.5–13036.5	F15	13090.5–13096.5
D	138–144	C16	12784.5–12790.5	D16	12843.7–12849.7	E16	13036.5–13042.5	F16	13096.5–13102.5
E	144–150	C17	12790.5–12796.5	D17	12849.7–12855.7	E17	13042.5–13048.5	F17	13102.5–13108.5
F	150–156	C18	12796.5–12802.5	D18	12855.7–12861.7	E18	13048.5–13054.5	F18	13108.5–13114.5
G	156–162	C19	12802.5–12808.5	D19	12861.7–12867.7	E19	13054.5–13060.5	F19	13114.5–13120.5
H	162–168	C20	12808.5–12814.5	D20	12867.7–12973.7	E20	13060.5–13066.5	F20	13120.5–13126.5
I	168–174	C21	12814.5–12820.5	D21	12873.7–12879.7	E21	13066.5–13072.5	F21	13126.5–13132.5
7	174–180	C22	12820.5–12826.5	D22	12879.7–12885.7	E22	13072.5–13078.5	F22	13132.5–13138.5
8	180–186	C23	12826.5–12832.5	D23	12885.7–12891.7	E23	13078.5–13084.5	F23	13138.5–13144.5
9	186–192	C24	12832.5–12838.5	D24	12891.7–12897.7	E24	13084.5–13090.5	F24	13144.5–13150.5

TABLE 3.4 CARS Band Frequencies (Continued)

		CARS group C		CARS group D		CARS group E		CARS group F	
CATV channel	VHF boundaries, MHz	FCC designation	Microwave boundaries, MHz	FCC designation	Microwave boundaries, MHz	FCC designation	Microwave boundaries, MHz	FCC designation	Microwave boundaries, MHz
10	192	198	C25	12838.5–12844.5	D25	12897.7–12903.7	E25	13090.5–13096.5	F25	13150.5–13156.5
11	198	204	C26	12844.5–12850.5	D26	12903.7–12909.7	E26	13096.5–13102.5	F26	13156.5–13162.5
12	204	210	C27	12850.5–12856.5	D27	12909.7–12915.7	E27	13102.5–13108.5	F27	13162.5–13168.5
13	210	216	C28	12856.5–12862.5	D28	12915.7–12921.7	E28	13108.5–13114.5	F28	13168.5–13174.5
J	216	222	C29	12862.5–12868.5	D29	12921.7–12927.7	E29	13114.5–13120.5	F29	13174.5–13180.5
K	222	228	C30	12868.5–12874.5	D30	12927.7–12933.7	E30	13120.5–13126.5	F30	13180.5–13186.5
L	228	234	C31	12874.5–12880.5	D31	12933.7–12939.7	E31	13126.5–13132.5	F31	13186.5–13192.5
M	234	240	C32	12880.5–12886.5	D32	12939.7–12945.7	E32	13132.5–13138.5	F32	13192.5–13196.5
N	240	246	C33	12886.5–12892.5	D33	12945.7–13951.7	E33	13138.5–13144.5		
O	246	252	C34	12892.5–12898.5	D34	12951.7–12957.7	E34	13144.5–13150.5		
P	252	258	C35	12898.5–12904.5	D35	12957.7–12963.7	E35	13150.1–53156.5		
Q	258	264	C36	12904.5–12910.5	D36	12963.7–12969.7	E36	13156.5–13162.5		
R	264	270	C37	12910.5–12916.5	D37	12969.7–12975.7	E37	13162.5–13168.5		
S	270	276	C38	12916.5–12922.5	D38	12975.7–12981.7	E38	13168.5–13174.5		
T	276	282	C39	12922.5–12928.5	D39	12981.7–12987.7	E39	13174.5–13180.5		
U	282	288	C40	12928.5–12934.5	D40	12987.7–12993.7	E40	13180.5–13186.5		
V	288	294	C41	12934.5–12940.5	D41	12993.7–12999.7	E41	13186.5–13192.5		
W	294	300	C42	12940.5–12946.5	D42	12999.7–13005.7	E42	13192.5–13198.5		

NOTE: Group C channels add 12,646.5 MHz to VHF Frequency; Group D channels add 12,705.7 MHz to VHF Frequency; Group E channels add 12,898.5 MHz to VHF Frequency; Group F channels add 12,958.5 MHz to VHF Frequency.

TABLE 3.4 CARS Band Frequencies (*Continued*)

FM Frequencies (Bandwidth = 25 MHz FM A, FM B; Bandwidth = 12.5 MHz FM K)

CARS group FM A		CARS group FM B		CARS group FM K			
Designation	Microwave boundaries, MHz	Designation	Microwave boundaries, MHz	Designation	Microwave boundaries, MHz	Designation	Microwave boundaries, MHz
A01	12700.0–12725.0	B01	12712.5–12737.5	K01	12700.0–12712.5	K21	12950.0–12962.5
A02	12725.0–12750.0	B02	12737.5–12762.5	K02	12712.5–12725.0	K22	12962.5–12975.0
A03	12750.0–12775.0	B03	12762.5–12787.5	K03	12725.0–12737.5	K23	12975.0–12987.5
A04	12775.0–12800.0	B04	12787.5–12812.5	K04	12737.5–12750.0	K24	12987.5–13000.0
A05	12800.0–12825.0	B05	12812.5–12837.5	K05	12750.0–12762.5	K25	13000.0–13012.5
A06	12825.0–12850.0	B06	12837.5–12862.5	K06	12762.5–12775.0	K26	13012.5–13025.0
A07	12850.0–12875.0	B07	12862.5–12887.5	K07	12775.0–12787.5	K27	13025.0–13037.5
A08	12875.0–12900.0	B08	12887.5–12912.5	K08	12787.5–12800.0	K28	13037.5–13050.0
A09	12900.0–12925.0	B09	12912.5–12937.5	K09	12800.0–12812.5	K29	13050.0–13062.5
A10	12925.0–12950.0	B10	12937.5–12962.5	K10	12812.5–12825.0	K30	13062.5–13075.0
A11	12950.0–12975.0	B11	12962.5–12987.5	K11	12825.0–12837.5	K31	13075.0–13087.5
A12	12975.0–13000.0	B12	12987.5–13012.5	K12	12837.5–12850.0	K32	13087.5–13100.0
A13	13000.0–13025.0	B13	13012.5–13037.5	K13	12850.0–12862.5	K33	13100.0–13112.5
A14	13025.0–13050.0	B14	13037.5–13062.5	K14	12862.5–12875.0	K34	13112.5–13125.0
A15	13050.0–13075.0	B15	13062.5–13087.5	K15	12875.0–12887.5	K35	13125.0–13137.5
A16	13075.0–13100.0	B16	13087.5–13112.5	K16	12887.5–12900.0	K36	13137.5–13150.0
A17	13100.0–13125.0	B17	13112.5–13137.5	K17	12900.0–12912.5	K37	13150.0–13162.5
A18	13125.0–13150.0	B18	13137.5–13162.5	K18	12912.5–12925.0	K38	13162.5–13175.0
A19	13150.0–13175.0	B19	13162.5–13187.5	K19	12925.0–12937.5	K39	13175.0–13187.5
A20	13175.0–13200.0			K20	12937.5–12950.0	K40	13187.5–13200.0

only 32 VHF carriers (2–M). This AML system used by cable operators is called *community-antenna radio service* (CARS) band and is in the frequency band of 12.7 to 12.95 gigahertz (GHz).

Of the groups in Table 3.4 from D33 to F32 use is specified for studio transmitter link (STL) service. This amplitude-modulated link (AML) service is used for connecting the studio facilities, usually in a city, to the transmitter or up-link site out of the city. The antenna systems used at these frequencies are usually dish type (parabolic) from 4 to 10 ft in diameter. The transmitting antenna may be 10 ft and the receiving antenna size may be 6 ft in diameter. As in satellite antennas, the larger the diameter the greater the antenna gain. A diagram of the basics is shown in Fig. 3.57. Now, so far nothing else has been taken into account. However, several other factors that add to this loss have to be considered.

3.11.2 Refraction of radio wave propagation

The transmitted microwave beam from the transmitting antenna does not travel in a straight line toward the receiving antenna. It may be bent by refraction and defraction by the density of the earth's atmosphere and the extraneous electromagnetic fields and the earth's magnetic field. We know that the density of the air is a function of altitude, so naturally the tower heights and the elevations between the transmitting and receiving antennas have to be taken into account.

Path loss / dB $= L_{dB} = 36.6 + 20 \log L + 20 \log f$

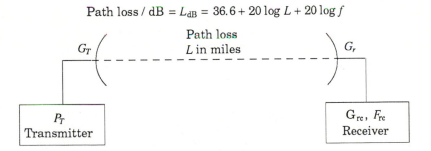

Example $L = 18$ mi $f = 12.95$ GHz   Find L_{dB}.

$L_{dB} = 36.6 + 20 \log 18 + 20 \log 12{,}950 = 36.6 + 25.1 + 82.2 = 143.9$
≈ 144 dB

Figure 3.57 Basic microwave link.

Obstacles in the line of sight path have to be considered as well. The earth's bulge is also a factor, as shown in Fig. 3.58a and b. Any obstruction in this area will be in the path. The amount of bulge in feet for any points in the path is given by the formula height of bulge $h = 0.667d_1d_2$.

3.11.3 *K* factor

So far any bending of the beam has not been taken into account. To do so the formula has to be modified by the K factor. For $K > 1$ the beam is bent toward the earth, which can allow possibly shorter towers. When $K < 1$, the beam is bent up, and for the same path distance the tower heights will have to be increased.

Essentially the value of K raises or lowers the bulge of the earth, and the formula is modified as follows:

$$h = \frac{0.667d_1d_2}{K} \qquad d_1, d_2 \text{ (mi)}$$

$$K = \frac{\text{effective earth radius}}{\text{true earth radius}}$$

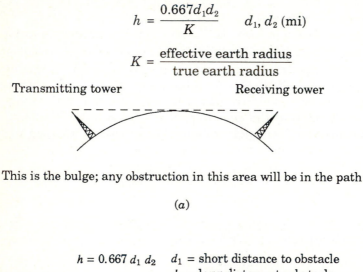

This is the bulge; any obstruction in this area will be in the path

(*a*)

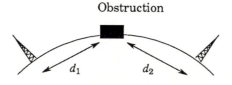

(*b*)

Figure 3.58 Earth's curvature and obstacle clearance.

Various methods are used to arrive at the K factor. The normal value often used is

$$K = \frac{4}{3} = 1.33$$

If one is actually engaged in making a path profile study, a more careful investigation of the correct K factor should be done for the area. For estimating purposes $K = \frac{4}{3}$ will not put one far off the mark.

3.11.4 Fresnel zone clearance

The *Fresnel zone clearance* is another factor to be taken into account for obstacles in the path of the beam. The Fresnel clearance accounts for the spreading of the beam, which now may be reflected off the obstacle returning to distort the beam pattern. Increasing the antenna heights solves the problem. Therefore, this phenomenon should be taken into account since raising the tower height later is costly. The additional clearance over an obstacle is given in Fresnel clearance zones. The zone is given as a radius in feet; 0.6 times this radius gives the optimum clearance of the obstacle. To calculate the Fresnel zone radius the following formula will be used:

$$\text{Fresnel radius } R_{mi} = 72.1 \sqrt{\frac{d_1 d_2}{fL}}$$

where f = frequency (GHz)
d_1 = distance from transmitter to obstacle (mi)
d_2 = distance from obstacle to receiver (mi)
L = path length $[d_1 + d_2]$ (mi)

Now since some of the basic properties of a line of site radio link have been discussed, we will apply them in a practical example called path profiling.

To check out our desired path, a set of 7.5-min maps covering the path should be obtained from the U.S. Geological Survey. Often these maps may be purchased for a few dollars from sporting goods stores or where camping, hunting, and fishing supplies are sold. The 7.5-min maps have a scale of 1:24,000 where 1 in covers 2000 ft. Elevation contour lines give the ground heights of the hills in the path. Once the main transmitter site and receiving site are plotted on the map a line drawn connecting these sites will be the line where obstacles should be investigated. The first step is to find on the map the geographical coordinates of the primary transmitting and receiving sites needed for the FCC licensing procedure. Then along the drawn line height points are picked off so Fresnel zone height corrections may be calculated. Often it is more convenient to transfer the information to linear graph paper so the path profile will be more clear. Ten divisions per inch is usually adequate, and the horizontal axis at 2 mi/in will cover 20 mi on a piece of paper, and the vertical axis at 100 ft/in will give 700 ft of vertical elevation. If more space is needed for more mountainous terrain and longer distances, pages can be taped or pasted together. All heights should be referred to mean sea level (MSL); however, the graphed profile only has to cover on scale the lowest and highest points.

The line drawn on the topographical map showing the proposed path should be analyzed for obstacles and should be marked for bodies of water, dry areas, etc.

Now plot the path of transmitter to receiver site in height and distance and determine the midpoint of the path. K will be found in one of the following ways. The first is to use the normal value of $K = 1.33$ for normal dry areas. For wet and water paths $K = 1$ or ½ can be used. A more accurate method is to refer to maps of sea level refractivity index contours N for the area. Usually for the continental United States August is the worst month because it is the hottest month. Now a graph of refractive index versus midpath elevation and K factor will give a possibly more correct value of K. See Figs. 3.59 and 3.60.

Example   Refer to Fig. 3.61. A microwave hop distance of 18 mi, a transmitter elevation of 500 ft, and a receiver elevation of 250 ft give

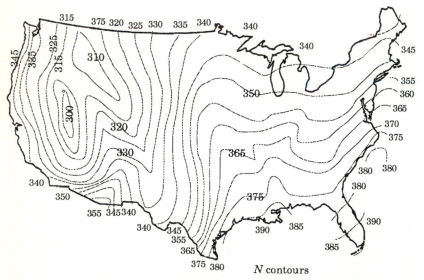

Figure 3.59   Sea level refractivity index for the continental United States: maximum for worst month (August).

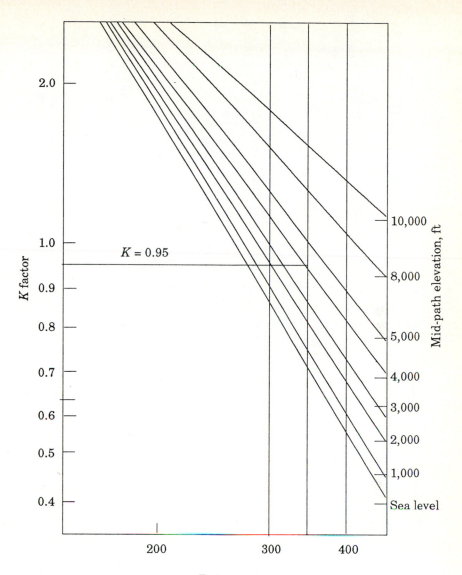

Figure 3.60 *K* factor scaled for midpath elevation above mean sea level.

a midbeam elevation of 375 ft. Consulting the map for area $N = 350$ and plotting on the chart of K give a K factor of 0.95.

Now earth's curvature EC for this midpoint, which is regarded as an obstruction, will be calculated and plotted, as well as other points of obstructions as obtained from the topographical maps.

$$\text{EC}_{\text{mid}} = \frac{0.667d_1^2}{0.95} = \frac{0.667 \times 9^2}{0.95} = \frac{0.667 \times 81}{0.95} = 56.8 \text{ ft} \approx 57 \text{ ft}$$

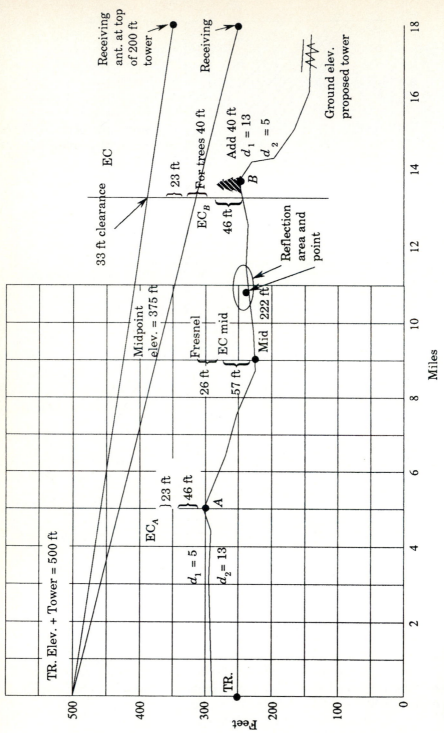

Figure 3.61 Microwave path profile.

$$EC_A = \frac{0.667 \times 5 \times 13}{0.95} = 45.6 \approx 46\text{ft}$$

$$EC_B = \frac{0.667 \times 5 \times 13}{0.95} = 45.6 \approx 46\text{ft}$$

The next task is to calculate the Fresnel height corrections. To do so, determine Fresnel zone radius, using the graph of frequencies versus path lengths or using the following formula for first zone midpath.

$$R_f = 1140 \sqrt{\frac{L}{f_{MHz}}}$$

where L = path length (mi)
$\quad\quad f$ = frequency (MHz)

$$R_f = 1140 \sqrt{\frac{18}{12,950}} = 42.5\text{ft} = 43\text{ft}$$

Note the previous formula $R_f = 72.1 \sqrt{\dfrac{d_1{}^2}{f_{GHz}L_{mi}}} = 72.1 \sqrt{\dfrac{81}{12.95 \times 18}}$

$$R_f = 42.5\text{ft} \approx 43\text{ft} \quad\quad \text{which is the same}$$

Now the height correction is 0.6 times 43 ft.

$$0.6 \times 42.5 = 25.5\text{ft} \approx 26\text{ft}$$

For obstacle points A and B the Fresnel zone clearance will be calculated. At point A, where the distance is 5 mi from the receiver and 13 mi to the transmitter, d_1 as a percentage of L is $5/18 \times 100$ or 27.7 percent; d_2 is $13/18 \times 100$ or 72.2 percent. This is corrected on the graph in Fig. 3.62, which actually corrects other path points as a percentage midpath value. So for our example 27.7 percent and 72.2 percent give 90 percent of midpoint value: 0.90×26 ft = 23.4 ft $\approx$ 23 ft.

At point B the ratio is the same, that is, 5 mi from the receiver and 13 mi from the transmitter.

The problem so far is that it can be seen that at the receiver site ground level is 150 ft and a 100-ft tower will not raise the path above the corrected B obstacle. We will certainly want to clear ground level.

Reflection points can range from about 10 to 50 percent of the distance from the shorter tower height. The following graph (Fig. 3.63) points to the possible reflection areas. If this area falls on very smooth damp terrain or water, then reflected energy can cause a problem. From the graph in Fig. 3.63 the reflection distance area is 0.42 times

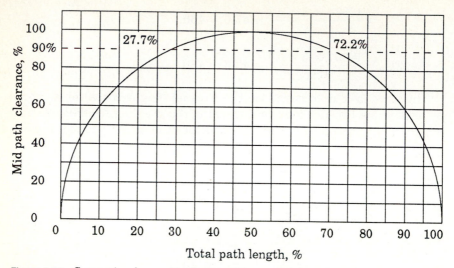

Figure 3.62 Conversion from midpath Fresnel zone clearance for other than midpath obstacles.

the total distance of 18 mi. So 0.42 × 18 = 7.56: 7.5 mi from the receiving antenna (the shorter tower height).

The foregoing path profile has just told us that we need certain tower heights to clear the obstacles and get a clear shot for the path

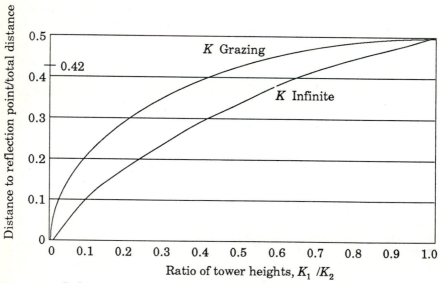

Figure 3.63 Reflection distance correction.

with a loss of 144.0 dB. The next task is to look at equipment, both transmitting and receiving, that can overcome this loss.

To calculate the signal level at the receive antenna we take the transmitted power level in dBm (decibel-milliwatts) minus the path loss, which in our example is 144.0 dB. Let us assume that the transmitter power is 1 Watt (W) (30 dBm) and the line loss is negligible and the transmitter antenna gain is 45 dB. The effective radiated power (ERP) is 75 dBm.

Recall that 0 dBm = 1 mW and in this case 50 Ω. Now at a transmitter power of 1 W

$$P_T = 10 \log \frac{1\text{ W}}{1 \times 10^{-3}\text{ W}} = 10 \log 1 \times 10^{-3} = 10 \times 3 = 30 \text{ dBm}$$

Now the gain of the antenna at 45 dB actually increases the transmitter power level of 30 dBm to 75 dBm.

$$P_o = 75 \text{ dBm}$$

The level at the receive antenna is this power less the path loss of 144.0 dB, which is

$$P_R = P_o - P_L = 75 - 144.0 = -69 \text{ dBm}$$

If the receive antenna has a gain of only 40 dB the receive carrier signal level becomes

$$C_R = P_R + G_R = 69 + 40 = -29 \text{ dBm}$$

Suppose the receiver manufacturer states that the noise power level at normal room temperature is -90 dBm; then with a C_R of -29 dBm we can calculate the receiver carrier-to-noise ratio.

Recall that C_R/N_R is in watts, milliwatts, etc., so

$$C_R/N_{R_{db}} = C_{R_{dB}} - N_{R_{dB}} = -29 - (-90) = 60 \text{ dB}$$

which means that the carrier level is greater than the noise level.

What we are after is a good video baseband signal-to-noise ratio. Practically with a carrier to noise level of 60 dB for our example we should expect to have a S/N ratio at baseband that will be adequate. Recall that a 52-dB S/N is considered a good-quality signal. Refer to the section on video signal measurements in Chapter 6.

The type of transmission AM or FM is important in that there are differences in the types of formulas used in calculating S/N ratios. It is extremely important when planning a microwave feed system to make some simple calculations and then consult the various manufacturers' field engineers for aid in making such calculations before choosing the type of equipment.

3.11.5 Path reliability, rain, and fade

Once the link is chosen and planned, the path reliability should be also investigated. For 0 percent reliability the system is out all of the time. A year of 365 days at 24 h/day equals 8760 h. Each month has 730 h however, use the average value of 720 h. Now, of course, 50 percent reliability means that the system operates half the time, which boils down to half the above figures. Now a reliability chart can be made, as shown in Table 3.5.

The formula given for *fade probability*, which is in "hours outage" due to multipath problems, is

$$F_P = F_{TR} \times F_{T/H} \times 2.5 \times 10^{-6} \times f \times L^3 \times 10^{-fm/10}$$

where F_P = fade probability (outage hours)

$\quad F_{TR}$ = terrain factor

$\qquad$ = 4 $\qquad$ for smooth terrain and water in path

$\qquad$ = 1 $\qquad$ for average terrain

$\qquad$ = 0.25 $\qquad$ for mountainous terrain

$\quad F_{T/H}$ = temperature-humidity factor

$\qquad$ = 0.5 $\qquad$ for hot-humid areas

$\qquad$ = 0.25 $\qquad$ for normal northern climate

$\qquad$ = 0.125 $\qquad$ for high, mountainous or dry desert

$\qquad f$ = operating frequency (GHz)

$\qquad L$ = path length (statute miles)

$\quad f_m$ = system fade margin; for AML C/N 35-and

$\qquad$ FML = threshold improvement C/N

TABLE 3.5 Reliability Chart

Reliability, percentage	Percentage	Outage time		
		Per year	Per month	Per day
0	100	8760 h	720 h	24 h
50	50	4380 h	360 h	12 h
80	20	1752 h	144 h	4 h, 48 min
90	10	876 h	72 h	2 h, 24 min
95	5	438 h	36 h	1 h, 12 min
98	2	175 h	14 h	29 min
99	1	88 h	7 h	14 min
99.9	0.1	8.8 h	43 min	1 min, 2 s
99.99	0.01	53 min	4.3 min	8.6 s

The outage hours due to rain may be calculated by analyzing the type of rain normally found in the path area, and its rate, and its hours. The following formulas are needed.

$$\text{Critical rain rate} = \left(0.465\frac{f}{L}\right)^{0.8}$$

$$\text{Slope of rain curve} = \log\left[\left(\frac{R_{max}/24 \text{ h}}{\text{hours of rain}}\right)^{0.72}\right]$$

where R_{max} = maximum rainfall (in) in any 1 h in last 5 yr. Now total hours outage due to rain is

$$\text{Rain outage hours} = \left(\frac{\text{critical rain rate}}{1 \text{ h rain}}\right)^{1/\text{slope}}$$

Let's go back and plug in some practical numbers.
For fade probability outage hours assume

$$F_{TR} = 1 \qquad \text{average terrain}$$

$$F_{T/N} = 0.25 \qquad \text{normal northern climate}$$

$$L = 18 \text{ mi}$$

$$f = 35 \text{ dB C/N AML}$$

Substituting

$$F_P = 1 \times 0.25 \times 2.5 \times 10^{-6} \times 12.95 \times 18^3 \times 10^{-35/10}$$

$$= 8.1 \times 10^{-6} \times 18^3 \times 10^{-3.5}$$

$$= 8.1 \times 10^{-6} \times 5832 \times 0.000316$$

$$= 8.1 \times 5832 \times 10^3 \times 10^{-6} \times 3.14 \times 10^{-4}$$

$$= 149.3 \times 10^{-7} \text{ h} \approx 0.15 \times 10^{-4} \text{ h}$$

Now let us calculate rain problem.

$$\text{Critical rain rate} = \left(0.465 \times \frac{35}{18}\right)^{0.8}$$

$$= (0.465 \times 1.94)^{0.8} = (0.9)^{0.8} = 0.92$$

Suppose that if for the 24-h period worst rain in last 5 yr was 8 in of rain. This is at a rate of ⅓ in/h for 24 h = 8 in. Also during the last 5 yr it was found that the worst rainfall for any 1-h period was 1 in/h.

$$\text{Therefore for the slope} = \left(\frac{8 \text{ in}/24 \text{ h}}{1 \text{ in/h}}\right)^{0.72} = \left(\frac{1}{3}\right)^{0.72} = (0.33)^{0.72} = 0.45$$

To find the hours of outage due to rain we combine the previous information as follows:

$$\text{Rain outage hours} = \left(\frac{\text{critical rain rate}}{\text{max rain rate/h}}\right)^{1/\text{slope}}$$

$$= \left(\frac{0.92}{1 \text{ in/h}}\right)^{1/0.45} = (0.92)^{2.22} = 0.83 \text{ h}$$

Taking into account the outages in hours due to fade (multipath) and rain (precipitation), we can calculate the percentage reliability.

$$\% \text{ Reliability} = \frac{8760 - (\text{multipath outage hours}) + (\text{rain outage hours})}{87.6}$$

$$= \frac{8760 - (0.15 \times 10 \text{ h} + 0.83 \text{ h})}{87.6}$$

$$= \frac{8760 - 0.83}{87.6} \approx \frac{8759}{87.6} = 99.99 \text{ percent}$$

For all practical purposes this is very good, because according to our chart 99.99 percent corresponds to 53 min/yr or 4.3 min/month or only 8.6 s/day outage time periods.

3.12 Optical Communications: Fiber Optics

It should be remembered from previous work that radio waves travel at the speed of light because light waves are also electromagnetic radiation. What makes light waves special is that they can be seen, i.e., the human eye is a detector of light waves or electromagnetic radiation in the visible region. In Fig. 3.64 is a spectrum of visible light and at the wavelengths shown for the various frequencies (colors) in the visible area of the spectrum

The angstrom (Å) is used by people working in optics and colorimetry, where 1 Å corresponds to a distance of 10^{-9} m = 1 nanometer (nm). Therefore, 5000 Å corresponds to

$$10^{-9} \frac{\text{m}}{\text{Å}} \times 5 \times 10^3 \text{ Å} = 5 \times 10^{-6} \text{ m} = \lambda \qquad \text{the wavelength}$$

Converting to frequency,

$$\lambda_{\text{m}} = \frac{300}{f_{\text{MHz}}} \qquad f_{\text{MHz}} = \frac{300}{\lambda_{\text{m}}}$$

and

$$f = \frac{300}{5 \times 10^{-6}} = 60 \times 10^6 \text{ MHz} = 60 \times 10^3 \text{ GHz}$$

If the preceding calculation is made for the ends of the visible spectrum, the frequency range of the visible spectrum turns out to be approximately 1×10^{14} Hz to 1×10^{15} Hz.

As a communication medium, light waves are extremely efficient.

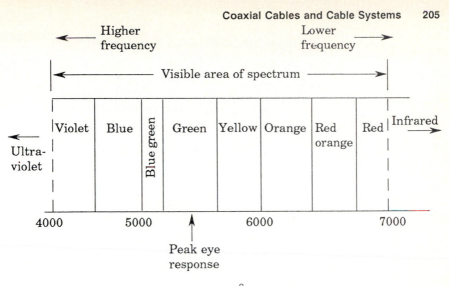

Figure 3.64 Visible spectrum.

Recall that radiation is given in microvolts/meter (μV/m), where "per meter" indicates the wavelength in meters. So for very short wavelengths, i.e., high frequencies, the denominator of the fraction gets smaller; thus the E (electric) field increases or less power is required to generate the E field.

After all our eye sees very well just using reflected light often at very low levels. In order to use such high-frequency electromagnetic waves, more study of light waves is necessary. At such high frequencies it is known that glass lenses can bend and focus the waves or rays and that such rays can be reflected from mirrors and shiny, smooth surfaces. It is also common knowledge that light waves can be polarized just as radio waves can and that such rays can also be absorbed by certain materials.

3.12.1 Reflection and refraction

The law that states the properties of reflection and refraction is *Snell's law*. To illustrate the principles, of Snell's law consider a boundary layer which divides two media of different refractive incidence. This may sound complicated, but it isn't because one medium can be glass and the other air. Therefore, light waves passing through a piece of glass is really a simple model. However, suppose we are talking about two sheets of different glass, one on top of the other, as shown in Fig. 3.65.

The incident ray traveling in medium 1 arrives at the boundary and

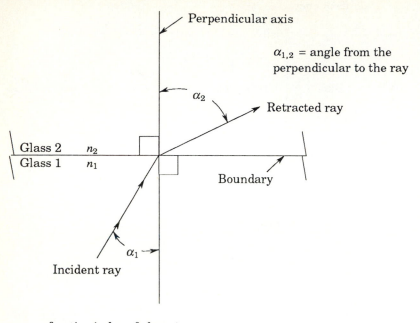

n_1 = refractive index of glass 1
n_2 = refractive index of glass 2

Figure 3.65 Snell's law.

passes through into medium 2 bent at a greater angle than the incident angle.

Now suppose that the angle of incidence is increased; then it is possible that the ray will never get into medium 2. The point where the ray just fails to enter medium 2, the incident angle, is called the *critical angle*. At any angle of incidence greater than the critical angle the ray will be reflected back into medium 1. The cases for the critical angle and angles of in a degree greater than the critical angle are shown in Fig. 3.66.

If medium 1 happens to be air, then the refractive index n = 1.0003 $\approx$ 1 and for a good grade of glass n = 1.50. Light waves are slowed as the refractive index increases. Therefore light waves travel more slowly in glass than in air. The *angle of incidence* is represented by α_1 (α is the Greek letter alpha); α_2 represents the *angle of refraction*. The relationship between these two angles is given by

$$\frac{\sin \alpha_1}{\sin \alpha_2} = \frac{\sin \alpha_i}{\sin \alpha_R} = \frac{n_2}{n_1} = K$$

where K = constant

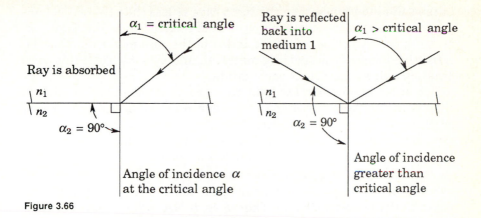

Figure 3.66

One of the first applications of optical communications appeared in very simple form in the light pipe head lamp indicators used in some automobiles. Here was an actual indication informing the driver that the bulbs were good and the lamp was actually "on." At first these light pipes were solid plastic bent rods covered with a plastic sleeve. Often mechanics working on the cars inadvertently caused a break in the light pipe. Even then some light did pass through, although the indication was dim. Since these plastic rods had to be custom bent in certain required shapes, manufacturing them was a costly and complicated procedure. Naturally it was found that a bundle of continuous fibers of filaments worked even better and was flexible, easy to install, and not subject to breaking. The efficiency of this product was good and uses were extended to passing light into difficult small areas for inspection of critical components. The medical profession used light pipes for diagnostic instruments not only for providing illumination but for optical viewing through the fiber-optic cable. Such a light pipe system is shown in Fig. 3.67.

Since the light rays travel in the glass medium, the refractive index should be as low as possible and the glass material should be pure and free of impurities that could harm the transmission. Such impurities can cause absorbance and refraction of the rays trying to pass through. The covering sleeve is also important. If the fiber is bare, as

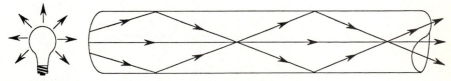

Figure 3.67   Light pipe: forerunner to fiber-optic systems.

in the solid light pipes, then the light rays at less than the critical angle will escape. This covering material is referred to as the *cladding*. Now suppose n_1 is the refractive index of the core material, n_2 is the refractive index of the cladding material, and $n_1 > n_2$. Then this combination will act as a light wave guide.

To investigate further, the light ray acceptance of an optical fiber is shown in Fig. 3.68. The angle of incidence has to fall in the cone of acceptance. The numerical aperture (NA) is related to the angle of acceptance. Mathematically NA = sin θ, where θ is an angle equal to half the acceptance cone angle, and NA is the numerical aperture. In terms of the refractive index NA = $\sqrt{n_1^2 - n_2^2}$.

Fairly normal commercial fiber-optic cable has an NA factor of about ½ (0.5) or slightly less. Therefore, if NA = 0.5 then sin θ = 0.5 from our calculator by entering decimal point and numeral 5 and pressing inv and sin to obtain θ = 30°. This means the acceptance cone angle is 60°. Therefore, any ray entering the fiber within this cone will pass through.

3.12.2   Types of fiber-optic cables

At present there are three categories of fiber-optic cable: (1) single mode, (2) step index, and (3) graded index. Both step and graded index are multimode types.

Single-mode fiber allows only one mode to propagate through the light wave guide. The step index fiber exhibits a sharp change in refractive index of the core and cladding material, whereas the graded index exhibits a smooth change. The diagram in Fig. 3.69 shows the difference between step index and graded index fiber.

The geometry of the cable as well as the refractive index of the cladding and core material allow the modes of operation. Essentially single-mode fiber accepts light waves parallel to its axis or focused

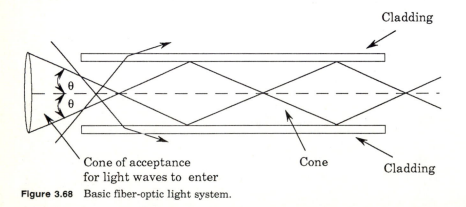

Cladding

θ
θ

Cone of acceptance
for light waves to enter

Cone

Cladding

Figure 3.68   Basic fiber-optic light system.

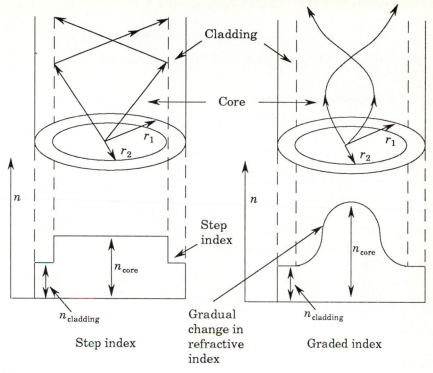

Figure 3.69

into the fiber. These light rays pass directly through the fiber to the end. The core has a very small diameter allowing only a few rays parallel to the core axis. Multimode-type fiber rays are reflected from the cladding back into the core so the many rays travel different paths hence arrive at the end at different phase relationships. The effects of multimode operation can cause a certain amount of signal distortion.

The relationship between the geometry of the fiber-optic cable and the composition of the cladding and core materials is called the *normalized frequency* F_n; where F_n is 2.4 or less, only one mode operation will result.

$$F_n = \frac{2\pi r}{\lambda} \text{NA}$$

For example, if the signal frequency of the light has a wavelength of 850 nm, NA = 0.5, and F_n = 2.40, then we can calculate the core radius.

$$2.40 = \frac{2\pi r}{850 \times 10^{-9}} \left(\frac{1}{2}\right)$$

$$2\pi r = (4.8)(850 \times 10^{-9})$$

$$r = (0.764)(850)(10^{-9})$$

$$= 650 \times 10^{-9} \, \text{m} = 0.65 \, \mu\text{m}$$

One can see that this is very small fiber-optic cable.

3.12.3 Fiber-optic cable losses

Again nothing is perfect. All the light that enters one end of the cable does not come out because of four basic losses: (1) absorption, (2) scattering, (3) connector loss, and (4) bending loss.

Impurities in the core material cause the absorption and scattering losses. Such impurities may be air bubbles in the plastic or glass core or foreign materials (specks) that interfere with transmission. The losses are lumped together so the measure of loss is in decibels per kilometer (dB/km). Attenuation on the order of 0.5 dB/km is fairly common. This is certainly far better than that of coaxial cable used in cable television systems.

Connection losses are still a problem today and more improved snap-together connectors are in various stages of development. Fusing through heat worked quite well, but it was precision work, required expensive equipment and highly trained technicians. The connector losses were light losses around the interface and scattering because of the boundary at the fused splice. Typical splicing losses should be 0.1 dB or less per splice.

Excessive bending by exceeding the prescribed bending radius distorted the cable, causing more modes of operation and light to exit the fiber and be absorbed in the cladding or jacket. Most manufacturers of optical fiber cable give the mechanical handling specifications, e.g., minimum bending radius, maximum pulling tension, and any other pertinent handling recommendations. Generally if cable is improperly handled, warranties and/or guarantees are usually voided by the manufacturer. Fiber-optic cable is usually much lighter in weight, quite flexible, and generally easier to work with as far as placing it, i.e., aerial or underground. Splicing, branching, and connecting to the optical amplifiers pose the problem.

3.12.4 Modulating techniques

Modulation is essentially varying the carrier in some intelligent manner so that at the receiving point demodulation can take place and the intelligence can be recovered. Light sources are usually single-frequency carriers (color) of high intensity. Most light sources used today are solid-state light-emitting diodes (LEDs) or injection laser diodes. LEDs are less efficient than lasers and the light energy is spread over the diode surface; thus, coupling to the fiber is a problem. Light energy is lost. The laser couples to the fiber core better because it is a high-energy concentrated beam with a narrow aperture. The LED generally lasts 10 times longer than the laser.

The light output of the optical transmitter can be varied (modulated) in frequency and amplitude. The FM technique, like any form of frequency modulation, has a better signal-to-noise ratio, hence is better for the long haul than AM technique. AM techniques are generally lower in cost and have wide use today for many applications. A fiber-optic transmission system in general terms is shown in the Fig. 3.70.

Recall that for AM broadcast radio service, e.g., operating at 1500 kHz with a top modulating frequency of 15 kHz corresponds to a ratio of 100 to 1. Therefore, using the same principle for a carrier with a wavelength of 850 nm gives

$$f_{\text{MHz}} = \frac{300}{850 \times 10^{-9}} = \frac{300}{850} \times 10^9 = 0.350 \times 10^9 \text{ MHz} = 350 \times 10^6 \text{ MHz}$$

for the carrier frequency. Now for the amplitude modulation case, two sidebands that are displaced from the carrier frequency (above and below) an amount equal to the modulating frequency for a bandwidth, in the case of a 1500-kHz carrier with 15-kHz modulation, equal to 30 kHz (30 kHz = 2 × 15 kHz), or twice the modulating frequency. For the carrier frequency of 350.0×10^6 MHz the signal bandwidth equals

$$\frac{350 \times 10^6 \text{ MHz}}{2 \times 100} = 1.75 \times 10^6 \text{ MHz}$$

This example is just posing theoretical limits. However, the limiting factors are the transmitter and receiver characteristics. The diagram

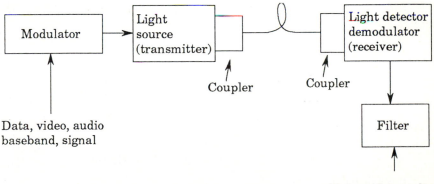

Figure 3.70   Schematic diagram of a fiber-optic communication system.

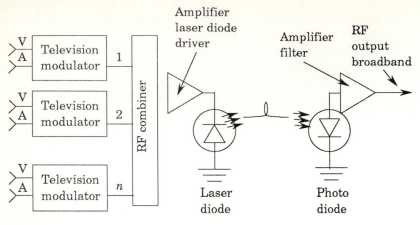

Figure 3.71   AM fiber-optic trunk system.

for the AM case carrying broadband RF carriers with AM video signal is shown in Fig. 3.71.

Since the transmitting laser has a transfer characteristic, operation has to be maintained within the proper limits of the linear portion of the transfer curves as shown in Fig. 3.72.

I_m should never be greater than $I_{B'}$ or serious distortion will result. If the driving signal is such that m (depth of modulation) is increased, C/N ratio will get better but distortion will increase. Thus there is a trade-off. An example should clear up any misunderstandings.

Example   A fiber-optic cable with a loss characteristic for the chosen carrier frequency of 2.5 dB/km is to be operated over a 9-km run. Two reels of fiber-optic cable with 4.5 km per reel will be spliced together with a splice loss of 0.5 dB. At the transmitter end and the receiver end the connection loss is 0.5 dB per connection. A list of typical values follows.

Transmitter power $P_{(+)}$: -2.5 dBm

Connector loss: 0.5 dB

Power into cable: -2.0 dBm

Power at receiver: -2.0 dBm $-$ 23 dBm $=$ -25 dBm

Receiver sensitivity: -30 dBm

Loss margin: 5 dB above minimum needed (workable)

Fiber-optic cable loss: 22.5 dB

Splice loss: 0.5 dB

Total cable loss: 23 dB

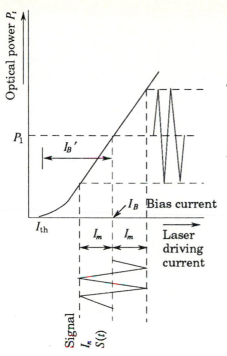

I_{th} = laser threshold current

I_B = laser bias current

$I_B' = I_B - I_{th}$ = maximum possible current swing

$S(t)$ = modulating signal (made up from many R.F. carriers)

$$m = \frac{I_m}{I_B'} = \text{depth of modulation}$$

$$P(t) = P_1\left[1 + m\,S(t)\right]$$

Figure 3.72 Laser transfer characteristic curve.

The preceding example is quite typical. The curve has a shape not so different from that of many electronic devices, one of which is the cable television trunk amplifier. Devices that have input/output transfer characteristic curves such as this one exhibit some second-order and third-order distortion components, mainly intermodulation beats. Such intermodulation beats, you will recall from the section in Chapter 3 on cable amplifiers, consists of $f_1 \pm f_2$ for the second-order case and $f_1 \pm f_2 \pm f_3$ for the third-order case. Usually the transmitting laser diode generates most of the distortion; hence caution must be used in keeping the bias stable and the driving signal amplitude in the linear portion of the transfer characteristic curves. Single-mode laser diodes operating with $m = 0.5$ or 50 percent depth of modulation have for the second-order case 35 to 45 dB down and 45 to 60 dB for the third-order intermod case. Circuits are used to measure the light output of the laser and the temperature, and through feedback control circuits stabilize the laser operation to limit the distortion. Having more RF channels modulating the laser causes beat stacking, i.e., composite triple beats.

From the diagram of the laser communications system (Fig. 3.70) the C/N ratio at the receiver is made up of the noise generated by the transmitter plus quantum noise at the receiver detector diode plus that contributed by the receiver electronics. If the laser power output is low, causing the receiver signal to be low, then the noise at the receiver will be the limiting factor for proper signal recovery. If the laser output is increased beyond the correct operating point, it may become the culprit by contributing more noise power. Therefore, it is a tradeoff to get the best C/N ratio.

Frequency division multiplexing (FDM) occurs when CATV carriers in their normal mode modulate the laser and at the receiver are detected and then amplified by normal cable television amplifiers to distribute the cable service to subscribers. To date, this system works well for short hauls (15 to 20 km).

3.12.5 Frequency modulation: trunking

For longer trunk runs, such as supertrunking, head end to HUB, or HUB to HUB, FM techniques are used. The block diagram in Fig. 3.73 illustrates this system.

This system intensity-modulates the laser according to the wideband FM broadband signal. Since video/audio FM systems are in-

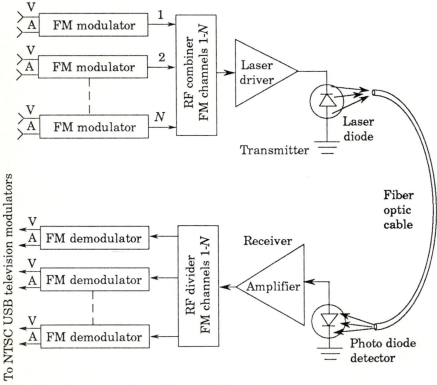

Figure 3.73   Fiber-optic FM trunking method.

corporated, the usual FM threshold improvement holds. For FM systems (see Section in Chapter 2 on FM) the modulation index is defined as

$$M = \frac{\Delta F}{F_m}$$

where ΔF is the peak frequency deviation and F_m is the highest modulating frequency. According to *Carson's rule of bandwidth*, which is accepted by the FCC and other agencies, $B_{IF} = 2(F_m + \Delta F)$: intermediate-frequency (IF) bandwidth B_{IF} equals two times the sum of the highest modulating frequency plus the peak frequency deviation. Carson's rule is a method of estimating the IF bandwidth necessary to reach the FM improvement threshold.

Example   If a video signal has as its highest frequency component 4.0 MHz (necessary to preserve the video and sync pulse information) and deviates an FM carrier 6.7 MHz find how much IF bandwidth is required. According to Carson's rule $B = 2(4 + 6.7) = 2(10.7) = 21.4$ MHz.

To be generous let us allow a B_{IF} to be 30 MHz. The relationship of the video signal-to-noise ratio to an FM modulated carrier is estimated by the following formula:

$$(S/N)_{WT} = \left[\frac{3M^2 B_{IF}}{2F_m}\right] + (C/N)_{IF} + W$$

The first two terms are given as a power ratio and not decibels. W is a combination of the International Radio Consultative Committee (CCIR) noise weighting function of 11.5 dB and the noise improvement ratio due to FM preemphasis, deemphasis of 3.6 dB, and the peak to peak/rms conversion factor of 9 dB.

Recall for a ± 1-V excursion of a sine wave, the ratio is 2 V pk to peak $\div 0.707 = 2.83$ and $20 \log 2.83 = 20 \times 0.45 = 9$ dB. Therefore W in decibels is $3.7 + 11.5 + 9 = 24.1$ dB. The first term can be calculated as follows:

$$M = \frac{\Delta F}{F_m} = \frac{6.7 \text{ MHz}}{4 \text{ MHz}} = 1.68 \qquad \text{the index of modulation}$$

Therefore;

$$\frac{3(1.68)^2(30 \times 10^6)}{2 \times 4 \times 10^6} = \frac{3(1.68)^2(30)}{8} = 31.75$$

For the second term $(C/N)_{IF}$ may be calculated by the formula

$$C/N_{IF} = \frac{P_{REC}}{K_B T_e B_{IF}}$$

This says that the receiver carrier-to-noise ratio is the received signal power divided by the received noise power at the IF bandwidth of 30 MHz. From before, recall that the receiver power was given as -26 dBm. Converting this corresponds to the power ratio

$$-26 \text{ dBm} = 10 \log P \qquad \log P = -2.6$$

Using the calculator, enter 2.6; press plus or minus ($\pm$); press inv, then log. This gives $0.0025 = 2.5 \times 10^{-3}$.

P_r corresponds to 2.5×10^{-3}, K_B is Boltzmann's constant of 1.38×10^{-23} J/°K. The missing quantity is T_e, which is the noise temperature of 300°K times the noise figure of the receiver. For a receiver with a worst-case noise figure of 20 dB this quantity has to be converted to a power ratio.

$$T_e = °\text{K}(\text{NF} - 1)$$

From memory 20 dB should give us the power ratio of 100 to 1.

$$T_e = 300°\text{K}(100) = 30,000 = 3 \times 10^4$$

$$\text{Now to calculate } (C/N)_{\text{IF}} = \frac{2.5 \times 10^{-3}}{(1.38 \times 10^{-23})(3 \times 10^4)(30 \times 10^6)}$$

$$= \frac{2.5 \times 10^{-3}}{(1.38)(10^{-13})(90)} = \frac{2.5 \times 10^{-3}}{124 \times 10^{-13}}$$

$$= \frac{2.5 \times 10^{-3}}{1.24 \times 10^{-11}} = 0.2 \times 10^8$$

$$= 20 \times 10^6$$

Therefore, the first term added to this number is quite insignificant. Now the 24.1 dB corresponds to a power ratio

$$2.41 = \log P_w \qquad P_w = 257$$

$$\text{S/N} = 31.75 + 20 \times 10^6 + 257 \approx 20 \times 10^6$$

$$= 10 \log 2 \times 10^6 = 73 \text{ dB}$$

If this is the optic fiber S/N, then if other equipment feeding the optical transmitter and receiver has, for example, a 62-dB S/N ratio then the 73 dB will not degrade this significantly.

3.12.6 Amplitude modulation

According to the theory of AM-type modulation, FM modulation with the same input bandwidth exhibits a 14 dB better signal-to-noise ratio than AM. Therefore, because of our calculation of 73 dB S/N for FM supertrunking, we can extrapolate to the AM case by subtracting 14 dB. This gives a value of S/N (AM) = 59 dB, which is still very good but would limit use of AM to shorter hops. Recall the work studied previously in combining C/N or S/N ratios. This example illustrates that one optical fiber hop of 40 km with a loss of 0.4 dB/km for the fiber-optic cable gives a loss for the 40-km run of 16 dB. If one can buy rolls of fiber-optic cable at 4 km/roll, then to cover 40 km there will be 11 splices. The splice connector with a loss of 0.25 dB/splice × 11 = 2.75 dB total connector loss. If an optical transmitter with signals modulated according to our FM trunk example has a power into the fiber-optic cable of −0.25 dBm, the receiver with a −26-dBm sensitivity should see −0.25 dBm −2.75 dB − 16 dB = −19-dBm signal. Therefore, a −26-dBm sensitivity gives a system margin of 7 dB. The link gives an S/N ratio of 73 dB. Recall that the noise builds up to the point that doubling the hops causes the C/N to decrease by 3 dB. Theoretically one could go to possibly six to seven times the 40-km distance; for example, 280 km at 0.61 mi/km corresponds to 170 mi. By now one may wonder about fiber-optic cable television applications. Great distances can be spanned by using FM and frequency division multiplexed (FDM) techniques such as previously discussed. Recall that in FDM there are many different signal frequencies (carriers), and in this case they are frequency-modulated. If the carriers are VSB National Television System Committee (NTSC) television carriers, then the method would be termed AM VSB (vestigial side bands) FDM. For shorter hauls such as 10 to 12 mi then AM-FDM can effectively be used as a fiber-optic trunk (backbone) feeding small minihubs (bridgers) using standard CATV techniques. Such a plan is shown in Fig. 3.74.

The diagram shows only four minihubs being fed. Of course, there could be more. The fiber-optic feed to the minihub has no electronics between the head end and the minihub. At the minihub location the receiver is usually a sensitive photo diode detector feeding an amplifier. The output of the amplifier is the AM_{VSB} signal carrier's "on channel," which drives the cable television distribution system. Since the signals arriving at the minihubs are high-quality signals, the short cascades of amplifiers can back-feed along the fiber-optic trunk route as well as the surrounding areas of the community. Visually short cascades of high-quality line extender CATV amplifiers and distribution cable feed the signals to the subscribers from the minihubs. Since there are no electronic elements contained in fiber-optic trunk runs, maintenance problems are very small. In addition, power outages affect only the areas where the outages occur.

3.12.7 Instruments and testing

Generally fiber-optic systems with the exception of the laser diode transmitters require very little maintenance. Since the fiber-optic ca-

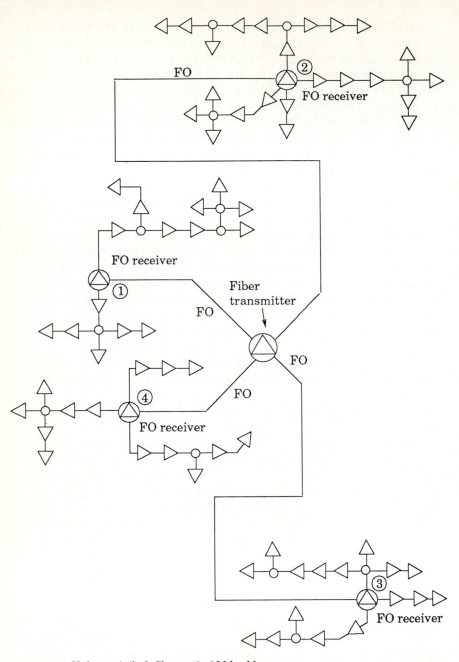

Figure 3.74   Hub to minihub fiber-optic AM backbone.

ble does not pick up lower-frequency nonoptic signals, it can be realized as radiation- and leakage-free. Electromagnetic interference (EMI) signals do not affect fiber-optic cable. A break in the cable due to a bad connector or splice can cause a failure.

3.12.8 Optical time domain reflectometer

An optical time domain reflectometer (TDR) is an essential instrument to have available. This instrument is made by several manufacturers and usually comes in a portable case, battery-operated. More expensive optical TDR instruments have such features as chart recording capabilities or digitized floppy disc storage. Both are useful features because the fiber-optic cable can be TDR-tested before being placed in service, thus maintaining a permanent record for comparison at a later date. These instruments are generally easy to use and the only problem is the device that connects the instrument to the fiber-optic cable.

To maintain the optical transmitter (laser diode) and the receiving components, the usual digital multimeter and oscilloscopes will suffice. As time goes on, more specialized testing equipment will appear.

3.12.9 Data transmission on fiber-optic systems

Fiber-optic communications of baseband video and data communications were earlier applications of fiber-optic cable. Essentially only one video channel was carried per fiber, which by today's standard was not very efficient. Digital data from file servers, computer to computer, using digital multiplexing techniques are still in use today. An example of such a system using what is known as T_1 carrier multiplexing is shown in Fig. 3.75.

Each computer has access to the bit stream for each direction. The T_1 multiplexer has a reference word indicating the start of the protocol and data words. Therefore, any computer can locate the digital data from any other computer for computer-to-computer communications. Again, single-mode fiber is used and the optical transmitters are lasers or laser diodes. Such a system is often referred to as an elementary local area network (LAN) using fiber-optic techniques.

The lasers used with most single-mode fiber-optic cables operate at longer wavelengths such as 1300 to 1500 nm. These wavelengths are out of the visible spectrum and located in the infrared areas. The chemical content of the laser diode material determines the wavelength of the emitted light. However, the wavelength of the light also governs whether the fiber-optic cable will operate in the single mode.

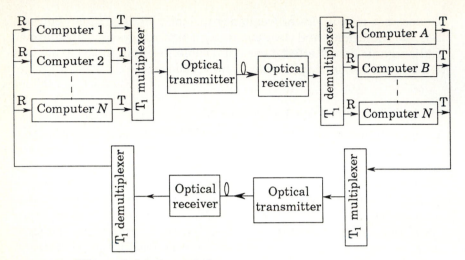

Figure 3.75   Fiber-optic data transmission.

Therefore, it is important that compatibility between components be considered when planning a cable communications system. Most manufacturers have a field or sales engineering staff available to consult with prospective customers. Most are quite knowledgeable about what other equipment works well with the equipment they are selling. Also the receiver diode, whether PIN (special type diode) type or APD (avalanche photo diode) type, should have a high sensitivity to the wavelength of the light emitted by the laser diode transmitter.

Many fiber-optic systems are in use today. Most of the earlier types operated in the pulsed (digital) or baseband video domain with one signal per fiber. As mentioned before, digital multiplexing was used to gain more data communications channels at the sacrifice of per-channel speed. Many channels of voice (telephone) and data are commercially carried on fiber-optic cables today. Cable television operators are now starting to use more and more fiber-optic communications in a variety of applications, such as supertrunking to replace head end-to-HUB and HUB-to-HUB microwave routes. Now the installation of head end to minihub fiber-optic distribution trunk "backbone" appears to be a cost-saving, low-maintenance approach to cable television systems.

3.12.10   Fiber-optic cable installations

Fiber-optic cable is placed on the utility pole plant in a similar fashion to coaxial cable. The minimum bending radius and the maximum pulling tension should be observed. For aerial plant the usual drip loops are placed at every pole and the cable is lashed up in the usual manner with lashing, wire bands, and spacers on steel messenger standard strand. Since the weight of the cable is less, often extra-high-strength or high-strength strand may not be needed. Before placing the cable on the pole plant it is often recommended that the cable be tested by using the optical TDR while it is still on the reel. Once the reel of cable is in place it should be TDR-tested before and after splicing. If a cable run is damaged, it is useful to know that before splicing is started. Once splicing is completed, the whole run can be TDR-tested. Of course, in a multifiber cable all fibers should be tested.

Installation of fiber optic-cable in buried systems should use a duct system. The cable has to be pulled through the duct, making sure the maximum allowable pulling tension is not exceeded or the minimum bending radius is not reached. Usually the maximum pulling tension for fiber-optic cable is about 600 lb. A dynamometer-type tension instrument should be used to set a capstan clutch on the pulling winch. The usual wire pulling eye made of wire mesh is used as the cable grip and coupled to the pulling line wound on the winch. Pulling lubrication compound applied to the cable and duct often lessens the pulling tension and makes the job easier. After the cable is placed in the duct it should be TDR-tested before splicing just as the aerial plant was. To date direct burial fiber-optic cable has not become available. However, most likely it will appear on the market as so much other equipment has.

3.12.11   Summary of fiber-optic systems

The foregoing material just scratches the surface of what is going to be a major communications medium. Laser diodes today are quite reliable and usually have a lifetime around 10 years. These diodes exhibit second- and third-order distortion components that have to be worked around. Since the second-order intermodulation beats fall in the frequency band, careful frequency planning should be done so the beats will fall in the spectrum areas between channels. This is very necessary for longer-haul supertrunk runs using FM-FDM techniques. For minihub standard NTSC-VSB signals only somewhere on the order of 12 to 16 channels per fiber should be considered. This method has one transmitting laser per fiber and one receiver for each 12 to 16 channels, thus limiting the distortion. For a 60-channel system then, the fiber-optic cable should have five separate fibers, one for each 12 to 16 channels. It would be a good idea to order six-fiber cable so one fiber could be reserved for reverse capabilities. This may seem to be good for now, but research indicates that by using wave frequency division multiplexing (WFDM) technique one fiber can carry many more channels of information. WFDM simply refers to more than one laser of different color (frequency) driving a fiber. Therefore, if two lasers of different wavelengths are driving a fiber with 12 television channels each, then each fiber will be carrying 24 channels. Now our six-fiber cable seems to be overkill.

As we have said before, for most communication systems, the single-mode fiber is the fiber of choice simply because it exhibits the widest bandwidth capabilities. This happens because all of the light is concentrated into a single ray that propagates down the so-called optical waveguide.

Selected Bibliography

Alley, Charles L., Atwood, Kenneth W. *Microelectronics*. Prentice-Hall, Englewood Cliffs, N.J., 1986.

Barker, Forrest. *Communications Electronics, Systems, Circuits, and Devices*. Prentice-Hall, Englewood Cliffs, N.J., 1987.

Bartlett, George W., editor. *Engineering Handbook*, ed. 6. National Association of Broadcasters, Washington, D.C., 1975.

Benson, K. Blair, editor in chief. *Television Engineering Handbook*. McGraw-Hill Book Co., New York, 1986.

Breed, Charles B. *Surveying*. John Wiley & Sons, New York, 1942.

Buchsbaum, Walter H. Sc.D. *Buchsbaum's Complete Handbook of Practical Electronic Reference Data*. ed. 2. Prentice-Hall, Inc., Englewood Cliffs, N.J., 1980.

CATV Cable Construction Manual. Comm/Scope Co./Pete Collins, Hickory, N.C., 1980. A company printed manual for customers.

Cunningham, John E. *Cable Television*. ed. 2. Howard W. Sams & Co., Indianapolis, 1985.

Design and Construction of CATV Systems. RCA Cablevision Systems, North Hollywood, Calif.

Freeman, Roger L. *Telecommunication Transmission Handbook*, ed. 2. John Wiley & Sons, New York, 1981.

Grant, William. *Cable Television*, ed. 2. GWG Associates, Fairfax, Va. 1988.

Johnson, Walter C. *Transmission Lines and Networks*. McGraw-Hill Book Co., New York, 1950.

Jordan, Edward C., editor in chief. *Reference Data for Engineers Radio, Electronics, Computers and Communications*. ed. 7. Howard W. Sams Co., Indianapolis, 1985.

Orr, William I. *Radio Handbook*, ed. 23. Howard W. Sams, Indianapolis, 1988.

Schrader, Robert L. *Electronic Communication*, ed. 2. McGraw-Hill Book Co., New York, 1980.

Schwartz, Mischa. *Information Transmission, Modulation and Noise*. McGraw-Hill Book Co., New York, 1959.

Taub, Herbert, Schilling, Donald L. McGraw-Hill Book Co., *Principles of Communication Systems*, ed. 7. New York, 1986.

Vergers, Charles A. *Handbook of Electrical Noise, Measurement and Technology*, ed. 2. TAB Books, Blue Ridge Summit, Pa., 1987.

4

Head Ends

4.1 Introduction

The head end of a cable television system is actually the beginning of the system which drives the distribution system serving the subscribers. The head end can take several forms; traditionally it consisted of a tall structure such as a tower strategically located on a hill or mountaintop with one or more antennas aimed at television broadcast stations. These antennas were connected via coaxial cables called *down leads* to electronic signal processing equipment housed in a climate-controlled building, usually located near the base of the tower. At this location the received off-air television signals were processed and combined on one cable needed to drive the cable television distribution system. The signal processing performed at the head-end site consisted of signal amplification, filtering, video and audio carrier level controls, and signal automatic level control. When satellite systems became a source of television programming, locating the satellite receiving antennas at the same site made sense. The only problem with this type of system is connecting in any local studio located away from the head-end tower site. Two ways commonly used to solve this problem were (1) trunking the head-end satellite receiving site to a hub location where such a studio would be situated, then driving the distribution system from that point and (2) activating a reverse channel feeding the local channel back to the head-end site, where the signal would be converted to a downstream channel for subscriber viewing. This method made necessary installation of reverse (upstream) amplifiers and diplex filters in the cable section from the studio location to the head-end site. These two schemes are diagrammed in Fig. 4.1.

Technically the method shown in Fig. 4.1a is superior mainly because the cable signals will not be degraded by merely adding in the

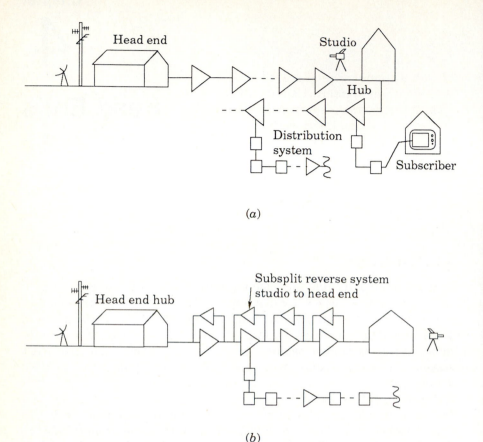

Figure 4.1   Supertrunking head end to studio.

studio signal to the trunk cable system, thus forming what is known as a hub. In the method shown in Fig. 4.1*b* a portion of the reverse system will have to be activated, increasing the system cost. The reverse amplifier equipment and the added head-end equipment are needed to convert from an upstream channel to a downstream channel, and usually more signal filtering is necessary to decrease the amount of upstream noise. It should be realized that an increase in equipment means an increase in maintenance, all adding to the cost. The downstream signal quality is most important because basically that controls quality of the pictures. High-definition television (HDTV) systems require extremely high signal quality.

4.2   Off-Air Broadcast Reception

Improving the reception of the television broadcast stations was the first purpose of the cable television business. Essentially, the cable system was just an extension of a remote high-elevation antenna system. Early cable systems were known as community-antenna television (CATV) systems. When other services, e.g., satellite reception, public access, local origination studios, and microwave pickup of remote programs, were added, the CATV systems became known as cable television systems. Usually off-air reception, or terrestrial broadcast as it is often known, accounts for less than half of the modern cable system channel capacity.

However, the FCC has always seemed to protect the interests of the television broadcaster by establishing picture quality standards for the cable operator. Even in this day of deregulation cable operators usually try to provide excellent off-air broadcast reception. The techniques needed to provide high-quality off-air reception should be thoroughly understood by the cable system technical staff.

4.3   Antenna Basics

Antenna, or aerial as it was once called, is essentially a tuned conductor placed above the surface of the earth. In the case of a transmitting antenna, it is desired that all the generated radio frequency energy be propagated through the atmosphere to be received by a tuned receiving antenna. The basic antenna is merely a conductor, i.e., a piece of wire suspended in space by insulators connected to end supports as shown in Fig. 4.2.

If current flows from negative to positive along the conductor, it causes the I (inductance) field to build up. The whole procedure reverses when the polarity of the transmitter sine wave reverses. Essentially the lower the transmitting frequency becomes, the longer the antenna wire becomes because the energy traveling down the wire and back has to return in time for the other half of the sine wave to be at the correct polarity. The length of the antenna used this way has to be an even multiple of the transmitting wavelength. However, an antenna that is exactly at a half-wavelength will also work. Shorter antennas essentially have to be loaded to make up the electrical difference for a half-wavelength. The energy on the antenna causes an electric field at the ends of the antenna and a magnetic field in the center. These fields build up and collapse, causing the voltage and currents to reverse. Some of the energy escapes into the atmosphere, never to return to the antenna. The magnetic and electrostatic fields that escape make up the signal to be received by receiving antennas, which can be constructed in the same manner as transmitting antennas. These fields propagate through the atmosphere and are attenuated by it. The farther these fields travel the weaker they become. Above the earth is a layer of air that is ionized by the sun's rays. Since

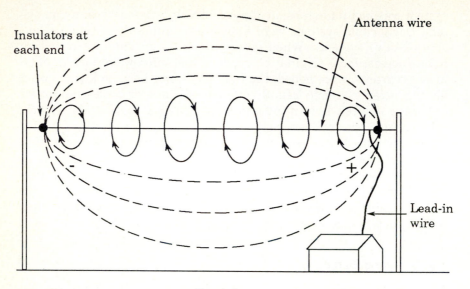

Figure 4.2   Basic antenna.

the air is dense near the earth's surface the fields are attenuated much more than in the thin air that becomes ionized and hence conductive. The transmitted electrostatic and magnetic fields can now travel easily through this conductive layer and hence travel faster than the waves traveling through the thick air near the earth's surface. At lower transmitting frequencies the fields have less of a penetrating effect on the ionized layer and hence are sent back to earth. Now, since the sun's rays have an ionizing effect on the earth's upper thin atmosphere, the time of day has an effect on how well the fields can travel through the ionized layer. The diagram in Fig. 4.3 illustrates this phenomenon.

High frequencies and very high frequencies that penetrate the ionosphere generally are lost as far as reception on earth is concerned. The choice of frequency and type of antenna greatly determines the propagation of the fields through space. Low-frequency signals are usually 200 kHz to 2 MHz, and these waves usually travel just above the earth's surface and hence have only a ground wave. For frequencies of 5 to 10 MHz the usable ground wave becomes weaker and the direct sky wave is stronger. Generally as the frequency increases the sky or direct wave is strongest. Frequencies from 10 to 30 MHz can penetrate the ionosphere while some is reflected back to earth.

Antennas can be placed in the vertical position, with the electrostatic fields at the top and bottom ends. These antennas are used for

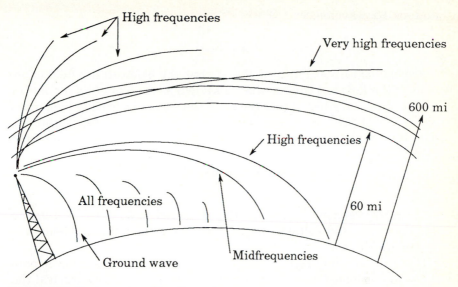

Figure 4.3 Radio energy traveling through space.

lower-frequency AM broadcast service at 540 to 1640 kHz. Television transmitting and receiving antennas are horizontally polarized.

Since an antenna has to be at a half-wavelength or even number of wavelengths to work, usually the half-wavelength type is chosen because it is the shortest. Essentially, a radio or light wave has a velocity or speed of 300,000,000 m/s. Therefore, a frequency of 1,000,000 Hz driving an antenna will cause a wave to travel out at 300,000,000 m/s. Since a hertz is one cycle of the signal sine wave per second so for 1,000,000 cycles per second one cycle will take (1/1,000,000) s, or 300 m in the time of one cycle, i.e., 1/1,000,000 s. Thus 300 m is the *wavelength*.

$$\text{Mathematically } \lambda = \frac{v}{f}$$

where λ = wavelength (m)
       v = 300,000,000 m/s
       f = frequency (Hz)

This relationship may be rewritten as

$$\lambda = \frac{300{,}000{,}000}{f_{Hz}} = \frac{300{,}000}{f_{kHz}} = \frac{300}{f_{MHz}}$$

Since 300,000,000 m/s = 186,400 mi/s, or nearly 984,000,000 ft/s,

$$\lambda_{ft} = \frac{984}{f_{MHz}}$$

and for 1/2 wavelength

$$\frac{\lambda}{2} = \frac{492}{f_{\text{MHz}}}$$

Experimentally it has been found that the electrostatic field at the ends of the antenna causes what is known as the *end effect* for the antenna. This end effect makes the antenna seem electrically longer and hence out of tune. So to correct the situation it has been found that only 95 percent of the total half-wave is needed. Therefore the correct formula is

$$\frac{\lambda}{2} = \frac{468}{f_{\text{MHz}}}$$

The half-wavelength antenna, often referred to as a *dipole antenna*, has a current and voltage distribution along the antenna element as shown in Fig. 4.4.

Since an antenna has a voltage and a current distribution it must have an associated resistance or impedance. From basic electrical theory

$$P = I^2 R$$

$$\text{so } R = \frac{P}{I^2}$$

where P = watts
I = amperes
R = ohms

This value is the *radiation resistance* of the antenna. The radiation re-

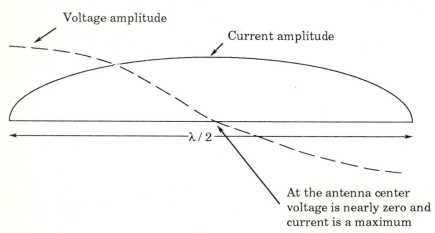

Voltage amplitude

Current amplitude

$\lambda / 2$

At the antenna center voltage is nearly zero and current is a maximum

Figure 4.4   Dipole antenna.

sistance at the center of the dipole antenna is at a minimum because the current is a maximum. At the ends of the antenna the voltage goes to the maximum and the current to a minimum; thus theoretically the resistance goes to infinity. The minimum resistance of a dipole antenna theoretically is 73 Ω at the center of the antenna. As the antenna is placed close to the ground, this resistance approaches zero; while it is moved away to one-quarter wavelength it is at 73 Ω. Increasing to one-third wavelength makes the resistance to 95 Ω, 73 Ω at one-half wavelength, 58 Ω at two-thirds wavelength, and 73 Ω at three-quarter wavelength, with variations becoming less and less as the antenna is elevated from the conducting earth.

Naturally the transmission equipment is connected to the antenna by means of a transmission line, which is often a coaxial cable of proper impedance needed to drive the antenna. As we have seen, the dipole antenna has an impedance of 73 Ω at the center of the antenna, which is nearly 75 Ω, so very little adjustment is needed to match to 75 Ω. To provide this connection, the dipole is cut at the center to give two quarter-wave sections, as shown in Fig. 4.5.

This type of antenna is used in such equipment as radiation detectors because the quarter-wave sections can be constructed of telescoping rods. The transformer can be mounted in a plastic box which also supports and insulates the rods.

Another type of useful antenna is the folded dipole. For television work, this antenna can be totally constructed of 300-Ω flat television wire. The configuration is shown in Fig. 4.6.

As one moves away from an antenna, the radiated energy is confined to a pattern according to the geometry of the antenna, the polarization, and the height of the antenna above ground. For example, consider the case of the simple dipole antenna one-half wavelength above ground, where the radiation resistance is 73 Ω at the center, as shown in Fig. 4.7.

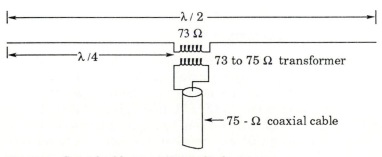

Figure 4.5   Coaxial cable connection to dipole antenna.

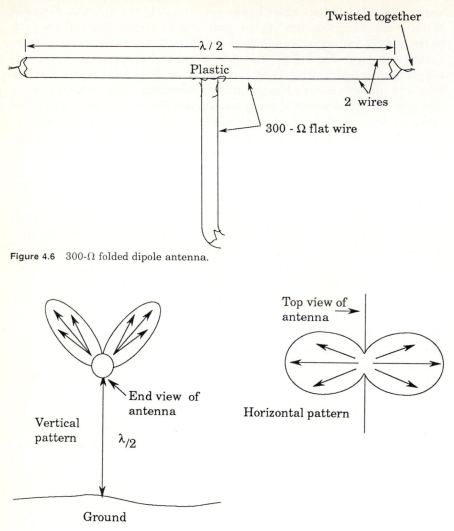

Figure 4.6   300-Ω folded dipole antenna.

Figure 4.7   Vertical and horizontal radiation patterns.

4.4   Television Antenna Systems

Television broadcast station receiving antennas fall into two categories: the broadband type and the single-channel type. The typical home receiving television antenna is the broadband type. This antenna is designed to receive low VHF, FM, high VHF, and UHF signals, none of them very well. Antennas of this type usually consist of two types of antennas mounted on the same bracket: the frequency band of 55 to 216 MHz on one set of elements and 470 to 890 MHz on the other, making up the *all-band antenna*. This type of home antenna was somewhat directional, and if stations were located more than 90° apart an electric rotation system could be installed to turn the antenna toward the station. Cable television systems usually employed only the heavy-duty version of this type of antenna for a search antenna, usually used to sample other television signals or act as a backup system for the primary antenna. The primary antenna of choice by cable television engineers is the single-channel narrow-band, high-gain, very directive antenna of either the log-periodic or Yagi-Uda array type. In general, the *log-periodic antenna* has a receiving pattern of one main front lobe and very small side lobes. The *yagi array* also has a large front lobe but two side lobes of significant size, one on each side. The front-to-back ratio of the yagi array is much lower than that of the log-periodic array. A diagram of a log-periodic antenna is shown in Fig. 4.8.

Antenna bandwidth can be made very large by selection of L and X (parameters shown in Fig. 4.8). However, the antenna becomes large.

Since a geometric progression of each element is required, α is the taper (angle) and τ is the common ratio.

Mathematically

$$\tau = \frac{L_n + 1}{L_n} = \frac{X_{n+1}}{X_n}$$

and

$$\alpha = \tan^{-1} \frac{L_n}{X_n}$$

where X_n and X_{n+1} indicate adjacent element spacing and L_n and L_{n+1} are adjacent element lengths.

One of the most important characteristics is the field intensity, or E plane pattern, which tells what the receiving pattern looks like. Such a pattern is shown in Fig. 4.9.

Since it should be seen that the receiving pattern is fairly broad in the frontal lobe, the question of what might happen if a signal on the same channel some distance away is received arises. The answer is that a condition called co-channel could result. The *co-channel* condition essentially results when two channels on the same frequency are being received, one of which is the desired one, the other the undesired or interfering frequency. When the received level of co-channel is 36 dB or less below the video carrier level of the desired channel, picture impairment of the desired channel results. To reduce the effects of co-channel, an antenna with a sharp narrow front lobe pattern is aimed at the desired station and away from the direction of the interfering station, improving the situation significantly. Such an antenna usu-

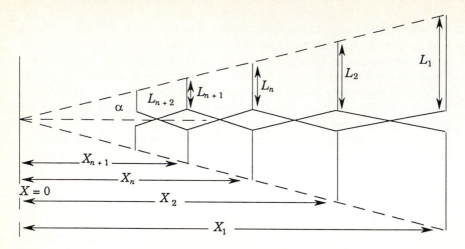

L_1 = 1/2 wavelength of lowest antenna frequency

L_{n+2} = 0.38 wavelength of highest antenna frequency

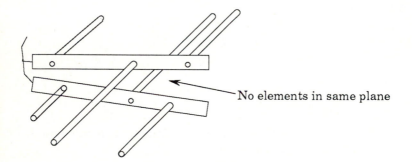

Figure 4.8 Log-periodic antenna.

ally consists of a stacked array of several log-periodic antennas mounted on a boom-type beam. Two antennas mounted in tandem are called a *dual array*; when four are used, a *quad array* results. Antenna stacking is always done in pairs. As might be suspected, the quad array is more effective than the dual- or single-antenna, but, of course, size, cost, and mounting difficulty are greater.

The mounting or placement of antennas on a tower is certainly not a matter of convenience. Proper spacing must be maintained so the tuned antenna elements are not affected by the adjacent antennas. For the same compass bearing, antenna separation above or below should be one wavelength of the lowest-frequency (largest) antenna. For example, consider the diagram in Fig. 4.10.

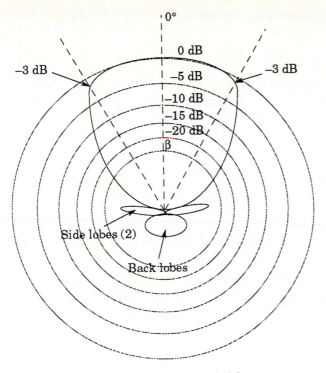

∡ β = amtenna beamwidth
in degrees (at –3-dB points)

Figure 4.9 Antenna radiation pattern.

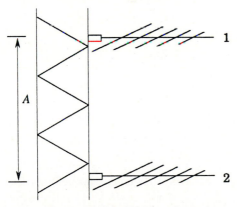

1   Channel 2 Lowest frequency

$$A \, (\text{ft}) = \frac{984}{\text{frequency (MHz)}}$$
$$= \frac{984}{55.25} = 17.8 \, \text{ft}$$

2   Channel 3

Figure 4.10 Vertical separation of antennas.

The difference in separation A for channel 10 (lowest frequency) and channel 12 is

$$A = \frac{984}{193.25} = 5.1 \text{ ft}$$

Naturally if the angular difference between the two antennas is due to a different station bearing, vertical spacing may be closer. For example, if the two stations are 180° apart (back to back) then the two antennas may be mounted back to back in the same plane, as shown in Fig. 4.11a.

Now if the angular difference is somewhere in between, e.g., the two antennas are 90° apart, as shown in Fig. 4.11b, and for channels 2,3 then: $A_{min} = A(1 - \theta/180°)$. Generally the rules for antenna placement are as follows:

1. The antenna for the weakest station should occupy the topmost section of the tower.

2. If the received signals are equal, place the highest frequency (smallest) on top so as not to weight the top unnecessarily.

3. Follow the procedures outlined for proper minimum allowed spacing.

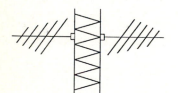

Antennas back to back
(180° apart)
(a)

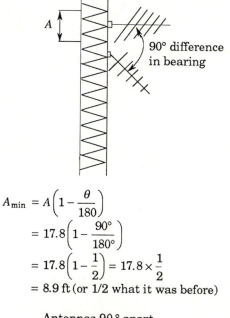

90° difference in bearing

$$A_{min} = A\left(1 - \frac{\theta}{180}\right)$$
$$= 17.8\left(1 - \frac{90°}{180°}\right)$$
$$= 17.8\left(1 - \frac{1}{2}\right) = 17.8 \times \frac{1}{2}$$
$$= 8.9 \text{ ft (or 1/2 what it was before)}$$

Antennas 90° apart
(b)

Figure 4.11 Antenna mounting location on tower.

In general, UHF-type antennas are broadband over the UHF television band. VHF log-periodic arrays almost always cover the low VHF band (channels 2 to 6) and high VHF band (channels 7 to 12), with two separate antennas covering both sections of the VHF television band.

Separate antennas are usually used for the FM radio band. When selecting antennas for a cable television system, the advisable first step is to get a computerized signal survey. Such a survey includes such parameters as the broadcast television transmitter output power, tower height and antenna gain, path loss between the transmitter site and the receiver site, and the receiver antenna height with the predicted signal level for a given receiving antenna type. This information is given for each television station within a specified radial distance from the receiving site; it is very useful in providing answers about antenna type, tower height, and any needed preamplification.

4.5 Antenna Preamplifiers

Once antennas are placed on the cable television receiving tower they have to be aligned, or pointed correctly, toward the transmitting television station for greatest signal strength. This procedure is performed by a technician on the tower with a battery-operated portable signal level meter connected to the antenna. The antenna is turned toward the station and set to its highest reading. If several stations are to be received by a common antenna, the antenna can be aligned to the weakest station. The length of down lead cable, of course, will attenuate some of the signal, and this loss has to be taken into account. The signal processors in the head-end electronic cabinets usually require a minimum of +10 dBmV of signal level. Anything less will degrade the carrier-to-noise ratio. In general, for every decibel less than +10 dBmV input to the signal processor will cause a 1-dB reduction in carrier-to-noise ratio. Since the head end should produce the best signal carrier-to-noise ratio before going down the cascade of amplifiers in the cable distribution system it is extremely important to allow antenna signal level to vary according to seasonal weather changes. For example, if it is known that a station signal level varies +5 dB, then for the signal processor to have a minimum level of +10 dBmV the median signal level should be +15 dBmV.

If the antenna produces only a +2-dBmV signal and this level has to be split for two stations to be received from the antenna, then a preamplifier should be considered. Whether the preamplifier should be located on the tower near the antenna or in the head end, where it would be convenient, is determined by the C/N ratio at the receiving antenna on the tower. The following example, illustrated in Fig. 4.12, will illustrate the procedure.

Now 46 dB is not really a good enough C/N to put through a cascade. If good results can be obtained by placing the preamplifier in the head-end building, then maintenance will be convenient. According to cascade theory, the noise factors have to be cascaded as follows.

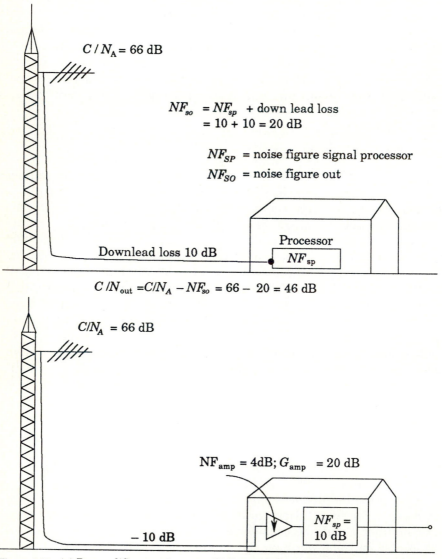

$C / N_A = 66$ dB

$$NF_{so} = NF_{sp} + \text{down lead loss}$$
$$= 10 + 10 = 20 \text{ dB}$$

NF_{SP} = noise figure signal processor
NF_{SO} = noise figure out

Processor
NF_{sp}

Downlead loss 10 dB

$C/N_{out} = C/N_A - NF_{so} = 66 - 20 = 46$ dB

$C/N_A = 66$ dB

$NF_{amp} = 4$dB; $G_{amp} = 20$ dB

$NF_{sp} = 10$ dB

-10 dB

Figure 4.12 (*a*) Preamplifier requirement. (*b*) Preamplifier in head end.

$$\text{Noise factor out} = F_o = F_c + \frac{F_a - 1}{G_c} + \frac{F_p - 1}{G_c G_{\text{amp}}}$$

where noise factor (NF) = $10^{\text{NF}/10}$

F_o = noise factor out
F_a = preamplifier noise factor
F_c = cable noise factor
F_p = signal processor noise factor
G_c = cable gain (a loss)
G_{amp} = preamplifier gain

$$F_o = \underset{\text{cable loss}}{10^{10/10}} + \frac{10^{4/10} - 1}{10^{-10/10}} + \frac{10^{10/10} - 1}{(10^{-10/10})(10^{20/20})}$$

$$= 10 + (2.5 - 1)10 + \frac{10 - 1}{(10^{-1})(10^2)} = 10 + 15 + \frac{9}{10} = 25.9$$

so   noise factor out in dB = $10 \log F_o = 10 \log 25.9 = 14.1$ dB

For this off-air channel C/N = (antenna C/N) – NF, so C/N off-air = 66 dB – 14.1 dB = 51.9 dB, which really is not very good.

Now we shall see how much improvement there is if the pre-amplifier is placed on the tower at the antenna, as shown in Fig. 4.13. Substituting

$$F_o = 10^{4/10} + \frac{10^{10/10} - 1}{10^{20/10}} + \frac{10^{10/10} - 1}{(10^{20/10})(10^{10/10})}$$

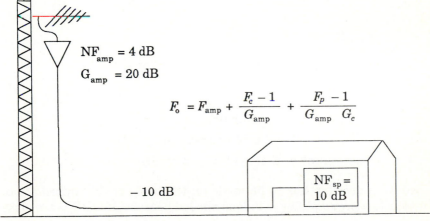

C/N = 66 dB

NF_{amp} = 4 dB

G_{amp} = 20 dB

$$F_o = F_{\text{amp}} + \frac{F_c - 1}{G_{\text{amp}}} + \frac{F_p - 1}{G_{\text{amp}}} \frac{1}{G_c}$$

– 10 dB

$NF_{\text{sp}} = 10$ dB

Figure 4.13   Preamplifier on the tower.

$$= 10^{0.4} + \frac{10 - 1}{100} + \frac{10^{-1}}{(100)(0.1)}$$

$$= 2.5 + 0.09 + 0.9 = 3.49$$

$$NF_o = 10 \log F_o = 10 \log 3.49 = 5.4 \text{ dB}$$

For this off-air channel the C/N ratio is now C/N = 66 − 5.4 = 60.6 dB.

Now to see what this does for an ending C/N ratio at the cascade end, where the cascade C/N is 48 dB, proceed as follows:

No preamp $\dfrac{C}{N_{SYS}}$ = − 10 log [$10^{-48/10} + 10^{-46/10}$]

$$= - 10 \log [10^{-4.8} + 10^{-4.6}]$$

$$= - 10 \log [1.58 \times 10^{-5} + 2.5 \times 10^{-5}]$$

$$= - 10 \log [0.0000408] = - 10 \times - 4.39 = 43.9 \text{ dB}$$

which is not very good.

For the preamplifier in the head-end building the system C/N becomes

$$\frac{C}{N_{SYS}} = - 10 \log [10^{-48/10} + 10^{-51.3/10}] = - 10 \log [10^{-4.8} + 10^{-5.13}]$$

$$= - 10 \log [1.58 \times 10^{-5} + 0.65 \times 10^{-5}]$$

$$= - 10 \log [2.23 \times 10^{-5}] = - 10(-4.65) = 46.5 \text{ dB}$$

which is acceptable.

For the preamplifier on the tower the system C/N ratio becomes

$$\frac{C}{N_{SYS}} = - 10 \log [10^{-48/10} + 10^{-60.6/10}] = - 10 \log [10^{-4.8} + 10^{-6.06}]$$

$$= - 10 \log [1.58 \times 10^{-5} + 0.09 \times 10^{-5}]$$

$$= - 10 \log [1.67 \times 10^{-5}] = - 10 [-4.78] = 47.8 \text{ dB}$$

To summarize:

No preamp C/N system = 43.9 dB

Preamplifier in head-end C/N system = 46.5 dB

Preamplifier on tower C/N system = 47.8 dB

Now the decision has to be made on the basis of this information. Obviously a preamplifier is indeed needed because 43.9 dB is too close to what is known as the minimum carrier-to-noise ratio for good television reception, and there is little margin for any signal variations.

The difference between 47.8 and 46.5 dB is not large enough to warrant placing the preamplifier on the tower instead of in the head end. If the preamplifier is placed on the tower and fails, then someone has to climb the tower and either replace the preamplifier or temporarily remove it; this may be difficult or impossible because of weather or lack of tower staff. If the failed unit cannot be replaced or removed immediately, then the channel will be unavailable for service, causing subscriber complaints. In this case the preamplifier should be placed in the head end and should provide excellent pictures, and if a failure occurs replacement or removal will be more convenient and can be performed immediately. It should also be remembered that the antenna problem started at the antenna and worked through to the cascade. Line extenders, amplifiers, and converters further degrade the carrier-to-noise ratio.

4.6   Television Signal Processing

Each of the antenna signals, which may consist of a single channel or a group of channels in nearly the same frequency range, has to be processed before being connected to the cable system amplifier cascades. This processing may be accomplished by a heterodyne signal processor or by a receiver that demodulates and remodulates (demod-remod) the signal. Earlier systems used an amplifier filter system called a *strip amplifier*, which filtered the channel and then amplified each television channel or group of channels. The heterodyne signal processor and the demod-remod method process each television channel individually. The demod-remod method is more expensive but provides an easier process that allows signal scrambling. Video-channel switching can be employed to enable a cable operator to change programming from one channel to the other easily when it is necessary to comply with nonduplication or syndicated exclusivity requirements. Many cable operators employ a mix of signal processors and demod-remod. Both methods allow automatic gain control and carrier substitution (on loss of signal) as well as control over the audio carrier and video carrier levels. A fairly typical method, showing the equipment connections for off-air signal reception, is illustrated in Fig. 4.14.

4.7   Heterodyne Signal Processor

The television signal processor receives its signal from the off-air antenna. This signal may be preamplified by a single-channel preamplifier, either tower-mounted or head-end-installed, to provide proper level to the RF input connector on the signal processor. Care

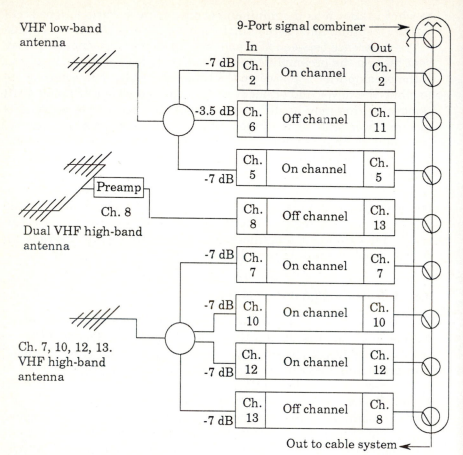

Figure 4.14 Off-air receiving method.

must be exercised not to overdrive the signal processor if a preamplifier is used. Care must also be used in determining whether a preamplifier is needed at all. If a preamplifier is overdriven, it will generate beats and noise that will be passed on through the processor. Often pads may have to be installed at the RF input connector of the signal processor on a seasonal basis to control the input signal level.

Referring to the block diagram for a signal processor in Fig. 4.15, the RF input terminal on the chassis feeds signal to the input converter section. A test point, often on the front panel of the input converter section, facilitates monitoring the input signal level by means of a signal level meter or spectrum analyzer. The input signal is amplified by amplifier 1 before being fed to the mixer. The local oscillator also provides its carrier frequency to the mixer, where the two signals are mixed, producing sum and difference frequencies, as well as input

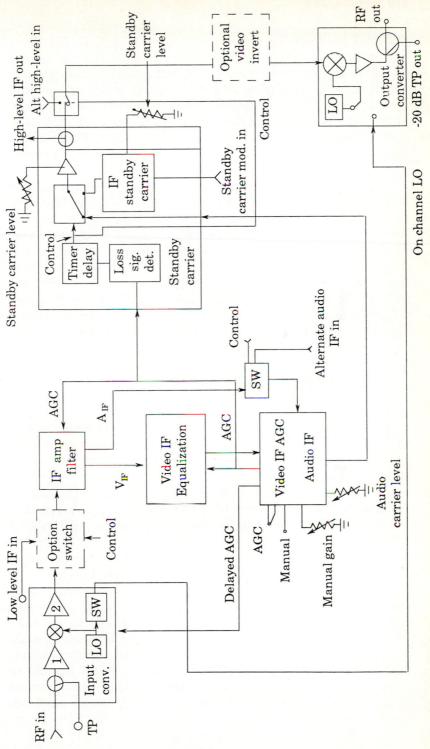

Figure 4.15 Signal processor block diagram.

and local oscillator signal frequencies. Amplifier 2 is a filter amplifier that selects the difference frequency and amplifies the signal. This signal is passed through a selector switch, which selects either the input IF signal or some other IF signal consisting of programming from some other source. The selected IF signal is passed through the IF amplifier filter system, which now separates the video IF signal from the audio IF signal. The video IF signal feeds the video IF amplifier equalization section, where compensation is provided for the video IF signal.

The corrected video IF signal from the video IF equalization system is fed to the video IF automatic gain control (AGC) section as well as the audio IF signal from the IF amplifier filter section. The video IF AGC circuit amplifies both video IF and audio IF signals separately where the audio IF carrier level can be independently adjusted for the prescribed 15 dB below video carrier level standard. The signals are then combined and passed to the standby carrier section. Also the AGC signal is developed in the video IF AGC section. Delayed AGC voltage is fed to the input converter, where the input level can be adjusted to close tolerances automatically so the input mixer will not be overdriven. This AGC signal is delayed so control is not so fast as to disturb the synchronization portion of the television signal. Normal fast AGC voltages are fed to the IF amplifier filter and the video IF equalization circuits to control the level of amplification automatically. The audio IF signal is passed through a switch before being connected to the video IF AGC circuit. This allows an alternate audio IF signal to be introduced at this point by the selector switch. Such a signal could be used as an audio override alert voice-over that could occur on all channels.

The combined audio and video IF signals are fed to the standby carrier circuit. This circuit monitors the AGC voltage. When the AGC voltage decreases significantly, indicating loss of input signal, the standby carrier module provides a substituted video IF signal, which is viewed as a blank, dark screen by subscribers. This prevents noise on this channel being fed into the cable system. This module has a timer that controls when the switch makes the selection to the alternate video IF or standby carrier. If the input signal keeps going on and off, the normal 15-s delay allows subscribers to see the short bursts of programming but prevents the longer-term outages. Also as an option in some manufacturer's processors the standby carrier module has an internal video modulator, which allows the standby carrier to be modulated by an alternate video source; this may consist of a message such as "Please Stand By." The timer can be deactivated in the standby carrier module, which allows an instant "on" condition.

This condition is a must when the television signal is used as a pilot carrier for the cascade of trunk amplifiers.

The combined, cleaned-up audio and video IF signals are fed to the output converter, which by virtue of its local oscillator mixer circuitry transfers the signals to a selected television RF carrier. As an option, if the input to output channels are the same as the received television channel the local oscillator from the input converter can also be used as the local oscillator for the output converter by a short piece of cable and a couple of switches inside the modules placed in the proper position.

The output signal appearing at the RF output connector contains the processed output signal on any of the usual CATV channel assignments. This signal is connected to a signal combiner network, where all of the cable television signals are combined to drive the cable system.

A trip through the signal processor for a standard television-received signal to be placed on a cable television harmonically related carrier (HRC) channel plan should clarify the process. For channel 7 input and HRC channel 11 output, channel 7 has a video carrier frequency of 175.250 MHz, a color carrier frequency of 178.83 MHz, and a sound carrier frequency of 179.75 MHz. Now the local oscillator is running at 221 MHz so when this frequency is mixed with the 175.250-MHz video carrier frequency, the *difference frequency*, which is the video IF carrier frequency, is 221.000 − 175.250 = 45.750 MHz. When the audio carrier of 179.750 MHz is mixed with 221.000 MHz the difference frequency is 221.000 − 179.750 = 41.250 MHz. For the color carrier, 221.00 − 178.83 = 42.17 MHz. Since the video IF bandwidth is 4.2 MHz this carrier is contained within the video bandwidth. However, since the video and chroma signals pass through the same IF filters, the difference in phase delay between these two signals must be compensated for by the video IF delay equalization circuits. This has to be done so the color signals will occur at the correct place in the video picture.

Notice that now the higher-frequency carrier at IF is the video IF carrier frequency, and the lower-frequency carrier is the sound IF carrier. As the IF signal is passed through the processor circuits and through the output converter, the reversal takes place again, correcting the situation. For channel 11 HRC the output converter local oscillator frequency becomes 243.750 MHz; when it is mixed with the video IF carrier of 45.750 MHz, the difference is 243.750 − 45.750 = 198.00 MHz channel 11 video carrier frequency. When 41.250-MHz audio IF carrier frequency is mixed with 243.750 MHz, the difference frequency is 243.750 − 41.250 = 202.50 MHz, which is channel 11 audio carrier frequency. For the color IF carrier 243.750 − 42.17 = 201.58 MHz.

Suppose the cable system was operated at a standard frequency channel plan and the input and output channels were on the same frequency; channel 7 in, channel 7 out. Then the switch could be placed so that the input converter local oscillator (LO) at a frequency of 221 MHz could also be the output converter local oscillator. The conversion would be as follows:

	IF	LO − IF	= carrier frequency
Video	45.750	221 − 45.75 = 175.250MHz	Channel 7 video carrier
Chroma	42.17	221 − 42.17 = 178.83 MHz	Channel 7 chroma carrier
Audio	41.250	221 − 41.250 = 179.75 MHz	Channel 7 audio carrier

4.8   Signal Processor Adjustments and Tests

To set up most processors, a good accurate signal level meter is a requirement. A spectrum analyzer will be most helpful. Most processors have a +10-dBmV minimum recommended RF input level. Any signal below this level will lower the C/N ratio for that channel on a decibel-for-decibel basis. For example, every 1-dBmV drop in input level below +10 dBmV will decrease the output carrier-to-noise ratio by 1 dB. It must be remembered that the C/N ratio is at its best at the head end so it is best to start with the best C/N because it will degrade the cascade C/N by the least amount. Even if the preamplifier adds noise it is better to use it and never allow the processor C/N ratio to decrease.

The signal level meter should be used to set the RF input level within the recommended limits, typically +10 to +30 dBmV; a reasonable choice would be +15 to +20 dBmV. If +30 dBmV is exceeded, then the processor input converter amplifier could be overdriven even with AGC, thus generating beats and noise and causing poor pictures and noisy sound. The output level should be adjusted to the proper level as measured at the signal combiner network output test point with the audio carrier level 15 to 16 dB lower than the video carrier level. It generally is not a good idea to operate the processors at maximum output level. Usually a +60 dBmV is the maximum level. Therefore, operation at +55 to +57 dBmV is recommended. One other adjustment is the standby carrier output level adjustment. The standby carrier should be activated by means of a switch if there is one or by simply removing the RF input signal and waiting the appropriate time, usually 15 s, until the front panel lamp indicates the standby carrier presence. Now the standby carrier level can be set to the same level as the video carrier by the appropriate control.

It is extremely important to adjust the standby carriers because when such stations go off the air and the standby carriers replace these signals, the cascade of amplifiers will not be overdriven and cause distortion. Even if the cascade AGC system works and maintains signal level the first amplifier will be overdriven, causing distortion. It is well known that the first few amplifiers in cascade have the most effect on the overall cascade performance and are most objectionable at the end of the cascade. Also, if any of the processors acts as a pilot carrier to the cascade and is at the incorrect standby carrier level, when the standby carriers become activated the whole cascade can become either too high or too low in gain and slope, causing disastrous results to the service.

A good test for the overall quality of a television signal processor is shown in Fig. 4.16. Unfortunately, this test requires a lot of equipment, which, of course, is expensive.

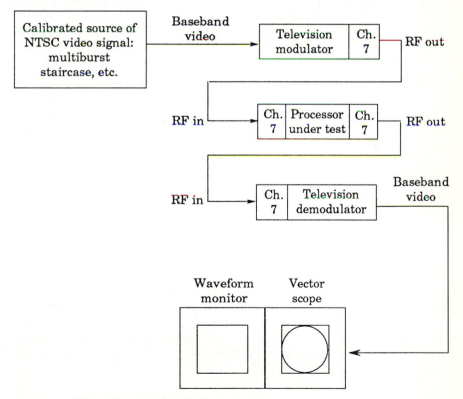

Figure 4.16 Heterodyne signal processor test setup.

Calibrated signals are fed to the television modulator, where they are placed on television channel 7 and operated on by the processor and demodulated by the demodulator. Essentially the video test equipment can be connected to the generating equipment to verify and photographically record the test signals. These signals are connected to the modulator and placed on a television RF carrier (channel 7). This modulator can be connected to the demodulator and the video signals verified and photographed. Now the processor is introduced into the circuit and any signal degradation due to the processor can be observed and photographed. The television modulator output channel has to be on the same channel frequency as the processor input converter. The processor output converter also has to agree with the demodulator input converter. If the equipment is all from the same manufacturer and the output converters, as well as the input converters, are interchangeable, then the calibration procedure will be simpler.

If the processors are used for a cable system using the harmonically related carrier (HRC) system the processor will require the output converters to be phase-locked to a reference comb generator. For the HRC system, this comb generator generates a carrier frequency starting at 54 MHz (Channel 2H) and proceeds with a carrier every 6 MHz until 450 MHz is obtained for a 450-MHz upper limit. Now it should be evident that 54 and 450 MHz are divisible by 6 and hence all video carriers in between since they are spaced 6 MHz apart. This is what constitutes the harmonic relationship. Therefore, the third-order distortion will produce beat components that fall on video carrier frequencies (zero beat) and do not interfere with the picture components in the video sidebands. To put this another way, the distortion components, although present, will not be seen in the pictures unless, of course, they become extremely large. The task at hand is to lock the output converter to the comb frequency. The block diagram to modify the previous diagram (i.e., Fig. 4.16) for phase locking is shown in Fig. 4.17.

The reference signal from the comb generator is fed to the phase lock local oscillator. The resulting IF goes to the servo, which compares the phase of the incoming signal IF to the phase lock IF. The control voltage then controls the phase lock local oscillator to keep the phase lock loop in lock. If the phase lock oscillator ceases, the output converter gets its local oscillator signal from the internal crystal.

Phase locking the local oscillator in the input and output converters to an interfering strong local channel can improve picture quality. However, it must be understood that one can only phase lock a processor to one reference.

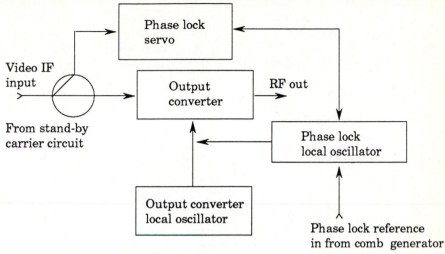

Figure 4.17   Signal processor phase-lock circuit.

4.9   Nonbroadcast Cable Programming

At the head end or HUB location local and satellite-delivered signals are processed and inserted on the cable system through the signal combiner. These signals are video and audio baseband voltages that are placed on a television video or audio carrier frequency by a device known as a *television modulator*. The modulator output is an RF carrier containing the television video and audio signals. A review of AM and FM basic modulation techniques will be helpful and will be achieved by consulting the references at the end of this chapter. Fig. 4.18 illustrates a typical equipment connection diagram for this portion of the head end.

4.9.1   Television modulator

The television modulator is essentially an extremely low-power television transmitter, i.e., a transmitter without the high-power output stages. The television modulator used in a cable television system takes baseband audio and video signals (voltages) and places them on video and audio television carrier frequencies which when combined with the other modulated carriers drive the cable system. The modulators produced by several manufacturers all operate in a similar manner with many options offered. A block diagram of a typical modulator is shown in Fig. 4.19. The video signal is connected to the video modulator through the optional video selector switch. If it senses the loss of the main video source this optional input switch automatically

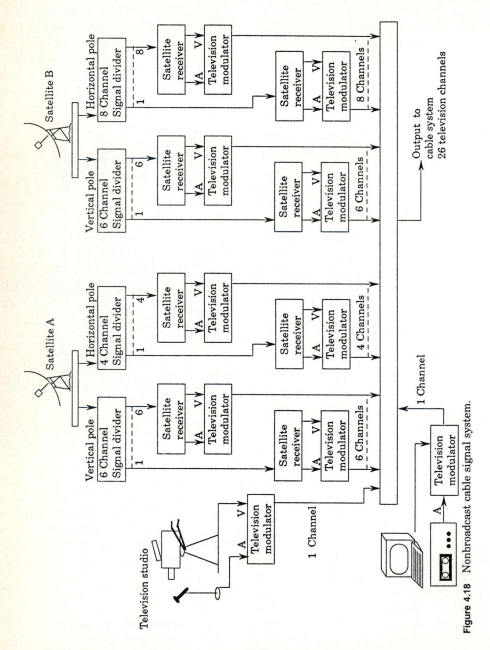

Figure 4.18 Nonbroadcast cable signal system.

248

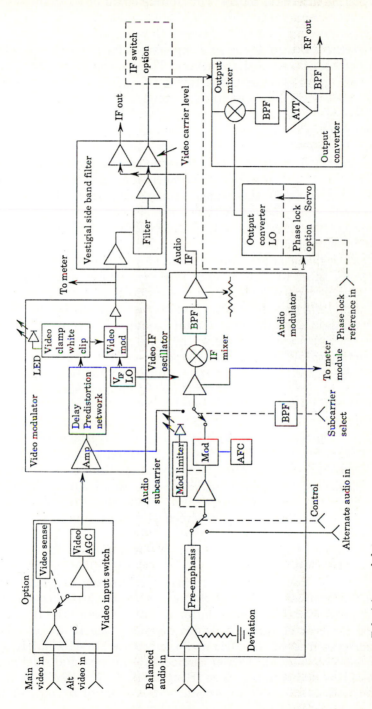

Figure 4.19 Television modulator.

selects the alternate source. This source could be as simple as "Please Stand By" or could be part of a video override message by switching off the main channel on command. Because one of two sources is selected, the video AGC circuit compensates for any difference between the two video source signal levels. The baseband video signal is fed through an isolation amplifier through a predistortion network which acts similarly to a preemphasis system. The signal is distorted in reverse to the distortion in the modulator. The video is clamped to a dc voltage level proportional to the white level (100 percent negative AM). The video IF oscillator provides the carrier frequency for the video signal. The signal from the video modulator consists of the carrier frequency and the upper and lower sidebands containing the video information. Since this video information is duplicated on both sidebands, only one set of sidebands is needed (the upper set). The lower sidebands are removed by the vestigial sideband filter; i.e., the lower vestige is filtered out.

The baseband audio voltage at an impedance of 600 Ω balanced is fed to the input amplifier in the audio modulator section. The level is adjusted at this point because the audio level is what changes the amount of FM deviation. After level adjustment the signal is fed to the preemphasis circuit. This circuit tends to accentuate the high-frequency components because they are usually much weaker. This means that the higher-frequency portion of the audio signal deviates the FM carrier more and in the receiver portion, the high-frequency components are deemphasized to be in proper perspective.

After the preemphasis the signal is fed to the audio FM modulators. A modulation limiter protects the modulator from overdeviating the FM audio subcarrier. When the threshold of overdeviation occurs and the limiter is working, an LED lamp indicates that the audio signal is too high in level. Before the modulation limiter an optional preemphasized alternate signal can be substituted. The audio modulator generates the audio subcarrier at 4.5 MHz, and the input audio level deviates the frequency of this subcarrier at an audio rate to perform the modulation function.

If the main baseband video signal is accompanied by an audio FM subcarrier with the program audio on it, it is called the *composite signal*. In this case the audio subcarrier is already in place, the audio modulator circuit is not needed, and the subcarrier can be injected through a level-adjusting isolation amplifier before the IF mixer. The video IF local oscillator provides drive for the IF mixer where the mixer converts the 4.5-MHz audio subcarrier to 41.25-MHz audio IF carrier. This audio IF carrier connects to the vestigial sideband filter, where it is added to the video IF signal at 45.75 MHz. This combined IF output is available to provide a drive signal to another piece of equipment, such as another modulator or processor. The combined IF signal also feeds the output mixer in the output converter section, where it is mixed with the local oscillator frequency to produce a television video and audio RF carrier frequency. This output signal is filtered and amplified before going to the RF output connector. The two filters remove any extraneous signals, e.g., the sum frequency of the mixed signals, and the amplifier provides an output level of +60 dBmV maximum.

In general, manufacturers of head-end equipment build both processors and modulators in modular form, which facilitates easy maintenance. Some manufacturers' equipment is more modular than others'. Usually the input and output converters are separate units. Some manufacturers of modulators then combine the video and audio sections with the IF modulator and vestigial sideband filter in one unit. Processors may appear with separate IF filters, standby carrier sections, and AGC systems. It may be said that the more modular the unit the rapid replacement is possible and makes for easier and cheaper maintenance. A system will only have to stock a few power supply modules and IF standby carrier modules, which are common between processors. The only modules are the input (receive) module of a processor or demodulator and the output converter module of a processor or modulator, which make the difference between the units. If a tuned input converter for a processor or demodulator fails, a good temporary solution is to use a reasonably good video cassette recorder (VCR) as a demodulator. The VCR should be tuned to the receive station frequency, and the video and audio outputs can be connected to a modulator. The output converter from the processor can usually be installed in the modulator so the same output frequency will be used.

4.9.2 Satellite communications for cable television

4.9.2.1 Introduction. With more channel capacity becoming available in modern cable television systems and the desire for more programming, the satellite method of providing multiple sources of television programming became available. In the earlier days only one satellite was available with only two or three channels containing programming for the cable television industry. At present, there are four satellites available with each satellite containing 24 transponder channels, thus providing 96 total channels, which is more than the whole channel capacity of the most modern cable television systems. Obviously a choice is now available among the programming available on these satellite transponders. Essentially, the satellite system works by the programming signal source radio's transmitting on a channel in the 6-GHz band up to the satellite, which receives this signal and retransmits it on a channel in the 4-GHz band in a pattern covering the continental United States and Central and South America. The signal path to the satellite is called the *up link* and from the satellite back down to earth, the *down link*. Each programming source will be contained in an up-down link pair of channels. The send and receive antennas are usually parabolic reflector types that concentrate the energy into a narrow beam for both transmitting and receiving. The larger the diameter of the parabolic antenna the more signal transmitting or receiving concentration is obtained, thus more antenna gain. A diagram of the simplified link for a single-channel transponder is shown in Fig. 4.20. The up link and down link can also include several transponders at each site.

The standard up-down link frequencies for the transponders on several satellites are shown in Table 4.1. Adjacent channel numbers for the transponders are 20 MHz apart and are transmitted on opposite poles by *polarization*, which reduces crosstalk to tolerable limits. The transmitting and receiving antenna feed systems that transmit or receive the energy sent or received by the parabolic (dish) antenna are spaced 90° (right angles) from each other to achieve this polarization of the adjacent signals.

To make such a system work correctly, it is essential to place the satellite into an orbital position to "hit" it with the up-link transmitting antenna. The technique that is employed uses an equatorial orbit. It should be recalled that the equator is an imaginary belt that is midway between the north and south poles and hence is midway on the earth's axis of rotation. When a satellite is placed in an equatorial orbit and high enough out of the drag of the earth's atmosphere, it orbits at the same speed as the earth's rotation, and hence will appear stationary

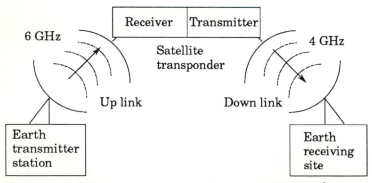

Figure 4.20   Satellite transponder link. 1 GHz = 1000 MHz = 10^9 Hz. The separation in frequency of 6 – 4 GHz = 2 GHz is more than adequate to separate the up and down signals.

TABLE 4.1   Commonly Used C Band Satellite Channel Information

Sat channel	Up link, MHz	Down link, MHz	Down-link polarization			
			Satcom 3R, 4	Galaxy 1, 2, 3	Westar 3	Westar 4, 5
1	5945	3720	V	H	H	H
2	5965	3740	H	V		V
3	5985	3760	V	H	H	H
4	6005	3780	H	V		V
5	6025	3800	V	H	H	H
6	6045	3820	H	V		V
7	6065	3840	V	H	H	H
8	6085	3860	H	V		V
9	6105	3880	V	H	H	H
10	6125	3900	H	V		V
11	6145	3920	V	H	H	H
12	6165	3940	H	V		V
13	6185	3960	V	H	H	H
14	6205	3980	H	V		V
15	6225	4000	V	H	H	H
16	6245	4020	H	V		V
17	6265	4040	V	H	H	H
18	6285	4060	H	V		V
19	6305	4080	V	H	H	H
20	6325	4100	H	V		V
21	6345	4120	V	H	H	H
22	6365	4140	H	V		V
23	6385	4160	V	H	H	H
24	6405	4180	H	V		V

20 MHz between channels	20 MHz between channels	12 transponders single-pole
2225 MHz difference between up and down links		

Note: For KU band satellites, see program supplier.

over the earth. Any earth transmitting or receiving antenna can now be stationary and will not have to track the satellite. Because even at nearly 23,000-mi altitude some atmosphere does exist, the satellite over a period of times drifts because it slows down slightly. An "on-board" fuel and jet system corrects for this slowdown, thus keeping the satellite "on station." When the fuel is exhausted, the useful life of the satellite is essentially over. Solar cells also charge the on-board batteries. Since these batteries have limited charge/discharge cycles before they in essence fail, the useful battery life is also limited, so that the useful satellite life is approximately 7 to 10 yr. It was expected that the space shuttle program could be used for communication satellite refueling and for battery replacement needed to increase useful satellite lifetime.

From the cable television operator's point of view, the receiving link is the only area under control. Further discussion will consider the down-link components, such as antenna parameters, low noise amplifiers, and down converters and receivers. Correct antenna pointing calculations and procedures will also be discussed.

4.9.2.2 Down link. At the satellite location in space, the receiver in the satellite receives the up-link signal on the 6-GHz band. The receiver drives the down converter, which converts the 6-GHz signal to a channel in the 4-GHz band, which in turn drives the power amplifier connected to the transmitting satellite antenna. This provides the down-link signal to the receiving stations placed throughout the country. Since the path is shortest directly under the satellite, at the equator the received signal strength is the strongest. As one moves around the various parts of the country the received signal strength varies. The output power as received on the ground is a function of the transmitted power from the satellite and the path loss. Both parameters are in effective isotropic radiated power (EIRP), which is usually in decibels in reference to the watt (dBW). As this energy is received on the ground a contour of equal received lines in EIRP in decibel-watts can be plotted on a map for future reference. Such a map is shown in Fig. 4.21. The earth receiving station can take several configurations. The most basic is shown in Fig. 4.22.

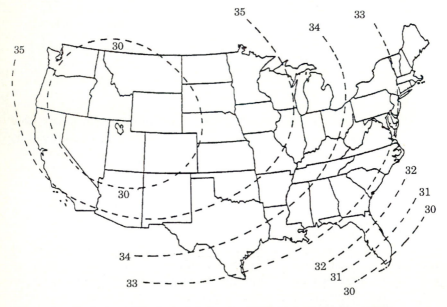

Figure 4.21 Typical satellite footprint showing EIRP in decibel-watts (dBW).

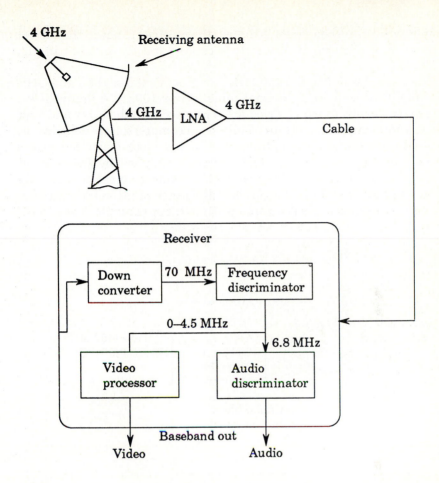

LNA: Low-noise amplifier

Figure 4.22   Basic satellite receiving system.

Since the cable loss at 4 GHz is large, a good idea is to amplify the signal at the antenna and then convert this frequency to a lower frequency which will lower the cable loss. This conversion process at the antenna is called *block down conversion* since all transponder frequencies are converted in a block to some lower band of frequencies. For example, at the 4-GHz frequencies transponder 1 is 3720 MHz and transponder 24 is 4180 MHz. These frequencies will be converted in a block to transponder 1 to 1450 MHz and transponder 24 to about 950 MHz. Notice that because of conversion the positions of the channels are reversed in frequency. The receivers that are made to work with the block-down converters take this into account when tuned. For example, when transponder 24 is dialed, the receiver tunes to the correct frequency. A diagram of this earth receiving station is shown in Fig. 4.23.

The cable loss at 950 to 1450 MHz is a great deal less than at 4 GHz. Usually if the connecting coaxial cable is over 100 ft, ⅞-in air dielectric coaxial cable, in which dried air is pumped into the cable and kept under constant pressure (2 lb usually) by a compressor air dehydrator, is used. The low noise amplifier (LNA) and block-down converters usually operate at $+15$ V_{dc}, which often is supplied by the receiver and sent along the connecting cable. The connecting cable is usually at an impedance of 50 Ω. The typical difference in attenuation for 50-Ω ½-in foam-filled cable is about 4 dB/100 ft. For example, loss of 50-Ω

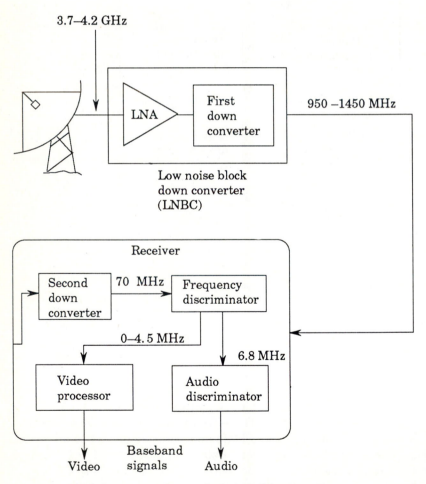

Figure 4.23 Block-down-conversion method of satellite receivers.

½-in foam-filled cable is 8 dB/100 ft at 4 GHz and 3.5 dB/100 ft at 1450 MHz. A discussion of C/N ratio and system performance follows.

4.9.2.3 Noise and temperature.   Since the impedance of free space at a specified temperature generates a specific amount of noise and the sun also acts as a generator of noise, it is more meaningful to calculate the noise power density (N_0) over the IF bandwidth of the receiver, taking into account all sources to calculate *carrier-to-noise* (C/N) *power density*, where N_0 = noise power per megahertz or watts/megahertz, and T = noise temperature (in degrees kelvin).

So mathematically $N_0 = K\,T\,(10^6)$

$$C/N_0 \text{ (dB – MHz)} = C/N \text{ (dB)} + 10 \log B_r \text{ (MHz)}$$

where B_r = receiver bandwidth (MHz).

Since the method of modulation for satellite communications is FM, the C/N_0 ratio over the receiver IF bandwidth produces the video signal when detected. The final baseband video signal-to-noise ratio determines picture quality. The signal-to-noise ratio S/N of a video signal with CCIR weighting corresponds to a peak deviation of 10.75 MHz and is given by

$$S/N = C/N_0 + 22.6 \text{ dB}$$

For the desired video signal-to-noise ratio of 54 dB then, from above, $54 = C/N_0 + 22.6$ and $C/N_0 = 54 - 22.6 = 31.4$ dB. This is the overall desired link C/N_0 that will give the overall video signal-to-noise ratio of 54 dB. To calculate further recall that noise for the system N_s at some point in the receiver is equal to the sum of the antenna noise and the receiver noise power in watts, microwatts, etc. This may be written as $N_s = N_A + N_R$. Also recall that noise power is a function of the bandwidth and temperature in degrees Kelvin (°K).

$$N = KTB_n$$

where K = Boltzmann's constant = 1.38×10^{-23} J/°K
   T = noise temperature (°K)
   B_n = bandwidth (Hz)

Noise for system = $N_s = N_A + N_R$

$$\uparrow \qquad \uparrow$$
$$\text{Antenna} \qquad \text{Receiver}$$

Substituting, $KT_sB_n = KT_AB_n + KT_RB_n$ (4.1)

Dividing by KB_n, $T_s = T_A + T_R$

T_A is the antenna temperature, which is not the actual temperature of the antenna on the ground. This is the sky temperature in degrees Kelvin the antenna is looking at. At low elevation angles the temperature is approximately 46°K and at straight up (zenith) 18°K. T_R is almost completely the noise temperature of the low noise amplifier (preamplifier).

This means that this amplifier will generate noise like a resistor at 120°K if that is the specified temperature of the LNA. Now if Eq. (4.1) is rewritten in decibel form for the system, $N_s = KTB_N$.

In decibel form N_s (dBW) $= T_s$(dB $-$ °K) $+ 10 \log B_N$

$$+ 10 \log 10^6 - 10 \log 1.38 \times 10^{-23}$$

where B_N is in megahertz.

To calculate $10 \log 10^6$ we should know that $10 \times 6 = 60$; then to calculate $10 \log 1.38 \times 10^{-23}$, enter 1.38, press EE, press 23, press plus or minus ($\pm$), press log, press inv, press EE, press equals ($=$). This gives -22.86. Press $\times$, press 10, press equal sign; this gives -228.6 dB. So N_s (dBW) $= T_s$ (dB $-$ °K) $+ 10 \log B_n$ $-228.6 + 60$.

$$N_s \text{(dBW)} = T_s \text{(dB} - \text{°K)} + 10 \log B_n - 168.6 \qquad (4.2)$$

From radio theory the received carrier power in decibels at a given point on earth is

$$C \text{ (dBW)} = \text{EIRP} - L_A - L_s - L_p + G_r \qquad (4.3)$$

$L_A = 10 \log K_A$ and $K_A = 1$, so $L_A = 0$. K_A is the atmospheric attenuation factor.

$L_p = -10 \log p$; in a well-designed system $p = 1$.

$$\text{So } L_p = -10 \log 1 = -10 \times 0 = 0$$

$$L_s = 20 \log \left(4\pi \frac{R}{\lambda} \right) \qquad \text{space loss (dB)}$$

Now calculating the C/N (in decibels) by combining Eqs. (4.2) and (4.3) gives

$$\text{C/N (dB)} = \text{EIRP} - L_s + G/T \text{ (dB/°K)} - 10 \log B_n + 168.6 \qquad (4.4)$$

where G/T (dB/°K) $= G_r$ (dB) $-$ (dB $-$ °K)

                    G = gain of satellite receiving antenna (dB)

                    T = effective system operating noise temperature at the same point where G is measured

The C/N_0 is the overall carrier-to-noise power density ratio where N_0 is in watts/megahertz so $N_0 = KB_n(10^6)$ for B_n in megahertz.

$$C/N_o \text{ (dB } - \text{MHz)} = C/N \text{ (dB)} + 10 \log B_n \text{ (MHz)}$$

where B_n is normally the IF bandwidth of the receiver.

Now Eq. (4.4) can be rewritten in terms of C/N_0 and solving for G/T

$$C/N_{dB} = \text{EIRP} - L_s + G/T \text{ (dB)} - 10 \log B_n + 186.6$$

Now the footprint map (a map contour of equal power–see Fig. 4.21) can be consulted for the receiving site. For Massachusetts, for example, the map gives 32.5-dBW EIRP.

To calculate L_s for a transponder frequency of 4000 MHz

$$\text{Recall } \lambda = \frac{v}{f} = \frac{300 \text{ m/s}}{f_{\text{MHz}}} = \frac{300}{4000} = \frac{3}{40} = 0.075 \text{ m}$$

The satellite is 22,200 mi above the equator; therefore, the distance from our point in Massachusetts to the satellite will be assumed to be 22,800 mi. Since 1 kilometer (km) = 0.6 mi,

$$R_{\text{km}} = \frac{22800}{0.6} = 38,000 \text{ km}$$

$$L_s = 20 \log \left[\frac{(4\pi)(3.8 \times 10^7) \text{ m}}{0.075 \text{ m}} \right] = 20 \log \frac{47.73 \times 10^7}{0.075}$$

$$= 20 \log 636.37 \times 10^7 = 20 \times 9.8 = 196.1 \text{ dB}$$

This is the space loss in decibels. Recall that a C/N_0 was needed at 31.4 for a 54-dB S/N_o. So

$$\frac{G}{T} = 31.4 - 32.5 + 196.1 - 168.6 = 26.4 \text{ dB}$$

This tells just what the expected G/T should be for the receiving system to provide a C/N_0 of 31.4 dB.

Since it is the ratio G/T that is important, increasing the size of the receiving antenna increases the gain and provides a lower temperature because the narrower receiving beam looks at a colder sky. Also the LNA may have more gain but mostly has lower noise temperature, thus making T smaller and the G/T larger. Recall that G/T (dB/°K) = G_R (dB) − T_s (dB − °K). Notice that the receiving system noise temperature is subtracted from the antenna gain. Therefore, for a given G/T such as 26.4 dB/°K a trade-off of lowering the temperature or increasing the gain will allow a selection of components.

Example A 4.5-m parabolic earth station receiving antenna which will be used to receive signals from satellite F_4 will have an elevation angle of 38°, and for transponder 15-vertical the frequency will be

4000 MHz. From published data T_A for the antenna is 22 (dB – °K) and gain is 43.6 dB.

$$G/T = 43.6 - 22 = 21.6 \text{ dB}$$

The conversion from decibels to degrees Kelvin may be necessary because some catalog data may give data only in degrees Kelvin.

$$T_s(°K) = 10^{22/10} = 10^{2.2} = 158.5°K$$

However, G/T of 21.6 dB is too much less than the required G/T of 26.4 dB.

A reasonable choice would be to go to a larger antenna, say a 7-m antenna with a gain of 47.5 dB and a 100°K LNA. The antenna noise temperature of 20°K for the link is 100°K + 20°K = 120°K, where 120°K corresponds to 10 log 120°K = 10 × 2.079 = 20.8 dB.

$$\text{Now } G/T = 47.5 - 20.8 = 26.7 \text{ dB/°K}$$

This is 26.7 – 26.4 = 0.3-dB/°K margin. Note also that 32.5-dB is pretty well down in the footprint. Usually modern satellites are hotter, so 33 to 35 dB is closer to reality.

The foregoing are some of the calculations one would want to crank through when selecting equipment. Normally most manufacturers supply a computer run for their customers or prospective customers. Such a computer run consists of performance figures of C/N_0, C/N, etc., for a customer's required geographical location (latitude and longitude to nearest second) for a selected satellite. Computer calculations are made for various choices of antenna size, LNA gain, and temperature. A good suggestion is to make a sample manual run with a pocket calculator or computer to test the figures.

4.9.2.4 Antenna orientation.

The installation of an earth receiving antenna used to be quite a job because LNAs did not have low enough noise temperatures, and the antennas had to be quite large. The feeds usually were prime focus: i.e., the waveguide feed looked into the reflector at the focal point. The geometry for a prime focus antenna is shown in Fig. 4.24a.

Another type of parabolic antenna is the Cassagrain type, which uses a secondary reflector which bounces the received energy into the end of the feed waveguide mounted through the antenna reflector vertex. This type of feed is usually more efficient in its energy gathering powers, thus giving approximately a 1-dB better C/N_0 of the receiving link. Fig. 4.24b shows this feed configuration.

Some antenna reflectors were not parabolic in shape but spherical. These antennas have a larger aperture angle. This type of reflector allows several prime focus feeds to receive the beamed signal from sev-

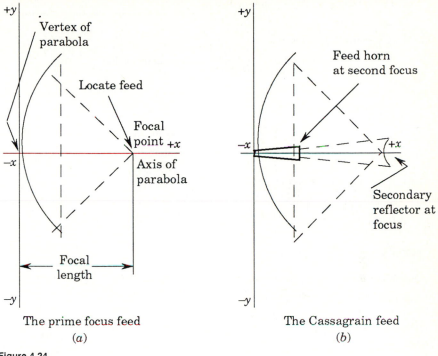

The prime focus feed
(a)

The Cassagrain feed
(b)

Figure 4.24

eral directions. This type of antenna can look at several different satellites with less performance than the parabolic type. Even though a parabolic type of reflector also can look at more than one satellite, the satellites in this case have to be adjacent to each other.

All of the types of satellite receiving antennas have to look at the satellite in space and hence have to be aimed precisely. The formulas based on the satellite-earth station geometry are given in Table 4.2. A pocket scientific calculator will suffice; however, a personal computer with the appropriate program that will provide all the performance calculations and look angles is available.

4.9.2.5 Earth station installation. In most cases the reflector mount requires a reinforced concrete base. The concrete base specifications are provided by the manufacturer. The wind loading criteria for the reflector are based on the specified mount and base. If the base is not adequate then excessive wind can cause sufficient forces to move the reflector, mount, and base, causing the signal to degrade. To restore service the antenna will have to be reaimed. All mounting structures can be moved in elevation and in azimuth. Some mounts can be moved over a greater angular distance, and some types are more easily

TABLE 4.2   Satellite Antenna Pointing Angles

The formulas are as follows:
Antenna azimuth angle (degrees) =

$$180 + \tan^{-1} \frac{(\tan D)}{(\sin X)} \quad \text{(from true north)}$$

Antenna elevation angle (degrees) =

$$\left(\frac{[\cos D \times \cos x] - 0.15126}{\sqrt{[\sin D]^2 + [\cos D \times \sin X]^2}} \right)$$

where X = earth station site latitude
      Y = earth station site longitude
      Z = satellite longitude
      $D = Z - Y$

Example

Earth station site latitude 41° 38′ 11.0″ N

Earth station site longitude 70° 27′ 42.0″ W

Satellite galaxy I longitude 93.5°W

Solution: Convert latitude and longitude of earth station site to decimal degrees.

Earth station latitude		Earth station longitude	
41°	41.0000	70°	70.0000
38′ = 38/60 =	0.6333	27′ = 27/60	0.4500
11.0″ = 11/3600 =	0.0031	42″ = 42/3600	0.0117
	41.6364 ≈ 41.64		70.4617 ≈ 70.46

where X = 41.64°
      Y = 70.46°
      Z = 93.5°
      D = 93.5° − 70.46° = 23.04°

Azimuth angle (degrees) $= 180 + \tan^{-1}\left(\dfrac{\tan D}{\sin X}\right)$

$$= 180 + \tan^{-1}\left(\frac{\tan 23.04°}{\sin 41.64°}\right) = 180 + \tan^{-1}\left(\frac{0.425}{0.66}\right)$$

$$= 180° + \tan^{-1}(0.644)$$

$$180° + 32.8° = 212.8° \text{ from true north}$$

Elevation angle (degrees) $= \tan^{-1}\left(\dfrac{[\cos D \times \cos X] - 0.15126}{\sqrt{[\sin D]^2 + [\cos D \times \sin X]^2}} \right)$

TABLE 4.2   Satellite Antenna Pointing Angles (*Continued*)

$$\tan^{-1}\left(\frac{[\cos 23.04 + \cos 41.64] - 0.15126}{\sqrt{(\sin 23.04)^2 + (\cos 23.04 \times \sin 41.64)^2}}\right)$$

$$= \tan^{-1}\left(\frac{(0.920 \times 0.747) - 0.15126}{\sqrt{(0.391)^2 + (0.920 \times 0.664)^2}}\right)$$

$$= \tan^{-1}\left(\frac{(0.6872) - 0.15126}{\sqrt{0.153 + 0.373}}\right)$$

$$= \tan^{-1}\left(\frac{0.5359}{0.723}\right) = \tan^{-1}(0.7389) = 36.5°$$

moved. When one is purchasing such equipment, the ease of installation and alignment is an important requirement.

Earth receiving systems require a good electrical grounding system. This is necessary to protect the LNA or LNA/block down converter at the feed point from lightning effects. One suggested layout for a grounding system is shown in Fig. 4.25.

If the coaxial feed cables are aerially connected to the head-end building on poles, the ground wire should be buried along the shortest path to the head-end building. Also, if a tower used for the off-air re-

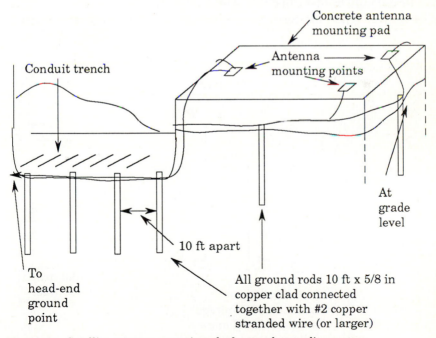

Figure 4.25 Satellite antenna mounting platform and grounding system.

ceiving system is at the same site, the tower grounding system should be connected to the earth station grounding system.

4.9.2.6 Antenna alignment procedures. After construction has been completed and the feed mounted, the antenna should be aimed, i.e., set to the calculated angles. The azimuth angle should be corrected for the magnetic declination of the site since an accurate compass will be used for aiming and the azimuth calculation is based on true north. A current topographical map will have this information. After the initial aiming has been completed, a receiver and a video monitor can be used to monitor the signal. The method is given in Fig. 4.26.

The transponder should be selected on the receiver and the equipment turned on. The power to LNA or low noise block converter (LNBC) usually will be provided by one of the receivers through either the power dividers or a power insertion device. With luck, a picture of sorts will be viewed on the monitor; if not, the antenna aiming should be rechecked. Usually it is more accurate to align the elevation angle with an inclinometer. This device has an angular scale and a level so the angle of elevation is measured from the level. The inclinometer should be placed near the base of the feed at the rear of the reflector for best results. Next the antenna should be slowly changed in azimuth back and forth over the specified setting. If any signal is received, the satellite should be identified so the relation to the desired satellite can be determined. When the correct satellite has been found and the antenna

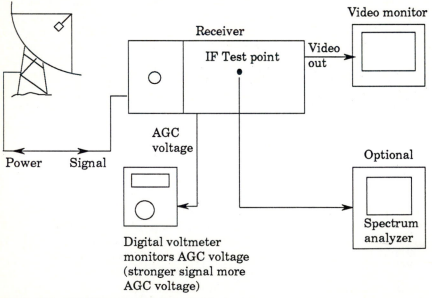

Figure 4.26 Satellite antenna alignment test setup.

aimed for the best possible signal, the feed mount with the LNA-LNBC system should be rotated to align the polarization.

Since adjacent transponders are separated in frequency by 20 MHz they are also transmitted by antennas placed at right angles to each other, thus achieving polarization. The receiving system also has to be aligned to correct for the polarization. The feed is manufactured at right angles so the whole feed should be rotated to align the axis. The procedure is to rotate the feed to maximize the received signal as viewed on the monitor and the digital voltmeter. As the desired signal peaks the adjacent signal should decrease. An incorrectly aligned feed will cause the adjacent channels to feed into one another, causing picture interference. The final peaking of the polarization can be performed by peaking the signal on the desired transponder while minimizing the adjacent transponder signal when the receiver is tuned to the adjacent transponder. After final polarization has been completed the azimuth and elevation can be retouched. The last step is to check all the other transponders to make certain the picture quality is good. If one or more transponders are marginal, small azimuth, elevation angle, and polarization adjustments may clear the picture adequately.

After the system has been used several months if any picture degradation is noticed, the concrete pad may have settled so the azimuth and elevation settings may have to be corrected. It must also be stressed that records of AGC voltage for the various receivers as well as any meter readings on the receiver should be kept in log form for future reference. Some receivers have LED bar graphs for signal strength; others have the usual meter movements.

4.9.2.7   Earth station siting.   The positioning and setting up of satellite receiving antennas on a head-end site can involve problems that a transit can solve. The siting problem is usually positioning the parabolic antenna with an unobstructed view of the satellites of interest. The procedure is as follows: (1) Obtain the azimuth and elevation angles (look angles) for the desired satellites by either a computer survey (supplied by many manufacturers) or calculations made on a scientific calculator. (2) Set up the transit over the prospective site and look toward the sky with the transit set for the elevation angle. (3) Slowly pan the telescope to the right and left of the calculated horizontal angle to check the field of view.

The above procedure should be repeated for each satellite needed, and if no trees or buildings block the field of view then the antenna should perform close to the expected performance values.

The transit is also used with a mirrored target or white target mounted on a similar tripod to test microwave link paths. Essentially a person with the target will mount it on a tall hill or building where a repeating antenna is desired. The transit setup at another prospective antenna site can check out the line of site. Accurate measurement of the horizontal azimuth angles can be made so the path can be plotted on maps. The magnetic declination angle should be taken into account for the area when relating the measured angles to compass (magnetic) and true north. The magnetic declination angle is usually obtained for the area from the U.S. Coast and Geological Survey topographical maps. Remember that accurate horizontal angle measurement requires use of the *vernier* (fine-adjustment) scales for the particular instrument. An example of a typical use of vernier scales is shown in Fig. 4.27.

Azimuth angles given by computer printouts provided by many manufacturers of earth station equipment are given in relation to true north. At locations in various parts of the continental United States the angular difference in azimuth from true north to magnetic (compass) north is called the *angle of declination*. If this angle is known, then a high-quality compass can give fairly accurate magnetic bearings ($\pm \frac{1}{4}°$). If a transit with a good compass is used, then accuracy can be on the order of $\frac{1}{10}°$. Then the bearing from true north can be calculated by correcting the magnetic bearing by taking into account the magnetic declination angle. In the northeastern part of the United

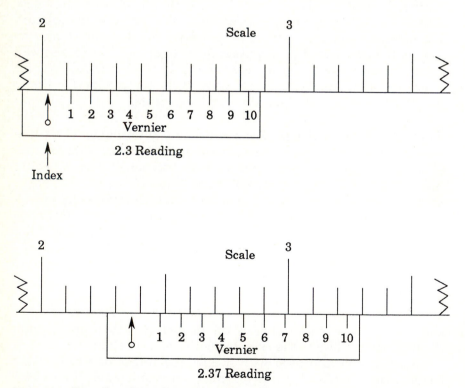

Figure 4.27   Examples of transit vernier scale readings.

States, e.g., the magnetic declination angle is about 15° to 15½° W. Therefore, this angle will be added to the magnetic reading to correct for true north as shown in Fig. 4.28a. The magnetic declination angle is given on every sectional topographical map published by the U.S. Coast and Geological Survey and is referenced to the center of the map. To set a look angle bearing line a transit can be set up to point the telescope along a compass site line and then corrected to true north by turning the transit scope to true north then turning from true north. The value of the bearing angle is as shown in Fig. 4.28b.

4.9.3 Frequency-modulated stereo service on cable systems

Providing high-quality FM stereo radio service can be as technically complicated as providing television signals on a cable system. Essen-

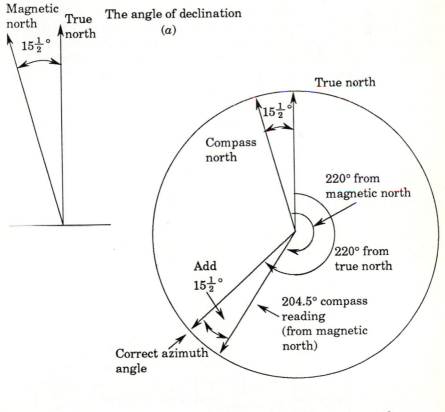

Figure 4.28

tially the best method is the same as that for television. However, for the FM signals, one or possibly two antennas are sufficient. If there is a strong high-power FM station operating near the cable television head end, traps or filters may be necessary to reduce the signal to a reasonable level. Some cable operators eliminate such a station, feeling that subscribers can easily receive this signal on the usual line power antenna clock radio or on the proverbial coat hanger. However, such a station has to be filtered completely enough so that other stations will not be affected. Some cable systems carry FM stereo radio in what is known as the *broadband* method. This is the way some early cable systems handled the television signals. Generally this method does not provide consistant high-quality service and requires constant adjustment, monitoring, and diddling. This rudimentary method is shown in Fig. 4.29.

The more elegant FM system individually processes each FM channel with a tuned FM signal processor. This processor is similar to a single-channel FM radio (crystal-controlled tuning) that has AGC and generates a 10.7-MHz IF carrier. This IF carrier is then mixed with the same crystal oscillator (on channel, in and out) or a different frequency (in on broadcast frequency, out on cable FM frequency). This is often referred to as *off-channel*. The reason cable operators sometimes change the output frequency is to space more evenly the FM channels to make tuning easier for the subscribers. This type of system is shown in Figs. 4.30 and 4.31. It is advisable to review FM theory in the references in the Selected Bibliography.

The level of the FM spectrum should be maintained to the same tolerances as the video signal. The output spectrum should be filtered by a bandpass filter (BPF) covering the whole 88- to 108-MHz spectrum.

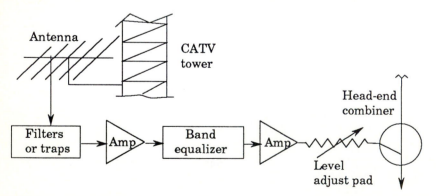

Figure 4.29 Broadband FM radio system.

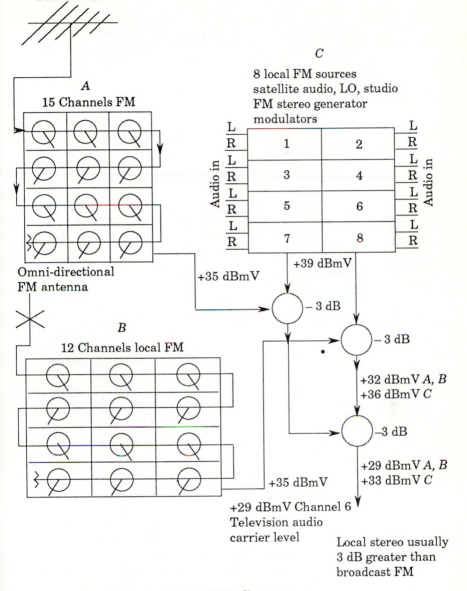

Directional
FM antenna

A

15 Channels FM

Omni-directional
FM antenna

B

12 Channels local FM

C

8 local FM sources
satellite audio, LO, studio
FM stereo generator
modulators

Audio in

L
R
L
R
L
R
L
R

1   2

3   4

5   6

7   8

L
R
L
R
L
R
L
R

Audio in

+35 dBmV

+39 dBmV

− 3 dB

− 3 dB

+32 dBmV *A, B*
+36 dBmV *C*

−3 dB

+29 dBmV *A, B*
+33 dBmV *C*

+35 dBmV

+29 dBmV Channel 6
Television audio
carrier level

Local stereo usually
3 dB greater than
broadcast FM

Figure 4.30   Individually processed FM radio system.

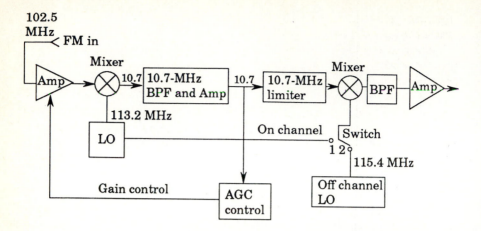

IF switch is in position 1. The 113.2-MHz
local oscillator converts the 10.7-MHz IF
signal to 102.5 MHz; with switch in position 2
station appears on 104.7 MHz.

Figure 4.31   FM signal processor.

Usually the processor has an output filter that filters either the output 200-kHz FM channel bandwidth or the FM 88- to 108-MHz band. The filtering is to attenuate extraneous signals, spurs, etc., and also any broadband noise the FM system may place on the cable system.

Testing FM signals can be more difficult than testing television signals simply because in the FM band there are 100 channels spaced 200 kHz apart. A number of commercially available signal level meters are too broadbanded to tune to individual FM carriers and give accurate level measurements. A reasonably good-quality spectrum analyzer that has a resolution bandwidth control can easily tune the FM band, giving the relative amplitude of the carriers at a glance. By narrowing the span and resolution bandwidth controls and carefully tuning through the carriers, measurements accurate within 1 dB of each carrier can usually be had. The point to remember is that an expensive piece of equipment is needed to maintain an FM system that produces only about 10 percent of the subscriber revenue. However, many cable operators feel they should provide FM stereo service and often it is a requirement of the license or franchise.

For many cable television operators a spectrum analyzer may not be available for FM testing. Since many signal level meters do not have a narrow enough bandwidth to differentiate between adjacent FM carriers, other means of testing need to be found. One piece of equipment that can be used is a high-quality FM stereo receiver. Of course, the amount of audio power output available is immaterial.

However, digital tuning and display, which make tuning through the FM spectrum of FM broadcast stations much easier, are preferred. Identification of these stations on the FM dial is much simpler. The next step is to test and calibrate the receiver for signal level. The diagram in Fig. 4.32 illustrates how this receiver can be used to make reasonably accurate level measurements on a per-channel basis.

The signal generator provides a strong unmodulated carrier on any of the FM frequencies. The level of this carrier into the receiver is controlled by the precision step attenuator B. Selecting step changes in signal level into the receiver will cause the AGC voltage to change in amplitude. Plotting a curve of signal level versus AGC voltage enables the receiver to become useful as an FM carrier signal level meter. This curve can be plotted on graph paper or a simple calibration chart of signal level versus corresponding AGC voltage can be made.

To locate the source of the AGC voltage in the receiver a service manual showing the schematic diagram is helpful. Usually the AGC

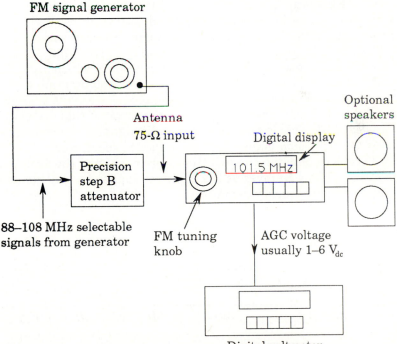

Figure 4.32 FM radio system testing method.

voltage can be found on the IF amplifier printed circuit board; it is often connected to a simple meter movement indicating received signal strength or an LED-type level meter. The signal level meter indicators found on most stereo receivers have insufficient resolution and range. Substituting a digital voltmeter solves the problem. The AGC voltage may be either positive or negative in polarity with respect to chassis ground and should not make any difference in the procedure. This whole setup should cost approximately $400 to $500, including the attenuator, receiver, and digital voltmeter. A speaker or speakers are not needed, and if they are not used the volume control should be kept off to prevent damage to the output amplifier. Use of speakers or stereo headphones can aid in identifying stations to be measured. Once the station is located, the digital voltmeter is read and compared to the calibration chart or curve for the corresponding signal level. For most purposes, this method is of sufficient accuracy and is well within most cable operators' budgets.

4.9.4   Broadcast stereo audio for television

Early in the 1980s several systems were proposed for using stereo audio in place of the standard television 4.5-MHz audio subcarrier system. Naturally, compatibility with the present system in use had to be part of the specification for any new system. One system was accepted by the FCC, the multichannel television sound (MTS). Many of the broadcast stations broadcast the audio portion as MTS, which provides broadcast stereo television (BTSC), as well as a second program audio (SAP) used for possible second-language audio, and a private communication channel (PRO). The layout of the spectrum of this system is shown in Fig. 4.33. This is the baseband audio signal that frequency-modulates the television audio 4.5-MHz standard subcarrier.

Fig. 4.33 tells us only the bandwidth requirements and the location. The diagram in Fig. 4.34 gives us the relative amplitudes that control the amount of FM frequency deviation.

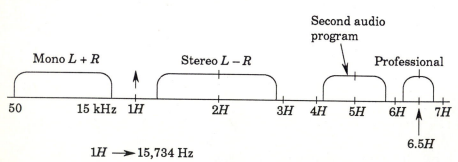

Figure 4.33   Spectrum location of MTS signals.

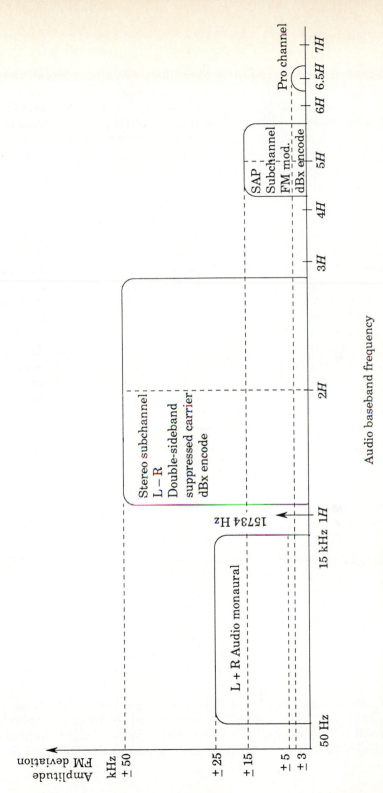

Figure 4.34 BTSC/MTS baseband audio signal.

273

There is a marked similarity to ordinary FM stereo. Recall that the normal FM carrier is frequency-modulated with the monoaural ($L + R$ audio signal) at ±75 kHz and the $L - R$ stereo subchannel modulated by the 38-kHz subcarrier. A 19-kHz pilot was transmitted to be used to manufacture the 38-kHz subcarrier, which was suppressed when transmitted. A subsidiary channel at 67 kHz was authorized for background music purposes sold by the broadcaster. The MTS system transmits the $L + R$ monoaural as before so compatibility with non-MTS television receivers is maintained. The stereo subchannel ($L - R$) modulates a subcarrier at twice the horizontal television line frequency ($2 \times 15,734 = 31,468$ Hz), referred to as $2H$. The 15,734 Hz carrier is also transmitted at low amplitude (±5 kHz) to prevent the sidebands from causing interference with the audio baseband signals. The second program audio channel, similar to the SCA, FM modulates a subcarrier at $5H$ ($5 \times 15,734 = 78,670$ Hz). The main difference is the narrow-band professional channel (PRO channel) operating on a subcarrier at 6 1/2H ($6.5 \times 15,734 = 102,271$ Hz). For the MTS system both the ($L - R$) stereo subchannel and the SAP channel have their audio baseband signal companded. *Companding* is compressing the dynamic range of the audio signal during transmission. Then expanding techniques employed in the receiver essentially restore the signal's dynamic range. Such techniques are used to improve the stereo separation and reduce noise. Twice the amount of deviation for the ($L - R$) portion is used to improve separation and reduce noise.

It should be noted from the previous figures that the baseband audio signal extends to about 110 kHz. Now care must be exercised by cable television technical people in handling television signals containing the MTS audio signal. The head-end heterodyne processors operating with normal television bandwidth essentially pass through the MTS audio signal. Possibly the sound trap in the processor may need tweaking if the quality of the stereo signal suffers. Also the sound carrier level should be maintained at 15 dB below the video carrier level, which is the normal level difference by CATV standards. The head-end treatment of what is known as demod/remod systems will require a different method of handling the MTS signal. If the demod only IF drives the modulator IF, then possibly the MTS signal will be passed through to the system. However, the benefits of the demod/remod scheme will not be realized. Subscriber converters of the baseband variety generally do not pass the MTS signal through. Essentially the MTS television set detects the MTS audio signal as shown in Fig. 4.35.

Another way MTS stereo can be handled by the cable television head end is to detect and decode the stereo left and right channel signals and place these audio program signals in the FM band in the

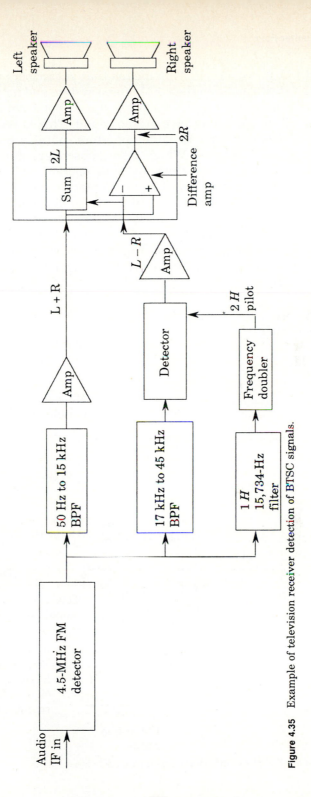

Figure 4.35 Example of television receiver detection of BTSC signals.

standard FM stereo format. This means that subscribers will have to receive the stereo audio on their FM stereo system and not turn up the normal television volume. Subscribers who do not have MTS television sets will be able to receive stereo sound. This is really not bad since the FM band for many cable operators contains premium program audio in FM stereo. When more subscribers purchase MTS-capable television receivers, then the cable operator will face the possibility of adding head-end equipment to place many of the satellite channels in the MTS audio format. Since the audio on a number of encoded satellite channels is also encoded, the output is a mono audio and left and right channels as well. The method to provide MTS audio to the television channel, shown in Fig. 4.36, incorporates a BTSC stereo generator, which is available from several high-quality audio manufacturers.

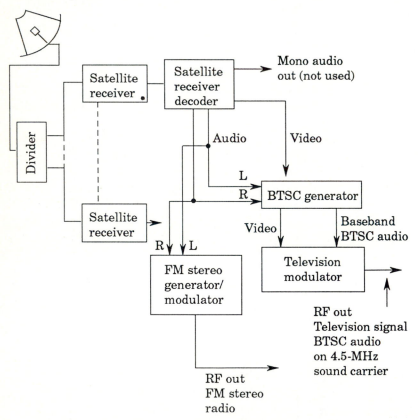

Figure 4.36   Method of satellite stereo audio to BTSC signal.

Earth station
antenna system

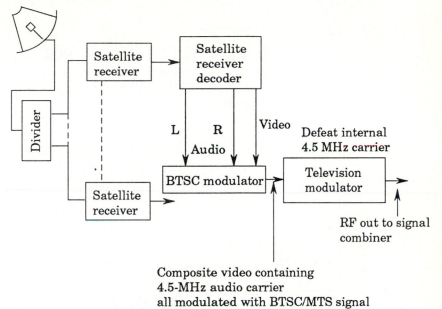

Figure 4.37   Composite BTSC method.

Some manufacturers have the MTS stereo signal available on a 4.5-MHz FM subcarrier all modulated. If it is available, it can be used in this manner with good results. The connections are shown in Fig. 4.37.

There are many techniques that can be employed in providing BTSC (MTS) signals to the cable system. New products are always coming out and should be investigated with regard to both technical and financial benefits. Most manufacturers publish a variety of information on how to use, test, and maintain their products. Most of the time all one has to do is ask.

Since a great many VCRs have BTSC capability some subscribers are electing to purchase a television that can act as a monitor for video as well as left and right audio inputs. The hookup is shown in Fig. 4.38.

4.9.5   Locally generated programming

One of the main attractions of cable television systems is locally originated programming. These channels can only be obtained by cable subscribers and not from any antenna system satellite or otherwise.

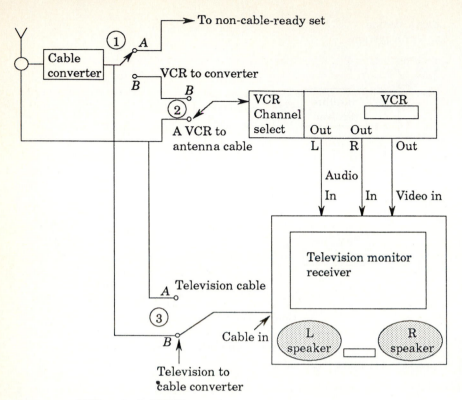

Figure 4.38 VCR method of BTSC reception.

Earlier cable systems usually designated one channel containing pro-gramming from a cable operator's own studio facility. This was simply known as *local origination* and programs consisted of local interest subjects. Later cable operators added alphanumeric channels contain-ing printed messages of community interest and/or classified ads. This equipment, known as a *character generator*, was controlled through a typewriter-type keyboard. Usually this equipment and the local studio were centrally situated in the community and often at the cable oper-ator's office. There, the customer service clerks could type in the char-acter generator pages as an added duty. The local origination staff could also schedule the camera work and studio appointments through the customer service clerks. Modern cable operators are often required to have other channels designated as public access, government, edu-cational, etc., as part of the licensing or franchising agreement with the community government. A second or third studio facility may be required to fulfill this obligation. These studio facilities may be quite expensive and complex, and the studio technical staff should be well versed in the technical aspects of studio construction and equipment. There are many good books covering television production and operations, some of which are listed in the Selected Bibliography.

A simple two-camera studio facility is shown in Fig. 4.39. The intercom system enables the control operator (producer-director) to communicate with the camera operators without disturbing the audio portion of the program. The tally light's "on" condition indicates which camera is being selected as active. The character generator provides the on-screen alphanumeric annotation. The audio tape recorder can provide background music for live cable-casting programming. It has been often said that people who try live cable programming subscribe to the theory that when you go live you're dead. Most cable operators

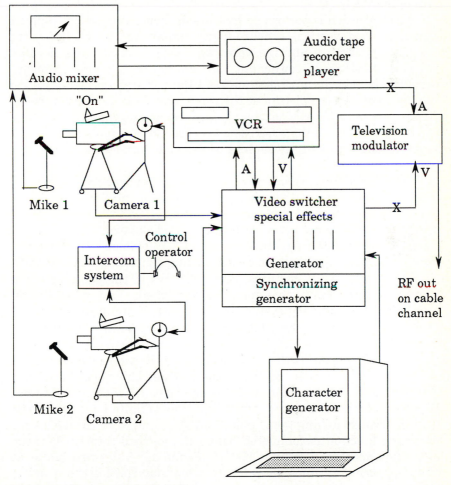

Figure 4.39   Two-camera local studio.

do their studio work at more convenient times and tape the results. Later the tape can be edited with all the mistakes removed, background music added, and graphics included, enabling the program to fit into a certain time slot. Smooth, professional programs result by this kind of programming. Unfortunately good editing equipment adds to the studio cost, but it opens a door by enabling a cable operator to insert local commercial advertising into the program material effectively since most cable studios record their programs on a VCR even if the programs are done live. This gives the studio a record of what went out on the cable. If a switching system is inserted at point *x-x* then the modulator can be switched to various sources, the studio being one. This enables the studio to be placed off-line for recording on the VCR while the modulator can be switched to another source of programming.

Synchronization is performed by the sync generator, which supplies horizontal and vertical synchronizing signals to the cameras, the character generator, and the video switch–special effects generator. Because all pieces of equipment that are generating television pictures receive synchronizing signals all the pictures are generated at the same point in the picture at the same time. This allows the special effects generator to provide the wipes, fades, enlargements, etc., from one picture source to the other. The synchronizing signals often are on separate conductors in the camera cables and character generator cables.

The synchronizing signals may be generated by a camera or a character generator internal synchronizing generator and are part of the video signal. A device known as a *gen-lock system* can use this video source as the master synchronizing signal and slave all other pieces of equipment together from this one source. Television broadcast stations usually use the separate synchronizing generator, which usually provides "rock stable" pictures. Since the cost of the equipment has been reduced in recent years, many cable operators have opted for this technique. If a switching system is incorporated, maintaining picture synchronization gives the added benefit of performing the switching during the time known as the *vertical interval*, which starts with the picture forming beam at the lower last line of the picture and the start of the visible portion at the top of the picture. This time duration is approximately the time for 21 lines or about 1.3 milliseconds (ms). Electronically speaking, this amount of time is quite long. Often vertical interval switching takes place just after line 9, which means the switching function is completed before the vertical interval test signals, often appearing on lines 17 and 18, and any video-text material, often appearing on lines 19 to 21. Caption material for the hearing-impaired and address converter control signals often use lines 12 to 16. As should be seen and understood, proper picture/video synchronization is a must for the up-to-date modern studio. Most synchronizing generators are able to operate very effectively in the gen-lock mode. This is particularly useful and necessary when the primary video signal is a video tape recorder, VCR, or laser video disc source. This primary video signal can be stripped of its sync signals and used as the primary sync source, thus slaving the other sources to the primary source.

In many cable television studios attention to the audio signal quality is deficient. Microphones, audio preamplifiers, and the audio control board are not given proper attention. Many times the audio levels of the various sources are not controlled properly so the cable channel programming may be too high or too low in volume, causing subscriber complaints. In extreme cases severe loud audio can overdeviate the audio carrier, causing video picture impairment as well. Audio limiting amplifiers can be employed to aid in balancing the audio signals. These amplifiers can provide a sort of audio automatic gain control (AGC) as well as automatic tone control and thus maintain audio fidelity during loud and soft audio passages. Also the television modulator can often be fitted with a deviation limiting circuit which does not allow the overdeviation (loud) condition to occur. This method takes care of only the extreme loud areas in programming, not the low-volume areas. A good review of some of the audio engineering books will aid in making the proper audio decisions. The criterion should be that the quality of the audio components be as good as that of the video components.

4.9.5.1 Lighting. Many of the new cameras use solid-state charge coupled device (CCD) image sensors instead of the vidicon or plumbicon tubes. The CCD-type image sensor works at low light levels and is less sensitive to the type of light. It is also more rugged. However, for studio use and for indoor telecasting artificial light is needed. Usually television studios have high enough ceilings to support a steel pipe grid system for hanging overhead light fixtures. These fixtures are made in several configurations where the uses and type of lamp dictate the use. Large incandescent flood lights often are used as general scene illumination types called "fill" lights and/or back lighting. Quartz-type lamps are used as the key lights to illuminate the subject. The studio lighting systems usually are all wired to a separate switch panel so the individual lamps may be controlled as needed. Certain dimming controls can be used with incandescent lamps. However, quartz-type lamps should not be connected to a dimmer because changing the brightness changes what is known as the *color temperature* of the lamp, i.e., its color content. Cable technical people should study television engineering, which is covered in some of the references listed in the Selected Bibliography.

4.9.5.2 Tape editing. Television production is often done in bits and pieces. Certain scenes are recorded at several locations and often require several "takes" before producing the desired results. The editing process enables the desired subject matter to be assembled to form a complete television program that will fit the desired time slot. The editing system often takes the form shown in Fig. 4.40. The tapes that contain the scenes to be selected are placed as needed on the source tape VCR and a blank tape cassette is placed on the editing VCR. The edit controller shuttles the source tape frame by frame to locate the scene to be selected. The controller backs up beyond the beginning of the scene a few frames and stops. This allows a preroll interval so both tape machines can be running at the proper speed before the editing process begins. When editing starts, the playback machine passes the signals to the editing VCR for recording on the output program tape. The editing operator stops the process when the desired scene is finished being recorded. It should be evident that the editing controller is the heart of the whole process. How elaborate and accurate this controller is determines the cost of the system. Some of the latest versions use a timing signal known as a *time code* that is added to the video tape systems, which aid in the editing process. Some system manufacturers have generated their own version of timing signals. The standard type is referred to as the Society of Motion Picture and Television Engineers (SMPTE) time code. The tape editing systems on the market today are vastly improved and the cost appears to be decreasing. These advanced types simplify the editing process, alleviating a lot of the intense operator control and producing professional results. Many cable operators are expecting HDTV systems soon, but they are not

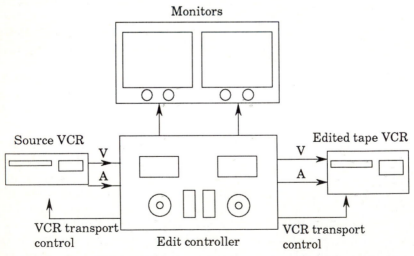

Figure 4.40   Tape editing system.

expecting that local programming will be in the HDTV format as yet. As soon as a few use it, the rest will follow.

4.9.5.3 Commercial advertisement insertion.

The cable television industry has finally realized that it can be competitive for the advertising dollars available. Many systems started placing local ads on the locally generated television channels to help offset the costs many system managers regarded as money "down the drain." Now many of the satellite program suppliers are allowing the vacant advertising slots to be available to cable operators to insert commercial ads. This point was often overlooked or scared off many cable operators, who considered advertising as "uncharted waters." Now many cable operators are placing local ads on several satellite channels on the available slots and are surprised by the money they are making. The technical aspects of ad insertion systems should be learned by cable technical personnel, who will inevitably have to live with the equipment. The technical staff responsible for maintenance should carefully investigate available equipment before purchases are made.

The ad spot that is designated as being available to a cable operator is marked at the beginning and end by start and stop signals. The program suppliers notify the cable operator of the schedule of locally available time slots, thus enabling the cable operator to sell a customer an ad, make the ad, and record it on tape for insertion during the available time slot. The start and stop signals are often dual tone multifrequency (DTMF) tones known as *cue tones*. The DTMF tones are the same as those used by touch tone telephone equipment and usually appear on the audio signal at a lower volume. Since these tones are heard by subscribers and are regarded by some as annoying, the start and stop signals are being placed on separate (unheard) subcarriers by some program suppliers. It is also being proposed that these signals be buried in the vertical interval. One of the simplest and most reliable ad insertion systems is known as the run of schedule (ROS) system. This technique is shown in Fig. 4.41. The inserted ads are shown as two 30-s ads in sequence (run of schedule). However, the inserted ad may be a 1-min ad occupying the two slots. Usually a start signal will appear at the beginning of the first 30-s spot and the stop signal at the end of the second 30-s spot. This allows the cable operator to sell two 30-s ads run as a 1-min pod. Commercially available ROS-type controllers often are able to provide control for up to four satellite channels with a VCR for each channel. The main problem with this type of system is the complexity of preparing the ad tapes with the ads placed in the order of appearance.

Other designs of commercial insertion equipment will enable ads to be recorded and played on more than one VCR per channel. This al-

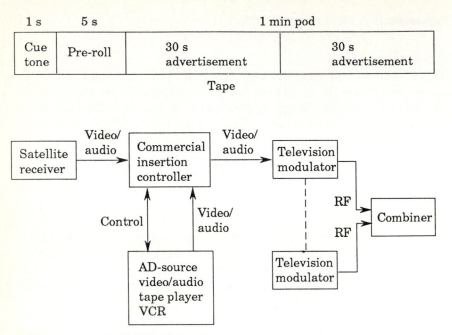

Figure 4.41   Run-of-schedule ad-insertion system.

lows, for example, ad number 1 to come from VCR 1, ad number 2 from VCR 2, etc. If each VCR has 2 h playing time and four VCRs are used, then the operator has to load only twice a day and often only once because most ads run during prime time plus a couple hours before and after (8 h). Commercial insertion equipment is often known as *random access* or *spot random access*, which means that the ad can be placed on any of the four VCRs, where each VCR has the same tape. Therefore, each ad will be fired off from any one of the four VCRs with the next one from one of the others. Each VCR is sharing the work, thus doing it all, as in the ROS-type system. The most up-to-date types are controlled by some type of computer system, such as the usual personal computer (PC). The schedule may be entered, the tape machines controlled, and logs of ads recorded under computer control, thus making operation easier and more professional. Such a system is shown in Fig. 4.42. The controlling computer receives network signals from the network interface unit (NIU) and selects the VCR through the matrix switches to the NIU for switching to the cable channel. Software has been developed by some manufacturers to enable the PC to do billing to customers, paying of invoices to ad sales people, and compiling of general sales reports. However, maintenance often falls to the technical staff. Therefore, proper documentation of installation and operating procedures, maintenance schedules, and schematic dia-

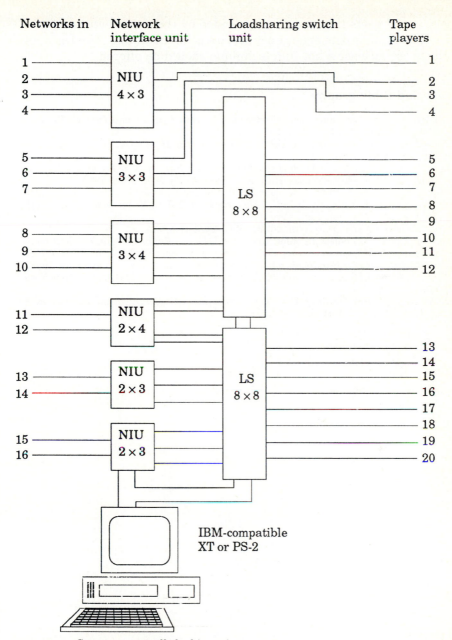

Figure 4.42   Computer-controlled ad-insertion system.

grams and parts lists should be provided by the manufacturer of such equipment. If on-site emergency service is not available from the manufacturer, at least a hot-line technique should be offered.

Selected Bibliography

Barker, Forrest. *Communications Electronics, Systems, Circuits and Devices*. Prentice-Hall, Englewood Cliffs, N.J., 1987.

Benson, K. Blair, editor in chief. *Television Engineering Handbook*. McGraw-Hill Book Co., New York, 1986.

Buchsbaum, Walter M., Sc.D. *Buchsbaum's Complete Handbook of Practical Electronic Reference Data*, ed. 2. Prentice-Hall, Englewood Cliffs, N.J., 1978, 1980.

Cunningham, John E. *Cable Television*, ed. 2. Howard Sams & Co., Indianapolis, 1985.

Freeman, Roger L. *Telecommunication Handbook*, ed. 2. John Wiley & Sons, New York, 1981.

Grant, William. *Cable Television*, ed. 2. GWG Associates, Fairfax, Va., 1988.

Jordan, Edward C., editor in chief. *Reference Data for Engineers: Radio, Electronics, Computers and Communications*, ed. 7. Howard W. Sams & Co., Indianapolis, 1985.

Orr, William I. *Radio Handbook*, ed. 23. Howard W. Sams & Co., Indianapolis, 1988.

Shrader, Robert L. *Electronic Communication*, ed. 2. McGraw-Hill Book Co., New York, 1967.

Taub, Herbert, and Schilling, Donald L. *Principles of Communication Systems*, ed. 2. McGraw-Hill Book Co., New York, 1986.

Chapter 5

Cable System Design
and Network Topology

Cable system design begins with the mapping procedure. The area to be designed, such as a neighborhood, a town, or a city, must be studied from a demographic standpoint as well as by a survey of the methods available to install the cable plant. Maps of the area will show the road layout and where the residential buildings are situated. Scaled maps obtained from the local water company, utility company, or department of public works (DPW) will be a good source of base maps, with which a cable television operator can plan where and how the cable plant can best be routed. For aerial cable plant construction on utility poles, the pole positions and locations will have to be plotted on such maps. To get this information a field survey has to be made: the distances between poles are measured and the pole numbers recorded, usually on what is known as a field survey form. This field survey is usually part of the *strand-mapping* procedure. When the field survey is completed and the pole locations and numbers are plotted on copies of the base maps along with the head-end location, the cable routing can then be determined. From the routing of the cable to the homes to be served, a list of pole numbers on a street-by-street basis has to be made so an application to the pole owners, i.e., the power company and telephone company, can be filed. This application with the street list of pole numbers is the usual requirement specified by the pole attachment agreement between the local electric power company, the telephone company, and the cable television company. During the initial field survey while the pole plant is being analyzed, it is also recommended to note on the records the relative positions of the prospective homes to be served and the utility poles that supply each home with

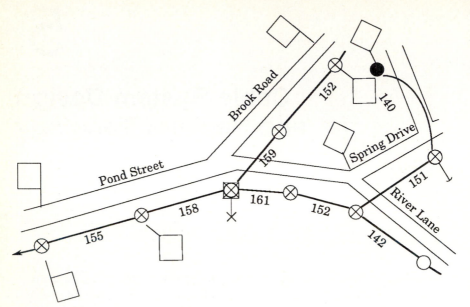

Figure 5.1 Strand map.

electric and telephone service. The cable service marketing department can then make a house count from the strand maps.

Underground construction routes also have to be plotted on the base maps with road crossings and available utility ducts noted. Serving both sides of a street with underground plant is always a problem. The choice is either to zigzag across the street or build plant on both sides or plant on one side and drop crossing ducts. Such problems cause the cost of underground plant to be much higher than that of aerial plant. An example of a section of a base strand map is shown in Fig. 5.1. A legend block usually found in the right-hand corner of the map shows the symbols for the various items marked on the map. An example is shown in Fig. 5.2.

The main part of the initial field survey effort is the measurement of the distances between the utility poles and the lengths along the underground cable route. These distances have to be measured accurately so the lengths of cable can be known precisely and the signal loss can be calculated. This cable loss is used to determine the location of the amplifiers. These distance measurements are entered on the map along with the pole numbers and other pertinent information.

5.1   Distance and Height Measurements

Distance and height measurements are often made during the construction of a cable television system and afterward as well. Before

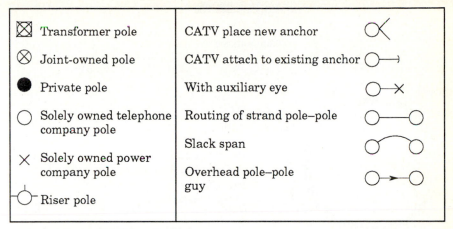

Figure 5.2 Legend block on strand map.

discussing such techniques, a short study on numerical accuracy is in order. As we have learned in high school and on the job as well, we currently use decimal numbers to express money values, time spans, and distance spans. For example, if we have measured a span of cable and it measures to 247 ft exactly and is written as such, then anyone reading notes of this measurement will assume the length to be measured to the nearest foot. If the value 247.5 ft is written, then the measurement has been made to the nearest 0.1 ft; if 247.18 ft, then measurement is to the nearest 0.01 ft. Since the actual on-site system measurements are made on pole-pole span lengths over the average terrain, which goes up and down, the accumulated errors should be investigated.

5.1.1 The measuring wheel

One of the more commonly used types of equipment is the measuring wheel. This is a wheel on an axle fitted with a handle that may be folded on a hinge to collapse for easy storage. On the rim the wheel is fitted with four horizontally protruding pins that trigger the counting lever of a mechanical counter. Each time the pin triggers the counter lever 1 ft is added to the counter. A knob on the end of the counter case allows the counter to be reset to 000; usually six digits are displayed so when the counter accumulates 999999 it will automatically reset and start again at 000000. Usually the distance measuring wheel will measure to the nearest foot. Consider the diagram of the measuring wheel in Fig. 5.3, which should clear up any questions.

Now, if this wheel is indiscriminately set to measure 10 pole spans, measured a span at a time, error will be accumulated. Suppose we had

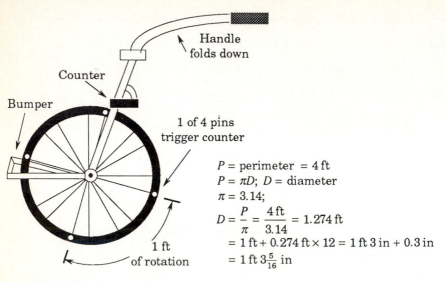

Handle
folds down

Counter

Bumper

1 of 4 pins
trigger counter

P = perimeter = 4 ft
$P = \pi D$; D = diameter
$\pi = 3.14$;
$$D = \frac{P}{\pi} = \frac{4\,\text{ft}}{3.14} = 1.274\,\text{ft}$$
= 1 ft + 0.274 ft × 12 = 1 ft 3 in + 0.3 in
= 1 ft $3\frac{5}{16}$ in

1 ft
of rotation

Figure 5.3   Measuring wheel.

the maximum error of 1 ft per span; then over 10 spans we would be 10 ft off. The question is, How bad is this error? Assume the cable loss at the upper frequency limit of 450 MHz to be 0.9 dB/100 ft. Then our error will be 0.09 dB or nearly 0.1 dB. If the amplifier cascade is spaced at 22 dB then 0.9 dB/100 ft will give a distance of

$$\frac{22\,\text{dB}}{0.9\,\text{dB}} \times 100\,\text{ft} = 2444\,\text{ft}$$

Now if there is an average of 40 poles per mile then

$$\frac{40}{5280\,\text{ft}} = \frac{x}{2444} = \frac{2444 \times 40}{5280} = 18\frac{1}{2} \qquad \text{poles between amplifiers}$$

Note

$$\frac{5280\,\text{ft}}{40} = 132\,\text{ft/pole span}$$

Therefore 1 ft in 132 ft, percent error/span = (1/132) × 100 = 0.75 percent. Since we can't handle one-half pole we could say a span of 19 poles with a maximum error of 1 ft/span will give 19 ft. Therefore, 19 ft × (0.9 dB/100 ft) = 0.17 dB error in loss.

A further consideration tells us that for a 20-amplifier cascade the accumulated error is 20 × 0.17 dB = 3.4 dB; of course, this error is spread over 20 amplifiers. Now if the wheel is reset to zero at each pole and the wheel is also set just behind a pin, then the accumulated error will be significantly less than the original not-so-bad error. To increase accuracy one could take repeated measurements and then take an average. This, of course, would be too time-consuming ever to be practical.

5.1.2   The transit and stadia rod

A steel tape is a favorite tool used by the building trades and by surveyors. Since the tape is laid out on the ground, over the terrain, some error will result. Often two persons stretching the tape between the points (poles) will give more accurate results. Many surveyors measure distance using a transit often called a *theodolite*. Essentially the *transit* is a telescope with a built-in graticule of vertical and horizontal cross hair lines. This is similar to a spotting scope used by rifle target shooters. The telescope is mounted in a holder that is free to turn in azimuth and in elevation. The elevation limit is from a level to nearly 90° straight up and in azimuth 360° completely around. Often a high-quality compass is installed inside the azimuth circular scales. Two bubble levels at right angles to each other aid in leveling the telescope by four leveling thumb screws. The transit instrument is mounted on a tripod with adjustable leg lengths. This allows use of the instrument on uneven ground. A photo of one of several varieties of instruments is shown in Fig. 5.4.

5.1.3   Transit and stadia rod method of distance measurement

The transit and stadia rod method, with care and practice, can produce pole span measurements of distance to a precision of 1 in 500 typically. Two persons are required and the work usually goes quite well with proper planning. Consider the diagram illustrating the measurement in Fig. 5.5.

The instrument person drops off the rod person at pole 1 and drives up to pole 4 and sets up the transit. The instrument person reads distance d_1 and records the results. The rod person moves up to pole 2, and the instrument person measures distance d_2. Then distance d_3 is measured in like manner. The rod person, after making the d_3 measurement, walks to pole 4 and moves the vehicle up to say pole 6 and walks to pole 7, where the d_6 distance is measured, and then walks back to pole 5 making the d_5 measurement followed by the d_4 measurement. Then the instrument person and the rod person get into the vehicle and the rod person is dropped off at pole 7 to repeat the procedure. Six spans are done at a crack. Subtraction of distances, for example, $d_5 - d_4$, will give span distance 5. To clarify the stadia principle of distance measurement a short discussion follows. The essential

Figure 5.4 Basic transit.

principle is that the vertical spread in field of view for a telescope as object (stadia rod) is increased in distance. The stadia rod has a variety of markings placed on one side of a square or flat rod approximately 12 ft long. Some stadia rods are 4 in wide and 1 in thick so they can be read at distances of 1000 ft. The procedure is as follows: The instrument is set over a point by use of a plumb bob hanging under the transit and then leveled by using the leveling screws and observing the two horizontal spirit levels. The rod person carries the rod to the distance to be measured, placing the bottom end of the rod on the point or next to the pole with the scale pointed toward the instrument.

The upper and lower stadia hairs in the transit telescope are observed. Usually the lower stadia hair is set over the nearest foot mark by tilting the scope upward or downward. The amount of rod observed between the lower and upper stadia hairs is a function of the distance of the instrument from the rod, i.e., the distance to be measured. The formula for the stadia distance is an application of the law of lenses.

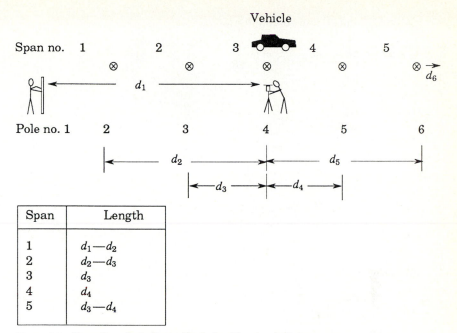

Figure 5.5   Transit and stadia method of cable stand distance.

$$\frac{f_2}{f_1} = \frac{AB}{ab}$$ (5.1)

$$\frac{1}{f_1} + \frac{1}{f_2} = \frac{1}{f}$$ f_1 and f_2 are conjugate foci and f is the principal focal distance   (5.2)

Consider the diagram in Fig. 5.6.

Therefore, for many instruments $D = 100\ AB$. D is the distance from the transit center (plumb line) to the stadia rod. An eyepiece view of a typical stadia rod measurement is shown in Fig. 5.7.

Newer instruments essentially work the same way, and reading through the operations manual should clarify the procedure. Usually the newer instruments are smaller, lighter in weight, and simpler to adjust.

5.1.4   Height measurements

The transit lends itself very well to the measurement of towers and tall structures. Measurement of tilt or bending of a tower structure can also be made with the transit. The procedure is first to set up the transit at a distance from the tower base. This distance should be nearly equal to the estimated height of the tower or greater. Once the

Combining ① and ② and solving for f_2

① $f_2 \times ab = f_1 AB$ so $f_1 = f_2 \dfrac{ab}{AB}$

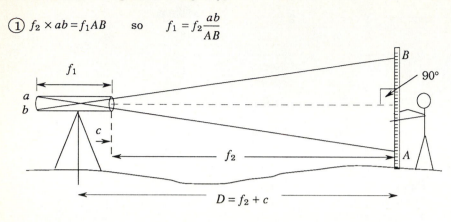

② Substituting for f_1 : $\dfrac{1}{f_2 \frac{ab}{AB}} + \dfrac{1}{f_2} = \dfrac{1}{f}$, $f(f_2) \dfrac{1}{f_2}\left(\dfrac{AB}{ab} + 1\right) = \dfrac{1}{f}(f_2)$

so $f_2 = f\left(\dfrac{ab}{AB} + 1\right) = f\dfrac{AB}{ab} + f$

$D = f\left(\dfrac{AB}{ab}\right) + f + c$

For most transits: $\dfrac{f}{ab} = 100 + f + c = 0.75$ to 1.50

For modern transits: $f + c = 0$

Figure 5.6 Principle of the transit.

instrument is leveled and plumbed over a point, the rod person carries the stadia rod to the tower and places it at the tower center. The instrument person measures and records the horizontal distance from the transit to the base of the tower. Then the instrument person raises the elevation angle of the telescope to center the top of the tower in the center of the telescope cross hairs. Now the elevation angle is read off the scale. Measurement to the nearest degree will be adequate.

However, a more accurate measurement can be made by use of the elevation angle vernier scale. The *vernier* is actually a fine-adjustment scale, which usually gives angular readings to either number of minutes and seconds or tenths of a degree. The operations manual for the transit will give the procedure for using the vernier scales for that particular instrument. The general procedure and a specific example are shown in Fig. 5.8.

Simple trigonometry indicates that if one side and one angle of a right triangle are known, the other angle and two sides may be calcu-

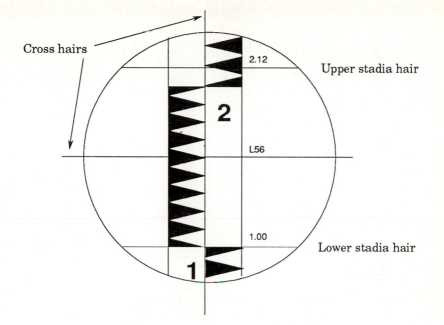

Cross hairs

2.12

Upper stadia hair

2

L56

1.00

Lower stadia hair

Read AB = 1.12 × 100 = 112 ft

Figure 5.7   Transit view of the stadia rod.

lated. Our trusty scientific calculator will easily make this calculation as follows:

Tower height H = 400 × tan 37° = 301.4 ft.

Press 3 and 7 to enter 37. Press tan.

Press ×. Press 4, 0, 0 to enter 400.

Now press = . Answer: 301.42 displayed.

One could say the tower was 300 ft; the height of the transit should be added.

To examine the accuracy needed for the measurement of the elevation angle, the calculator can give us an indication. Let us assume that the horizontal measurement is accurate for this example, and the elevation angle error is 30 min, or ½ degree. We can calculate the tower height and take note of the difference.

$$H = 400 \tan 37.5° = 306.9 \text{ ft}$$

Error 306.9 − 301.4 = 5.5   5½ ft error

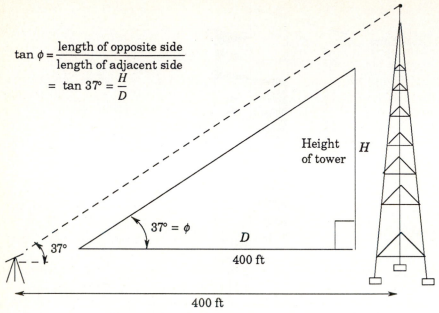

$$\tan \phi = \frac{\text{length of opposite side}}{\text{length of adjacent side}}$$

$$= \tan 37° = \frac{H}{D}$$

Figure 5.8 Transit measurement of tower height.

Suppose an error on 1 ft was made in the 400 measurement and the elevation angle was at 37°. Then $H = 399 \times \tan 37° = 300.7$ ft.

The foregoing facts tell us that most error can be attributed to the elevation angle, so if the readings in angles are not near a full degree mark then use of the elevation angle vernier can help. Of course, the horizontal measurement should be made with the stadia (line of sight) or a steel tape stretched over between the tower base center and the transit tripod center. The measuring wheel usually gives the least accurate results because of the unevenness of the terrain and the difficulty of making a long, straight, accurate measurement path.

If the plumb of the tower, or straightness, is desired, the transit can easily make this observation. Consider that once the instrument has been set for the tower height measurement and the tower is observed through the instrument telescope as the elevation angle is slowly lowered toward the ground (horizontal), any departure from the vertical hairs will indicate any change from tower straightness. Of course, this works best for straight-guyed towers. The steady transit-mounted telescope is also useful in observing any tower problems, e.g., loose ladder, cracked light globes, chipping paint rust.

5.2   Elements of Cable System Design

Since the selection of cable, amplifiers, and couplers/splitters is at best difficult, it will be assumed that these decisions have been made using prudent design specifications and cost considerations.

Normally the cable system is designed from the head end to the system ends, keeping track of the system losses and gains (signal levels) so each subscriber tap has a signal level within proper limits. Such a tap port signal level for a 450-MHz system might be +17 dBmV (max) and +10 dBmV (min) at either 54 MHz (channel 2) or 450 MHz. Tables of loss versus frequency for both trunk and feeder cable should be readily available. Loss for the passive devices (splitters and directional couplers) at the lowest and highest frequencies should also be known. Amplifier input levels and output levels for normal operation are also design requirements. All of these parameters at 68°F (20°C) should be known, as well as any variation as a function of temperature. For example, cable has less loss as the temperature decreases and usually varies about 1 percent per 10°F. Cable manufacturers provide detailed charts and/or graphs describing their products' characteristics and the way cable varies as a function of temperature.

5.2.1   Trunk system design

The design philosophy will be demonstrated by an example of a portion of a system shown in Fig. 5.9. Notice that the output of the head end into the trunk cable is at the same level as the desired amplifier output. Also notice that as a result of the high loss of 2500 ft of cable at 220 MHz the +32-dBmV signal level decreases to +12 dBmV and at channel 2 only to +19-dBmV. Obviously the amplifier needs to have an equalizer installed to equalize the difference. Such an equalizer would decrease the +19-dBmV signal level at channel 2 to about +12 dBmV while reducing the +12-dBmV signal at 220 MHz only about 0.5 dB. This equalizer would be most likely labeled 20 dB/220 MHz. Since the cable lengths to amplifiers 2 and 3 are the same, they too would have 20-dB/220-MHz equalizers installed. The span between amplifiers 3 and 4 has a total cable length of 1500 ft; 1000 ft. down this span is a directional coupler that feeds the system with television channels 13 through 62 (450 MHz). The equalizer in amplifier 4 needs to equalize 1500 ft of cable. The cable distance from the studio and satellite earth station site is only 1000 ft so the equalizer in amplifier 4 will overcompensate for the satellite and studio signals. The options are to equalize with a line equalizer the studio/satellite channels before the directional coupler point or, more often, to adjust the signal levels at the studio/satellite location to agree with

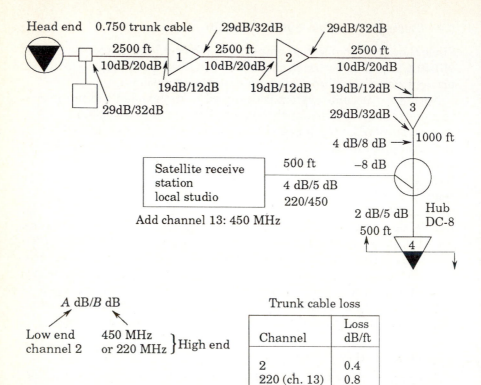

Figure 5.9   (a) Head end to hub system.

the correct signal level driving amplifier 4. Doing it this way will also compensate for the 8-dB insertion loss of the directional coupler. Usually 450-MHz systems have a trunk slope of 6 dB. The equalizer in amplifier 4 will be on the order of 10 to 12 dB at 450 MHz. A level of +19 dBmV is far too high and would overdrive amplifier 4, causing signal distortion. A plug-in pad value of 7 dB will drop the signal level to +12 dBmV, the nominal input signal for a number of amplifiers in use today. After the pad and equalizer are installed the input signal to amplifier 4 should be measured and verified by using a signal level meter or spectrum analyzer connected to the input test point. If the upper channels need to be adjusted, the satellite-studio hub modulators can be set so the correct signal level appears at amplifier 4 input. Now the gain and slope controls can be adjusted for +34 dBmV at 450 MHz and +28 dBmV at channel 2, giving a 6-dB slope. If the alignment is done at 68°F then these values will be valid. However, if the temperature is different, as shown in the chart in Fig. 5.10, the correct level will be found by consulting the amplifier manufacturer's specifi-

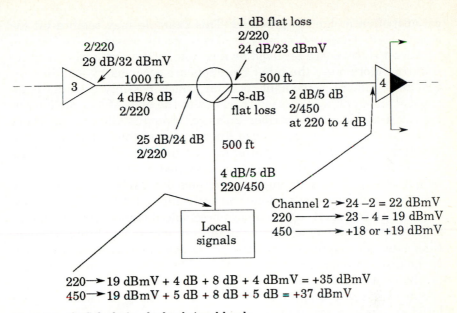

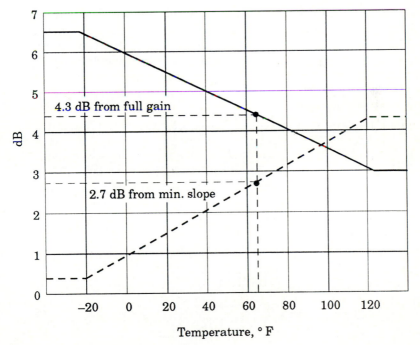

Figure 5.9   (b) Calculation for local signal levels.

Figure 5.10   Temperature alignment.

cations; often a chart is provided. This example may seem a bit too complicated. However, this type of system in which the television broadcast stations are received by a remote mountaintop site, is quite common. More often the other channels are received by satellite antennas at a more convenient location, where possibly a studio is located. This trunk system is carried out to the ends of the cable plant transporting signal. Subscriber taps are never installed in the trunk cable system. Feeder cable, also known as distribution cable, is fed signal from the bridging amplifier in the trunk amplifier station housing. This cable has subscriber taps installed with tap values in descending order. For most modern systems 450-MHz and above, only two line extenders in cascade are usually allowed. Line extenders are not used to transport signal any appreciable distance and operate at high levels to overcome the cable and tap loss of the feeder system. Line extender amplifiers with two in cascade do not contribute much to the distortion and noise problems, and hence can operate at higher levels.

Since the trunk amplifiers are spaced according to the unity gain building block theory their positions are basically fixed according to the accumulated losses. Therefore, the feeder cable is lashed to the same strand as the trunk cable, and often several feeder cables designated for different areas have to be bundled together. Some manufacturers supply an intermediate bridging amplifier, which is placed between the normal trunk amplifier locations and drives a bridging amplifier to supply signal to a feeder cable. Use of this type of amplifier is shown in Fig. 5.11. Amplifiers A and B have bridging amplifi-

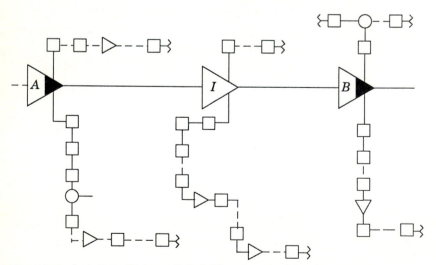

Figure 5.11   Insertion of intermediate bridger in normal trunk span.

ers, each feeding two ports. If a new area requiring service opens up, then a three-way feeder maker will have to be placed in either amplifier A or B and cable will have to be overlashed along the trunk route to the new area. If an intermediate bridging amplifier is available, it can be spliced directly into the trunk cable at the point needed. Its housing contains the intermediate amplifier and a normal bridging module to supply the feeder cables. The intermediate bridging amplifier has only a small amount of loss and thus does not interfere with the signal level into amplifier B. The only feeder cables required are those actually feeding the new area.

5.2.2 Feeder system design

Feeder design may offer even more choices to the system designer. When doing design work on feeder systems the main goal is to provide the proper signal level at the subscriber's tap port. Most designs should allow a tap port for each dwelling unit with the exception of large apartment houses. For large multiple-dwelling units, such as high-rise apartments or large spread-out complexes, the cable feeder (distribution) system essentially has to be wired through the premises. If amplifiers are used to increase the signal level inside the apartment complex, these amplifiers should be considered part of the system when performance tests are made.

For the example in Fig. 5.9a, amplifiers 1, 2, and 3 supply only the off-air signals and amplifier 4 carries all the signals. Amplifier 4 also has a bridging amplifier installed in the same housing, which supplies signal to two feeder cables. The symbols used for the various devices shown on cable system design maps are given in Fig. 5.12.

After leaving the bridger amplifier, the feeder cable usually has more signal slope than the trunk system. Typical values for a 450-MHz system would be 9 dB with output levels on the order of +48 dBmV at 450 MHz and +39 dBmV at channel 2. This signal level feeds the cable spans with the subscriber taps installed. The basic design philosophy of feeding signal to a subscriber tap is illustrated in Fig. 5.13. Notice that the signal level from the lowest channel (2) and the highest channel at 450 MHz is sloped upward by 9 dB at the bridger output. At the end of the cable the signal level has lost a great deal of this slope (39.5 − 36.2 = 3.3 dB) as a result of the cable loss. Taps also exhibit some loss versus frequency that should be taken into account in feeder system design. To illustrate the design technique further, a cable span of taps will be designed from the bridger output to the first line extender amplifier, as shown in Fig. 5.14; the calculations are shown in Table 5.1.

From the calculations made in Table 5.1 the cable lengths with

TABLE 5.1   System Points versus Signal Level or Signal Loss

Point in system	Channel 2, dB/450 MHz	Point in system	Channel 2, dB/450 MHz
Bridger output level	39/48	Signal in tap 4	32.12/29.9
Cable loss (500 ft)	2.8/8.5	Tap 4 value	17 dB
Signal in tap 1	36.2/39.5	Tap 4 port signal level	15.12/12.9
Tap 1 value	23 dB	17-dB Tap through loss	0.0/1.3
Tap 1 port signal	13.2/16.5	Tap 4 output level	31.22/28.6
23-dB Tap through loss	0.5/0.8	Cable loss (120 ft)	0.66/2.0
Tap output signal	35.5/38.7	Signal in taps	30.56/26.6
Cable loss (150 ft)	0.83/2.55	Tap 5 value	14 dB
Signal in tap 2	34.67/36.15	Tap 5 port signal	16.56/12.6
Tap 2 value	20 dB	14-dB Tap through loss	1.4/2.0
Tap 2 port signal level	14.67/16.15	Signal out tap 5	29.16/24.6
20-dB Tap through loss	0.6/1.0	Cable loss (150 ft)	0.83/2.55
Tap 2 output signal	34.1/35.15	Signal in tap 6	28.33/22.05
Cable loss (150 ft)	0.83/2.55	Tap 6 value	11 dB
Signal in tap 3	33.27/32.6	Tap 6 port signal	17.33/11.05
Tap 3 value	20 dB	11-dB tap through loss[a]	2.9/3.5
Tap 3 port signal	13.27/12.6	Signal out tap 6	25.43/18.55
20-dB Tap through loss	0.6/1.0	Cable loss (180 ft)	1/3
Tap 3 output signal	32.67/31.6[b]	Signal at end (180 ft)	24.43/15.55
Cable loss (100 ft)	0.55/1.7		

[a]0.33 dB over: not too bad.
[b]Nearly flat.
A line extender amplifier. Equalizer has to be equivalent to the cable (1350 ft of cable) with a corresponding loss of 23 dB. Therefore, a 23-dB equalizer is needed. However, notice that the tap through loss also tilts the system so a 23-dB equalizer would be the starting value.

their losses decrease the signal into the subscriber taps. The signal into the tap is divided equally by the tap value to the four subscriber ports, with the signal being passed through to the output connector to the next tap through the following span of cable. Notice that the tap values decrease in descending order, where the first tap is a 23-dB tap, followed by two 20-dB taps, a 17-dB tap, a 14-dB tap, and on the end an 11-dB tap. The signal starts with a 9-dB slope at the bridging amplifier, and at the input to the first line extender there is a tilt (reverse slope) of 9 dB. If the bridger output were flat, the tilt at the first line extender would be 18 dB. This fact can be confirmed by adding up all the losses at channel 2 and all the losses at 450 MHz and subtracting the totals. The main criterion is to provide a good tap port signal for the subscriber. An examination of Table 5.1 will indicate that the signal is indeed within the range of +17 dBmV maximum and 10 dBmV minimum. This signal specification should be adhered to, especially since high-definition television receivers are around the corner, so to speak, and any noise due to weak signal will be very objectional. This method is repeated to the second line extender amplifier and then through the last string of taps, where the system terminates with a

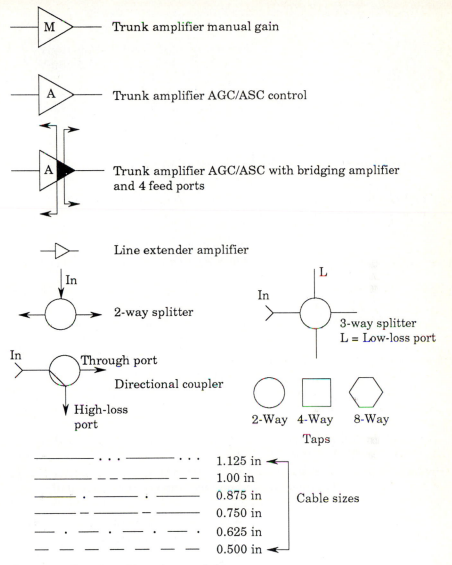

Figure 5.12 Common cable system symbols.

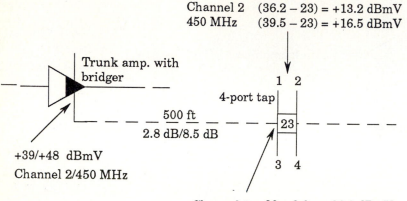

Signal level at each of 4 ports
is 23 dB less than input level

Channel 2   (36.2 − 23) = +13.2 dBmV
450 MHz   (39.5 − 23) = +16.5 dBmV

Trunk amp. with bridger

4-port tap

+39/+48 dBmV
Channel 2/450 MHz

500 ft
2.8 dB/8.5 dB

1   2
23
3   4

Channel 2 +39 − 2.8 = +36.2 dBmV
450 MHz   +48 − 8.5 = +39.5 dBmV

0.500 Cable loss: Channel 2/450 MHz ⟶ 0.55 dB/100 ft
1.7 dB/100 ft

The subscriber signal level is +13.2 dBmV at Channel 2
and +16.5 dBmV at Channel 62 (450 MHz) which is within
the signal specifications of +17 dBmV (max) and +10 dBmV (min)

Figure 5.13   Subscriber tap levels.

75-Ω line terminator connected to the last tap. Usually 11-dB eight-port taps, 7-dB four-port taps, and 4-dB two-port taps have built-in terminators. Since subscriber taps are directional, the signal has to enter on the input connector, and on most taps connector and direction of signal are labeled.

The previous example showed a straight span of cable taps serving a street to the end. However, the usual case may also involve side streets that branch off; this arrangement requires the use of either signal line splitters or directional couplers. An example of the use of these devices is shown in Fig. 5.15.

$$\text{Bridger output level} \left(\frac{39}{48}\right) - \left(\frac{1.65}{5.1}\right) = \frac{37.35}{42.9} \quad \text{into tap 1}$$

A 26-dB tap will give a port signal level of

Channel 2 (37.35 − 26) = 11.35 dBmV

and   450 MHz (42.9 − 26) = 16.9 dBmV

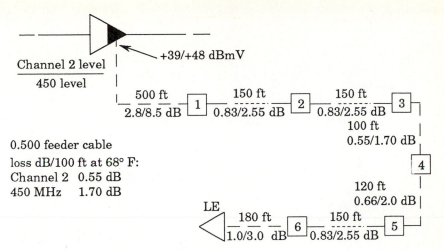

Figure 5.14   Span of taps example.

The tap through loss for a 26-dB tap is 0.5/0.8 dB. The signal leaving this tap is channel 2 (37.85 − 0.5) = 36.85 dBmV, 450 MHz (42.9 − 0.8) = 42.1 dBmV. The next cable span decreases this signal to channel 2 (36.85 − 0.083) = 36.02 dBmV and 450 MHz (42.1 − 2.55) = 39.55 dBmV, which enters tap 2. To provide a minimum signal of 10 dBmV, a 26-dB tap will again do the job, providing a tap signal of channel 2 (36.02 − 26) = 10.02 dBmV, 450 MHz (39.55 − 26) = 13.55 dBmV. Again this 26-dB tap will decrease the input signal channel 2 (36.02 − 0.5) = 35.52 dBmV and 450 MHz (39.55 − 0.8) = 38.75 dBmV. The 100-ft span decreases the signal level even more: channel 2 (35.52 − 0.55) = 35 dBmV, 450 MHz (38.75 − 1.7) = 37.05 dBmV. This level is now decreased 3.5 dB by the splitter channel 2 (35 − 3.5) = 31.5 dBmV and 450 MHz (37.05 − 3.5) = 33.55 dBmV. The 100-ft cable span will drop this level by channel 2 (31.5 − 0.55) = 31 dBmV, 450 MHz (33.55 − 1.7) = 31.85 dBmV. Notice at this point the signal at channel 2 is nearly the same as at 450 MHz. Obviously tap 3 and 5 will be the same value, which is 20 dB. The tap port signal level for each of these taps will be channel 2 (31 − 20) = 11 dBmV and 450 MHz (31.85 − 20) = 11.85 dBmV. The 20-dB taps decrease the signal to channel 2 (31 − 0.6) = 30.4 dBmV, 450 MHz (31.85 − 1.0) = 30.85 dBmV.

The next cable span decreases the signal level channel 2 (30.4 − 0.55) = 29.85 dBmV and 450 MHz (30.85 − 1.7) = 29.15 dBmV. Now, taps 6 and 4, because they are fed equal signal level through equal lengths of cables, will be the same value at 17 dB. The tap port signal level for taps 4 and 6 will be channel 2 (29.85 − 17) = 12.85 dBmV, 450 MHz (29.15 − 17) = 12.15 dBmV,

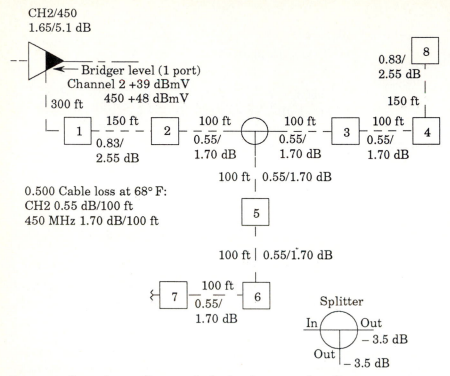

Figure 5.15   Span of taps split to two feeder legs by means of a splitter.

which is again nearly equal. The 17-dB tap decreases the input signal by the tap loss. Channel 2 (29.85 − 0.9) = 28.95 dBmV, 450 MHz (29.15 − 1.3) = 27.85 dBmV. The next cable span decreases the signal level to channel 2 (28.95 − 0.55) = 28.4 dBmV, 450 MHz (29.15 − 1.7) = 27.45 dBmV. Tap 7 should be a 17-dB tap to provide a tap port signal of channel 2 (28.4 − 17) = 11.4 dBmV and 450 MHz (27.45 − 17) = 10.45 dBmV. The terminated signal level across a 75-Ω terminator connected to the output terminal of this tap will be channel 2 (28.4 − 0.9) = 27.5 dBmV and 450 MHz (27.45 − 1.3) = 26.15 dBmV. If more signal level is desired at this last tap, the choice would be to make tap 7 a 14-dB tap. However, if the leg consisting of taps 3, 4, etc., is to be extended, a more prudent design would be to change the splitter to a dc eight-directional coupler.

If the −8-dB high-loss port is placed in the tap 5, 6, 7 leg, the tap values will have to be changed 4.5 to 5 dB less to keep the subscriber port signal essentially the same. By following this procedure the leg consisting of taps 3, 4, etc., will have the signal level increased by 3 dB, thus increasing the tap values by 3 dB. Fig. 5.16 illustrates this more correct design method.

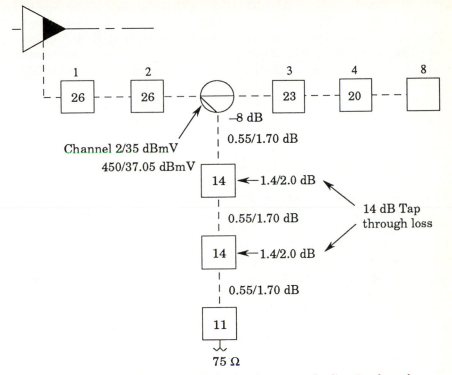

Figure 5.16   Span of taps split to two feeder legs by means of a directional coupler.

However, since the lower-value tap through loss is greater, the levels should be checked especially on the tap legs consisting of taps 5, 6, etc. The signal output on the −8-dB tap leg will be channel 2 (35 − 8) = 27 dBmV, 450 MHz (37.05 − 8) = 29.05 dBmV, which is further reduced by the 100-ft cable loss channel 2 (27 − 0.55) = 26.45 dBmV, 450 MHz (29.05 − 1.7) = 27.35 dBmV. Now tap 5 will become a 14-dB tap giving a port signal level of channel 2 (26.45 − 14) = 12.45 dBmV, 450 MHz (27.35 − 13.25) dBmV, which is within the desired tap signal level window of 17 dBmV (max) to 10 dBmV (min). Another 100-ft span of cable further reduces the signal level: ch 2 (26.45 − 0.55) = 25.9 dBmV, 450 MHz (27.35 − 1.7) = 25.65 dBmV. The next tap will be chosen as 14 dB with a port signal level of channel 2 (25.9 − 14) = 11.9 dBmV, 450 MHz (25.65 − 14) = 11.65 dBmV.

Tap 6 could be an 11-dB tap, but the through loss is greater than 14 dB. It is more conservative to design for the maximum tap value so additions can be made if the street is extended in length. The signal leaving tap 6 is reduced by the last 100 ft of cable channel 2 (25.9 − 0.55) = 25.35 dBmV, 450 MHz (25.65 − 1.7) = 23.95 dBmV.

The last tap value is 11 dB with a port signal of channel 2 (25.35 − 11) = 14.35 dBmV, 450 MHz (23.95 − 11) = 12.95 dBmV. This leg could be extended another 100 ft or so with an 8-dB four-port tap if the need arose. The other leg can be extended to a terminated tap or a line extender amplifier. When the signal level on the cable gets to +15 dBmV at 450 MHz a line extender is needed. This signal level is quite common for most manufacturers' line extender amplifiers. Most varieties of line extenders have no internal equalization so the plug-in equalizer has to do it all. Most trunk amplifier types have some built-in equalization as specified by the manufacturer. For example, if a 20-dB equalizer has been calculated for 20 dB of cable and the amplifier has 5-dB built-in equalization then the actual value of the plug-in equalizer will be 20 − 5 = 15 dB. Caution must be observed when using equalizers. For example, a 20-dB equalizer at 300 MHz is much different from a 20-dB at 450 MHz type.

The actual design method, as previously discussed, is manual, usually performed in pencil on blue-line map copies using a pocket calculator. This method is low in cost and efficient for small extensions or special cases. When the extension has been constructed, tested, and proofed-out, the extension should be drafted on the system's permanent Mylar maps and new blue-line maps made for everyday use.

The design choices are many and the experience of the designer determines the efficiency of design. Good design provides the proper signal level at the tap port, is low in cost, and can be easily extended. This is a tall order.

5.2.3 Rural design methods

When serving rural areas, the transportation of signal by the trunk cable accounts for most of the cost of construction. When the feeder to trunk ratio approaches 1:1 then a more efficient and cost-effective method should be found. Rural systems usually do not have the desired heavy concentration of subscribers, and the system is often spread out. The trick is to minimize the trunk and extend the feeder as far as practical. Some manufacturers have developed high-gain, lower-distortion amplifiers that permit more than two amplifiers in cascade and at 450 MHz as many as six. A skeletal view of this technique is shown in Fig. 5.17.

It should be noted that actually these amplifiers are fed from the trunk and have built-in feeder ports. The in-town areas can be served by the standard trunk-feeder system and the outlying areas fed signal by the special high-gain, high-level amplifiers using a single cable.

A technique called back feeding is often needed and used in the more rural areas. Back feeding is the method of placing the signal

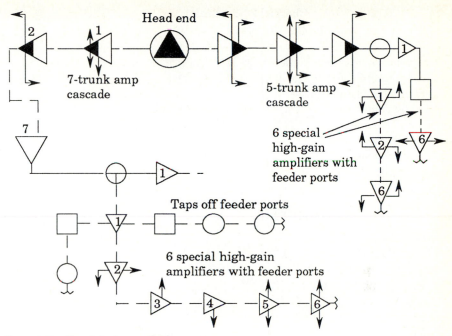

Head end

2

1

7-trunk amp
cascade

7

6 special
high-gain
amplifiers with
feeder ports

1

2

6

6

5-trunk amp
cascade

1

1

Taps off feeder ports

1

6 special high-gain
amplifiers with feeder ports

2

3   4   5   6

Figure 5.17 Rural design method.

point ahead of the desired distribution point and feeding the signal back to this point with a separate section of distribution cable. This technique is shown in Fig. 5.18.

The system is operating correctly and does not extend much after the line extender. Now a tap is needed at point A. It should be evident that any tap or device inserted at this point will affect the input signal level to the line extender, causing its output to drop. A better plan would be to install a splitter at the output of the amplifier; since the system does not extend appreciably beyond the amplifier, the tap plates can be changed (effectively changing the tap value) to restore the tap port signal level to its previous value. Now, designing back from the amplifier feeding the splitter through 500 ft of cable, channel 2 ($39 - 3.5 - 2.75 = 32.75$ dBmV), 450 MHz ($48 - 3.5 - 8.5 = 36$ dBmV) gives this signal level input to the tap. This tap should have a value of 20 dB, giving a tap port signal level of channel 2 ($32.75 - 20 = 12.75$ dBmV); 450 MHz ($36 - 20 = 16$ dBmV). If one were to cut a tap into the feeder 500 ft downstream, the signal level at this point would be channel 2/25.75 dBmV and 450 MHz/23.5 dBmV, calling for a tap value of 11 dB. The through loss for even a two-port 11-dB tap is 2 dB at 450 MHz, causing the 15-dB normal signal to be reduced to 13 dBmV at the line extender input. Therefore, this should

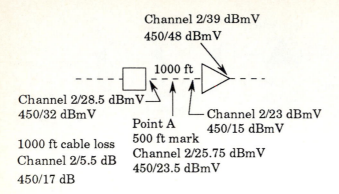

Channel 2/39 dBmV
450/48 dBmV

1000 ft

Channel 2/28.5 dBmV
450/32 dBmV

Channel 2/23 dBmV
450/15 dBmV

Point A
500 ft mark
Channel 2/25.75 dBmV
450/23.5 dBmV

1000 ft cable loss
Channel 2/5.5 dB
450/17 dB

(*a*) The problem

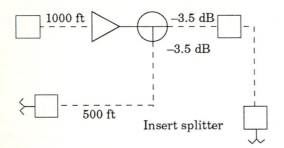

1000 ft   −3.5 dB

−3.5 dB

500 ft
Insert splitter

(*b*) The solution

Figure 5.18   Back feeding technique.

not be done and the solution with the splitter and 500 ft of cable will work just fine.

To extend rural plant farther, two port taps should be used where only a couple of houses exist with little chance of further subscriber expansion. Two port taps have less through loss than eight-port and four-port taps, which permit the cable system to proceed a greater distance.

5.2.4   Computer-aided design

5.2.4.1   Methods.   It should be noted that the cable television design process is a lot like bookkeeping. Keeping track of the signal level at the system branches, all the tap levels, as well as the carrier-to-noise and intermodulation figures is a difficult job at best. All of the system component parts, e.g., actives, passives, and pole line routes, should be plotted on the system maps. When the programmable printing calculator was put on the market, system designers quickly started using this device and design programs emerged. Development of the personal computer (PC) and the software (programs) for computer-aided design (CAD) of cable systems was a great benefit to designers faced with massive rebuild projects. Such system design depends on the method used by the computer program. When choosing a program, care must be exercised so the program is not too restrictive or, on the other hand, does not provide bells and whistles not actually needed. Some programs enable an operator to catalog all the chosen tap, passive, and active (amplifier) inventory in the files along with the specifications. Most programs are *menu-driven*, asking questions of the designer, who then enters his or her choices. A skeletal type of diagram similar to a tree diagram is shown on the screen to allow the designer to keep track of the signal level as an ongoing process. Distortion figures or carrier-to-noise ratios can be calculated at any amplifier location, and at the end of most design programs a bill of materials can be printed out. If the cost of materials has been cataloged into the data base, the cost per mile can be indicated. Such a system and software are not beyond reach of some smaller systems. The equipment and software for some systems are less than $10,000.

Larger systems are carrying this concept to another level, with a computer-aided mapping system that is really an automated drafting system. Building the data base with the mapping procedure is the main labor-intensive phase, which must be done initially. One system uses a special grid table on which a background street map is placed. A cursor device with a cross hair pointing indicator is traced over the street layout, reducing the points on the map to computer coordinates. At this time the field survey notes from the strand mapping procedure are entered as well. The computer system driving an x-y plotting system can then plot (draft) the strand maps in pen and ink on Mylar or paper. During design, areas of such maps can be presented on the computer monitor screen; when the design is completed for a specified area, a complete design map can be plotted on Mylar. Blue-line copies of the Mylar maps can provide the drawings used by the construction crews. However, some companies are plotting on paper and using large photocopy machines to make the construction prints. Construction drawings made this way do not fade and withstand moisture much better than blue-line copies.

The cost and the trained personnel for such equipment are beyond the means of medium and smaller cable operators. Also, the so-called learning curve for the operators of such systems is long and hence too expensive. Several service companies are offering this service on a contract basis. This could make it attractive for smaller systems. Design maps containing the large amounts of information can be stored on computer diskettes for archive storage. For rebuilds, upgrades, or system expansion projects, use of the computer stored data can simplify the procedure and lower the costs.

5.2.5   Reverse cable systems

Since the frequency bands on a cable system can be separated by using the appropriate filters, one band of frequencies can be assigned as the downstream or forward system and the other band designated the upstream or reverse system. This capacity makes the cable system bidirectional. Amplifier modules are incorporated in the same housing and also use the same power supply module. Such an amplifier configuration is depicted in Fig. 5.19, which shows the forward and reverse signal paths.

The diplex filter separates the forward channels from the reverse channels, which are then routed to the respective amplifier modules. The commonly used subsplit reverse uses the 5- to 30-MHz band as the reverse band, giving 25 MHz ÷ 6 MHz television channel = 4 television channels. The frequency assignments for the reverse system are given in Table 5.2.

The subsplit reverse system can only use approximately 4½ channels T7 to T10 and ½ T11. Use of the band between the upper end of T11 and channel 2, the first forward channel, is prohibited by the diplex filter. Another system split is the *midsplit*, which designates channels 5 to 112 MHz typically as the reverse channels and 150 MHz on up as the forward systems channels. For a 450-MHz system, 5 to 112 MHz gives nearly 18 channels reverse and (450 − 150) =

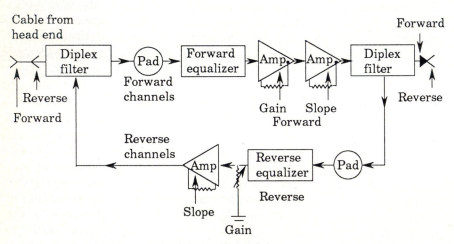

Figure 5.19   Forward-reverse amplifier configuration.

TABLE 5.2   **Subsplit Reverse Channel Frequencies**

Channel	Video Carrier, MHz	Color, MHz	Audio, MHz	Bandwidth, MHz
T7	7.00	10.58	11.5	5.75–11.25
T8	13	16.58	17.5	11.75–17.75
T9	19	22.58	23.5	17.75–23.75
T10	25	28.58	29.5	23.75–29.75
T11	31	34.58	35.5	29.75–35.75
T12	37	40.58	41.5	35.75–41.25
T13	43	46.58	47.5	41.75–47.55

$300 \div 6 = 50$ channels forward. The band between 150 and 112 MHz is prohibited by the diplex filter.

The high-split system is also available with the diplex filter cross-over band moved up in frequency. The forward band for a 550-MHz system would be 234 to 550 MHz, giving approximately 52 channels, and the reverse band would be typically 5 to 174 MHz, giving about 28 television channels. It should be realized that a true cable-ready television set or a converter is needed. It may seem that use of the reverse system should be attractive. However, for the usual cable television system, the delivery of high-quality downstream signals is most important. The reverse system is used to provide pickup of a remote television studio site or home alarm signal, for which the 4½ reverse channels are sufficient.

At first glance it may not be quite clear what reverse network topology implies. Recall from the discussion in Chapter 3 that in a reverse system noise combines in splitters. The following example discusses this process for the subsplit reverse system.

The 5- to 30-MHz reverse bandwidth of the subsplit type of two-way cable system operates fairly well for the 4½ television channels in the reverse direction. The cable loss at these low frequencies chiefly produces either low-gain, low-noise amplifiers in the reverse direction or fewer high-gain units, in which every amplifier station does not use a reverse amplifier module. Limiting the number of amplifiers decreases the amount of reverse noise. In midsplit or high-split cable systems, which are often used in broadband local area networks (LANs), the reverse bandwidth is increased so the number of channels in the forward direction (bandwidth) is more nearly equal to the number in the reverse direction. Usually the reverse direction is the lower-frequency band and the forward the upper-frequency band. Of course, the reason for this arrangement is the nature of the network topology. Consider the diagram of the forward and reverse topology in Fig. 5.20.

From this diagram the noise at point X from the HUB is the noise at

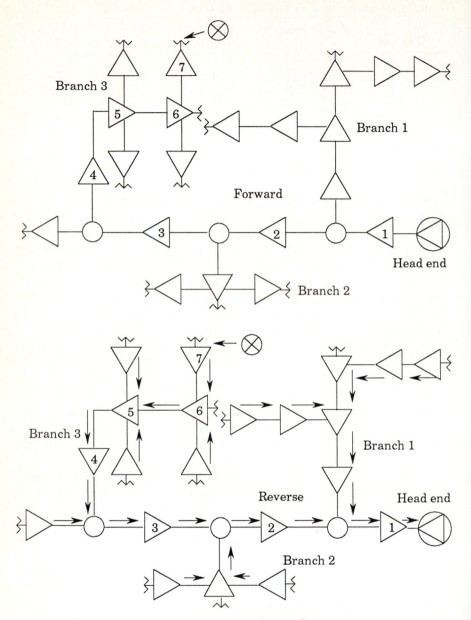

Figure 5.20   Forward and reverse systems shown separately.

the HUB with the noise contributed to it by amplifiers 1 to 7 in the conventional cable television manner. Now in the reverse direction any signal generated at point X to the HUB is going to be degraded by all the contributed noise buildup according to the arrows. To determine the noise at the HUB, meticulously calculate the noise as it is combined in the bridging amplifiers from the line extenders and at each trunk split all the way back to the HUB. If two-way service is not needed from every branch, it is far better to activate only the needed branches with reverse amplifiers and diplex filters. If interactive service is desired, then the whole cable system has to be reverse-activated. To limit noise buildup, addressable bridger leg switching has to be used so the interactive reverse signals are not buried in noise. Many cable television system operators emphasize two-way cable service in franchising and licensing applications only to find later that it is not used or needed. Often it is better to reverse-activate only the trunk system, which in many cases passes the locations needing a reverse channel anyway, thereby eliminating the expense and most of the noise, which lessens the usefulness of the reverse system. Now it should be evident that as the reverse bandwidth is increased, the noise problem increases, even though the upstream port uses the lower bandwidth, mainly because of the topology of the network. If a completely instituted reverse system is needed, then careful design of the whole system, both forward and reverse, should be made to limit the number of amplifier stations, reverse amplifiers, etc. Bridger-leg switching may also have to be used to turn on the areas feeding upstream signals only on command as necessary by a controlling computer system.

Designing for a two-way system is usually done in the following manner. First the forward system is designed and mapped out. Then the designated locations for the desired upstream channels should be located on the map, with the design proceeding from these locations back to the head end using the same basic techniques of losses and gains. For a subsplit (5- to 30-MHz) reverse system the cable loss is quite small, so a reverse amplifier may not be needed, and many manufacturers allow or make use of a simple bypass jumper installed in the amplifier housing. Manufacturers of most reverse amplifiers, particularly the mid- and high-split systems, have available AGC/ASC type amplifiers. Mid- and high-split systems usually have a reverse amplifier module in each housing. Cable equalizers are available from manufacturers for the sub-, mid-, and high-split systems and are so labeled. Caution should be observed in storing these parts so they do not get mixed up with the forward system components. Labels do fall off and become illegible, making it possible for the wrong parts to get into the wrong places.

5.2.6 Status monitoring systems

Use of the reverse system has been proposed, and some equipment presently on the market allows downstream (forward) system performance to be measured. The measurement results are then transmitted to the head end or HUB location through the reverse system for immediate observation or recording for later analysis. Status monitoring systems employ a downstream data receiver that receives the data carrier, demodulates the data signal, and looks for its particular address and instructions. The data signal in the form of a digital pulse train is decoded by the control module, which makes such parameter measurements as high- and low-pilot carrier levels (AGC voltages), power supply voltages, and housing temperature. These measurements are put in digital form with a built-in analog-to-digital converter (ADC) usually in the control circuitry and sent to the upstream data transmitter. Generally the upstream channels are only active when an amplifier is sending back data. The upstream data carrier is demodulated and decoded at the HUB/head end, where the data are analyzed as to the amplifier that sent the data and the value of the measured parameters. The parameters are compared against the standard values stored in the controlling computer-based data files. Some status monitoring systems employ switchable pad values in both downstream and upstream signal paths in the trunk amplifiers. These switchable pads allow tests to aid in noise source location as well as bridger-leg switching. By connecting the bridger leg containing the desired return signal and keeping the rest disconnected the upstream noise caused by the reverse network noise funneling phenomenon is controlled. It must be remembered that adding more equipment, subject to problems and maintenance to ensure reliability, must be carefully considered in terms of maintenance value and costs.

5.2.7 Reverse alignment

Alignment of the reverse system involves a different technique than the forward system. Now signals have to be injected into each amplifier's reverse input port (forward output port) and the reverse amplifier balanced on the output. One proposed scheme positioned a television camera to observe a spectrum analyzer connected to the upstream test port. When a signal was injected (by a 5- to 30-MHz comb generator) into the upstream path, the amplifier was adjusted while observing a television receiver connected to a downstream channel observing of the camera's view of the spectrum analyzer. Downstream sweeping can take place simultaneously, making the forward and reverse alignment to each amplifier. Most systems perform the forward alignment first, thus enabling subscribers to be connected as soon as possible. Reverse alignment follows, immediately before proof of performance testing.

5.3   System Powering

Another topic important to the distribution system is system powering. As previously discussed in Chapter 3, 60-V alternating current (V_{ac}) is inserted on the cable and carried on the cable to provide power to the amplifiers. Necessary filter chokes prevent the signal from becoming connected (short-circuited) to the power system. In practice a power supply powers only a segment of the system, and power blocks prevent more than one power supply from becoming connected. If this happens, then serious plant damage, e.g., burned choke coils and blown fuses, and power supplies, can result. Proper fusing is necessary to prevent such disasters. Geographically the cable power supply should be connected at the middle of the segment so the currents through both output sides of the power inserter are nearly equal.

A power inserter essentially places an ac voltage source on the cable without causing any disturbances to the television signal carried by the cable. The power inserter, therefore, becomes a low-pass filter for 60 Hz, 60 V, and a band rejection filter to the signal on the cable. This filter network is placed in a usual type of cable housing with weatherproof and RFI signal-proof gaskets similar to line splitters and directional couplers. The ports are usually labeled as ac input (from power supply) and RF (to cable system); there are two RF ports, for cable in and cable out. The diagram in Fig. 5.21 illustrates the concept.

Cable television power supplies are usually rated as (720-960-VA) outputs. This means that the 720-VA power supply can provide 60 V at 12 A and the 960-VA power supply can provide 60 V at 15 A. Power supply output is 60 V_{ac} and the power is given as the apparent power in volt-amperes. The power inserter's current-carrying capacity is also given, for example, 15 A/port. However, if each cable port is carrying 15 A, the input port will be required to carry 30 A and the power supply will blow its fuse or open its circuit overload breaker. The load on the output should be split in half for each port so that the sum does not exceed the current-carrying capacity of the input port or the capabilities of the power supply. Power supplies can become overloaded if amplifiers are indiscriminately added to the system. Most amplifiers have a maximum current passing value of 10 A. The diagram in Fig. 5.22 illustrates a potential problem.

An illustration of cable power insertion design is provided by the example of a segment of cable plant in Fig. 5.23. The parameters for this example are as follows:

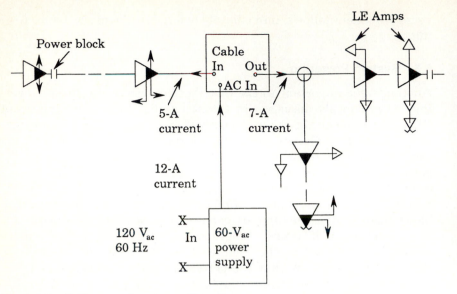

Figure 5.21   Cable power inserter configuration.

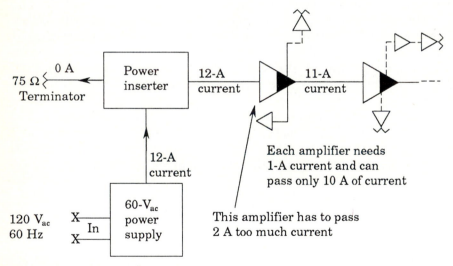

Figure 5.22   Unbalanced power feed.

1. Each trunk amplifier draws 0.8 A.

2. Each line extender draws 0.5 A.

3. The resistance of the trunk cable is 0.75 Ω/1000 ft.

4. The resistance of the feeder cable is 1.70 Ω/1000 ft.

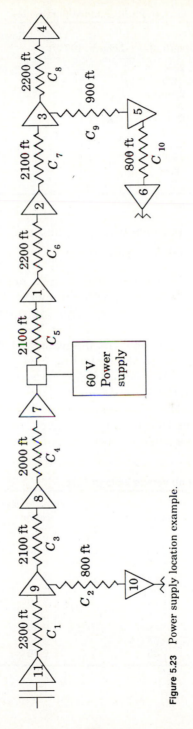

Figure 5.23 Power supply location example.

319

TABLE 5.3   Cable Section Length Resistance Parameters

Cable section	Length, ft	Length, 1000 ft	Resistance, 1000 ft	Cable section resistance
C_1	2300	2.3	0.75	1.73
C_2	800	0.8	1.70	1.36
C_3	2100	2.1	0.75	1.58
C_4	2000	2.0	0.75	1.50
C_5	2100	2.1	0.75	1.58
C_6	2200	2.2	0.75	1.65
C_7	2100	2.1	0.75	1.58
C_8	2200	2.2	0.75	1.65
C_9	900	0.9	1.70	1.53
C_{10}	800	0.8	1.70	1.36

5. The minimum allowable voltage to any amplifier is 40 V.

6. The bridging amplifier draws 0.3 A.

The idea is to determine whether any amplifier does not have enough voltage to allow it to function properly. The first step is to find the resistance of all cable sections. This information is given in Table 5.3.

The second step is to label the diagram with the resistance values and the current values as shown in Fig. 5.24.

Third, calculate all the voltage drops caused by the resistance of the cable. These calculations are shown in Table 5.4.

Fourth, make a table starting at the 60-V power supply and calculate all the voltages feeding the amplifiers. At the power inserter 60 V_{ac} is applied to the cable system. Table 5.5 illustrates the process.

TABLE 5.4   Cable Section versus Voltage Drop

Cable section	Resistance, Ω	×	Current in cable section	=	Voltage drop for cable section
C_1	1.73		0.80		1.4
C_2	1.36		0.50		0.7
C_3	1.58		2.4		3.8
C_4	1.50		3.2		4.8
C_5	1.58		4.5		7.1
C_6	1.65		3.70		6.1
C_7	1.58		2.9		4.6
C_8	1.65		0.80		1.3
C_9	1.53		1.00		1.5
C_{10}	1.36		0.50		0.7

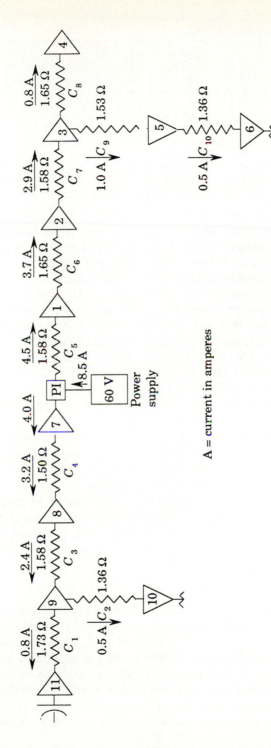

Figure 5.24 Current magnitude and direction.

321

TABLE 5.5   Amplifier Input Voltages

Amplifier		
	1	$60\text{ V} - C_5$ drop $= 60 - 7.1 = 52.9$ V
	2	$52.9\text{ V} - C_6$ drop $= 52.9 - 6.1 = 46.8$ V
	3	$46.8\text{ V} - C_7$ drop $= 46.8 - 4.6 = 42.2$ V
	4	$42.7\text{ V} - C_8$ drop $= 42.2 - 1.3 = 40.9$ V
	5	$42.2\text{ V} - C_9$ drop $= 42.2 - 1.5 = 40.9$ V
	6	$40.7\text{ V} - C_{10}$ drop $= 40.7 - 0.7 = 40.0$ V
	7	60 V (60 V inserted at amplifier 7)
	8	$60\text{ V} - C_4$ drop $= 60 - 4.8 = 55.2$ V
	9	$55.2\text{ V} - C_3$ drop $= 55.2 - 3.8 = 51.4$ V
	10	$51.4\text{ V} - C_2$ drop $= 51.4 - 0.7 = 50.7$ V
	11	$51.4\text{ V} - C_1$ drop $= 51.4 - 1.4 = 50.0$ V

This example illustrates poor power supply planning. However, there is a 10-V difference between amplifier low-voltage points at the ends of each branch. Notice that the current drawn by the left leg is nearly equal to that of the right, which is good. Also, most power inserters state that they will pass, for example, 10 A. If all the connector contacts in the power inserter are the same and will carry 10 A in each leg, then the feed leg will carry 20 A and hence will be overloaded. Overloaded contacts generate heat, which in turn chars the printed circuit inside the housing, melts the solder, and in general destroys the unit, thus causing a system outage. In this example the feed current is 8.76 A, which is well below 10 A. If it is desired to use power supplies capable of delivering more current, then power inserters able to withstand the high current must be used.

If more of the system is to be powered from a power supply then perhaps larger-size cable with a lower loop resistance should be used or a solid copper center conductor of the same size of cable can be used. The name of the game is loop resistance decrease.

When designing for system powering, care should be taken in choosing the location for the power supply. It should be on a pole that is easy to service, i.e., not a corner or junction pole or one on a hillside. If the pole chosen has a street light, then night servicing will be simpler.

5.3.1   Grounding and bonding

The cable television industry as a whole has to conform to many federal, state, and local electrical codes, rules, and regulations, which essentially deal with safety of personnel and property. The National Electric Code deals mainly with the wiring of buildings, both exterior and interior, such as the power, telephone, and the cable television drop to the interior wiring. The National Electric Safety Code involves the pole and underground plant in the public way or in areas where the public is concerned. By now it should be evident that the two codes will essentially have to cross because the homes, apartments, or buildings are connected to the utility or cable television exterior plant through the service drop.

The National Electric Code specifies that the cable television drop be grounded at the entrance point on the building and that this ground be connected to the same electrode as the power ground. For example, if the power ground is an exterior ground rod, then the cable television ground block will be wired to a ground clamp placed around the power ground rod. The ground rod clamp with the power company ground wire should never be disturbed by the cable television installer because power service to the subscriber may be interrupted and the installer will be exposed to a shock hazard. In some cases it may be permissible to connect the cable television ground wire to the electrical meter metallic housing. However, some electrical utility companies do not permit this practice. In the past the electrical service grounding electrode was the water service pipe at the entrance point. The problem with this practice was that this pipe usually came into the building basement, so that using this as the cable television entrance ground meant passing the ground wire along with the cable television drop wire into the basement, so that the ground point was not on the exterior of the building. To solve these problems, a conference with the local wiring inspector should be in order.

The National Electric Safety Code has set the standards for utility pole plant and underground plant. As most of us know, before cable television, only the electrical and telephone companies occupied space on the poles. Usually the poles are jointly owned, and each company has a one-half share. The cable television industry as a whole leases space on the poles and agrees to occupy that space and pay accordingly. The cable television industry by virtue of the pole attachment agreement also agrees to observe and obey the applicable sections of the National Electric Safety Code.

The cable television messenger strand is usually bonded, that is, connected, to the telephone company messenger strand at the first, last, and every tenth pole in a cable run. This connection is made with bare number 6 soft drawn copper wire with weaver-type bronze connecting clamps. Thus the cable television strand is connected to the telephone strand, which is in turn grounded to driven ground rods placed by the telephone company. The power company distributes electric power on a three-phase wye-type circuit with a grounded neutral. Usually at every power company distribution transformer, or pot, the neutral and the transformer case are grounded to a driven ground rod at the base of the pole. The connecting wire is covered for the bottom 8 to 10 ft of the pole by either wooden or plastic pole molding. Now it should be evident that the cable television plant, telephone plant, and power plant are connected by the grounding system. The cable sheath of the coaxial cable is the return conductor for the cable television 60-V powering system. The cable sheath is grounded through the coaxial connectors to the device housings to the messenger strand to the telephone company strand and ultimately to the power company grounds. Therefore, the cable television company, along with the telephone company, are integral parts of the power company network. Fig. 5.25 illustrates a fairly typical pole plant situation.

If at any time the power ground is lost, then the accompanying unbalanced wye currents flowing through the neutral wire will also flow through the parallel grounding systems of the telephone and cable television strands and the cable television coaxial cable sheath. The amount of current flowing through each wire depends on the resistance of the wire: hence more current will flow through the lowest-

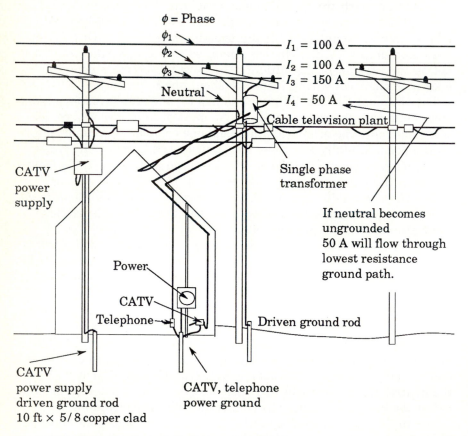

ϕ = Phase

ϕ_1

ϕ_2

I_1 = 100 A

ϕ_3

I_2 = 100 A

I_3 = 150 A

Neutral

I_4 = 50 A

Cable television plant

CATV power supply

Single phase transformer

If neutral becomes ungrounded 50 A will flow through lowest resistance ground path.

Power

CATV

Telephone

Driven ground rod

CATV power supply driven ground rod 10 ft × 5/8 copper clad

CATV, telephone power ground

Figure 5.25   Example of a typical plant arrangement.

resistance conductor. The coaxial cable aluminum outer conductor has much lower resistance than the galvanized steel messenger strand. One problem that arises from this arrangement involves longitudinal sheath currents. If such currents build to high enough values, the cable television plant can become damaged in the form of blown fuses and failed power supplies and amplifiers. Testing between strands (telephone and cable) with a voltmeter will indicate whether there is a difference in voltage, which in turn will cause longitudinal sheath currents.

Now, if there is a problem with the power company or telephone company grounds, then their plant will act as an antenna or floating conductor, which will attract lightning. If a lightning strike actually hits the plant, then a loss of power, telephone, and cable television service usually results. With the resumption of electrical service failed fuses, amplifiers, etc., will be revealed. Some cable television engineers feel that their plant will be better off by carrying their own grounds. This, of course, would be a violation of the National Electronic Safety Code. Technically, the cable television plant may not be subjected to the power and telephone company grounding problems, but a difference of potential (voltage) between the plant bonding systems could result, constituting a shock hazard. Hence the reason for the National Electronic Safety Code.

Therefore, in order for all three pole plant systems to coexist on the same poles, careful observation of the applicable codes and practices in bonding and grounding is necessary.

5.3.2 Lightning

Lightning is a result of the difference in electrical potential (voltage) of the charge buildup from cloud to cloud or cloud to earth. Many of us have observed the flash caused by the electric discharge from cloud to cloud or cloud to earth. It is the discharge from cloud to earth that causes most of the lightning damage to the electrical plant. Consider the diagram shown in Fig. 5.26. When this discharge occurs the difference of potential (voltage) disappears, and the arc conducts heavy currents with the accompanying magnetic fields, which will induce voltage in surrounding electrical or communications pole plants. Because the discharge is essentially to ground, then if any of the pole plants is not grounded, voltage will be induced. It is this induced voltage caused by the arc discharge that places voltage spikes on the power, telephone, and cable television plant. The spikes, if not absorbed by gas surge protector metal oxide varistors (MOVs), will cause fuses and circuit breakers to open. If the transient energy spike is very short, and the surge protectors do not do their job and the fuses and

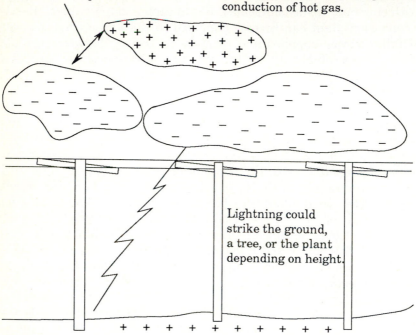

Difference of potential

When the difference in potential is large enough, the air breaks down (is ionized) and becomes a path of conduction of hot gas.

Lightning could strike the ground, a tree, or the plant depending on height.

Figure 5.26 Lightning formation.

circuit breakers do not have time to open, failure of some of the amplifiers will result. A plot of several types of transients is shown in Fig. 5.27.

The *A* curve maximum amplitude may be the same as that of the *B* curve. However, the *A* curve decreases to 50 percent in less time than the *B* curve. Hence, the *B* curve type will do more damage because it lasts longer. The *C* curve lasts a lot longer, never gets as large as the *A* and *B* curves, but can become large enough to cause damage. There is no way to predict the type of transient; hence the protective devices may use several types of suppressors in combination to cover the short high-energy types as well as the lower-energy, longer-duration types of transients.

We will now consider the problem of obtaining a good ground.

5.3.3 Grounding techniques

The basic common grounding technique is to drive a low-resistance metallic rod into the earth. Since copper is a good conductor, it is a

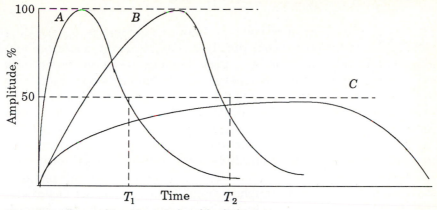

Figure 5.27   Types of transients induced by lightning.

good choice. However, because it is fairly expensive, and copper on the outside of the rod has contact with the earth, and copper is soft and pliable compared to steel, copper clad steel provides the best rod. Such rods may be obtained in ⅜-in diameter, increasing in ⅛-in increments to about ¾ in. Rod lengths vary from 4 to 10 ft, usually in 2-ft increments. Naturally the larger rods have more surface area with the earth than the smaller rods and hence a lower-resistance ground.

To improve the ground, i.e., to lower resistance, larger rods driven deeper may be effective. Adding chemicals, e.g., calcium or sodium chloride, around the ground rod is also helpful. The larger ground rods may be purchased with screw tops so another rod may be fastened to the end and driven deeper on top of the other until the desired result is obtained.

Testing a ground is usually conducted by an instrument called a *megger* and a test rod. The test rod is driven in close proximity to the desired ground rod and the megger is connected between the two ground rods. The megger essentially measures the resistance of the earth between the two rods. Usually the electric utility companies want a ground of 25 Ω or less. In normal soil conditions the 10 ft, ⅝-in copper clad ground rod properly driven to the end will do the job. In sandy, dry soils several 10-ft rods may be driven end to end with only 35 Ω attainable. Proper grounding and bonding of the cable plant are necessary to protect the plant from power company transients, sheath currents, and lightning.

Several manufacturers have been experimenting with a scheme to isolate the cable television drop from the pole plant. Such isolation is designed to reduce the effects of electric transients due to power company switching and/or lightning on the home equipment. As of this writing, such isolation equipment is not available.

5.4   Subscriber Installation

The installation procedure used by many cable system operators is usually not given proper consideration. Far too many operators employ installation contractors who provide many crews and essentially overrun a system. Improper supervision by the cable operators, who quickly lose control because of the volume of new subscribers, results in quick and often shoddy installations, a very common cable system start-up problem. The technical staff entry-level position is usually the installer job. Performing this task requires more mechanical aptitude and athletic ability to hustle ladders or climb poles than technical expertise. It must be stressed from the beginning that all installation personnel, whether company employees or contract installers, represent the cable operator and all work must be performed according to certain required standards. These standards should be enforced by a trained company technician who examines, usually on a spot inspection basis, the installations. Keeping proper records of the work quality should indicate responsibility for any poor subscriber installations. Usually the first subscriber contact the cable operator has is the customer service representative, who initially takes the order for service or provides service information. Some cable operators employ contract sales personnel, who then make the first subscriber contact. Then the installer technician performs the installation work on the subscriber's property or apartment. This early subscriber contract creates a lasting impression and dispelling bad impressions takes a long time.

5.4.1   Installation materials and tools

It is often commonly agreed that the weakest link in any cable system is the "F" connector, most likely because of the great number of them used in a system. F fittings, as they are often called, are incorporated in the head-end connecting cables since many cable operators purchase prepackaged head ends. These head-end connecting cables are often machine-made and are of high quality. However, they do have to be installed tightly to the equipment to control signal level and return loss. Unfortunately, the F fittings used on subscriber installations are not made with as much care and attention. Most of the problem is in the preparation of the cable ends, and many manufacturers have drop cable preparation tools. New and improved F fittings that when incorporating the proper cable preparation and connector crimp tooling can provide tight weatherproof connections are now appearing on the market. The chief technical person of any cable company should be constantly on the lookout for new and improved subscriber installation products and tooling for testing and examination before system incorporation.

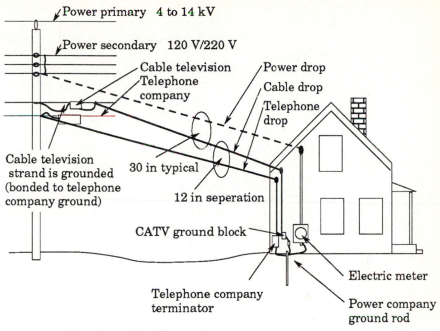

Power primary 4 to 14 kV

Power secondary 120 V/220 V

Cable television
Telephone
company

Power drop

Cable drop

Telephone
drop

Cable television
strand is grounded
(bonded to telephone
company ground)

30 in typical

12 in seperation

CATV ground block

Electric meter

Telephone company
terminator

Power company
ground rod

Figure 5.28 Subscriber cable drop.

5.4.2 Drop grounding and bonding

For safety reasons the subscriber drop cable is grounded at the point of
entry to the dwelling. This point is a grounding block with double female F fittings and a connecting point for a grounding wire. This wire
should be connected to a ground point at the same location as the commercial power grounding point. The commercial power ground clamp
must never be disturbed by any cable technician. The ground clamp
for the cable ground wire should be connected to the same ground rod
or water pipe as the power ground. Municipal water supplies often use
plastic pipe to provide domestic water service; these are unusable for
grounding purposes. Commercial power companies are driving their
own ground rods where plastic pipe is used. The local telephone companies either drive their own ground rods or connect to the power company ground. The ground wire from the ground point is connected to
the cable ground block before entering the building. Once inside the
building the drop cable connects to various splitters and ultimately to
the cable converter or the (cable-ready) television set. After the installation has been completed, the installing technician should take signal level readings at least on a few high-frequency and low-frequency
channels usually just before connecting the cable to the converter.
These signal level readings should be either filed or entered into a
computer data base for analysis or troubleshooting procedures if the

need arises. Many chief technicians do not have a good idea of the various actual signal levels floating around a system, possibly because they do not listen to the installing technicians. The electrical diagram for a drop is shown in Fig. 5.28.

Once inside the subscriber's dwelling the drop cable to the cable converters, television sets, and VCRs should be routed by the shortest route practical and fastened to the structure by cable clips or staples. Caution should be observed when using cable staple guns so the staple, particularly when exposed to weather, does not enter the cable, causing severe signal loss either immediately or later. Maintenance of proper signal level is important for good subscriber service and will be even more critical to HDTV reception.

Connections to subscriber equipment can be complicated and leave much to the imagination. Many pamphlets and instructional materials illustrating the many methods and options of connections are available to the cable technician. It must be stressed that any chosen method should not cause signal degradation, which will mean a service call.

Selected Bibliography

CATV Cable Construction Manual, Comm/Scope Company/P. Collins, Hickory, N.C., 1980.

Cunningham, John E. *Cable Television*, ed. 2. Howard W. Sams & Co., Indianapolis, 1985.

Grant, William. *Cable Television*, ed. 2. GWG Associates, Fairfax, Va., 1988.

Handbook for Distribution Products Training School, rev. ed. Scientific-Atlanta, Atlanta, 1984.

Institute of Electrical and Electronics Engineers. *National Electrical Safety Code*. New York, 1984.

Chapter 6

Measurement Instruments and Test Procedures

6.1 Introduction

As in measurement of any alternating current system, three measurement parameters of extreme importance are magnitude, phase, and frequency. Since we are dealing with a cable television system that includes a variety of signals, from time to time these three parameters will be measured and usually with a high degree of accuracy. Each of these parameters has several degrees of importance with reference to the many points in the system. In particular, signal level may be measured as voltage, current, and power. However, in CATV work the power ratio on a logarithmic scale is often used, as previously discussed. Where RF carriers are concerned, signal amplitude, i.e., level, is determined by a signal level meter. This instrument is in essence a frequency selective voltmeter. Baseband video signals are measured for amplitude and phase by an oscilloscope and a vectorscope, respectively. The basic oscilloscope is an amplitude versus time graph made on a cathode-ray tube (CRT) type of display. The vectorscope provides a phase amplitude plot on a cathode-ray-type polar display. Measurement of frequency is usually performed on the various carrier frequencies and is made with an instrument called a *frequency counter*, which counts cycles in a 1-s time frame and displays the information in digital form using LCD or LED numeric displays.

6.2 Signal Level Measurements of RF Carriers

The basic signal level method of measurement selects a carrier and measures the level. A simple diagram of this technique is shown in Fig. 6.1a.

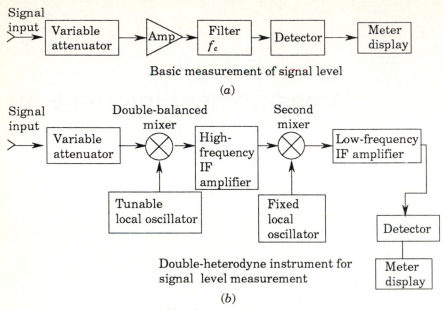

Basic measurement of signal level

(a)

Double-heterodyne instrument for signal level measurement

(b)

Figure 6.1 Techniques for signal level measurements.

The signal source at the input may contain many signals, including the one to be measured. The attenuator adjusts the signal level to prevent overdrive of the amplifier. The steps of the variable attenuator are logarithmic, i.e., in 1-dB or 10-dB steps. The amplifier has a known amount of gain. The filter selects the carrier to be measured and the detector actually measures the amount of signal and displays the value on the meter display. When using such a rudimentary instrument, the signal source is connected to the instrument and the instrument turned on. The input attenuator is adjusted to maintain the measured signal level in the range of the meter display. The filter that selects f_C is usually of the fixed type. Therefore, signal level meters of this type are often used for only a single frequency or at most only two or three frequencies. The level displayed on the meter usually has to be added to the step attenuator value to measure the signal amplitude correctly.

Meters of this type are usually small ruggedized battery-operated devices used by cable television installation people. Usual accuracy of this type of instrument is ± 2 dB. Therefore, to obtain accurate measurements when many carriers are involved a different type of instrument is required. Such an instrument will have to be tunable, i.e., frequency-selectable, with good sensitivity and an accurate display. *Sensitivity* is the ability of the instrument to measure a wide range of level values accurately. The method used in the design of a tunable signal level meter is a technique used in radio receivers called superheterodyne. *Heterodyne* means that signals are mixed, developing sum and difference frequencies plus the original frequencies. In practice, the incoming signals are mixed and a selected reference frequency produces predicted sum and difference frequencies. Now if the reference frequency is varied so the difference between the incoming desired frequency and the reference frequency is the same value for all cases, then this difference frequency can be filtered and measured. A block diagram of an instrument using double heterodyning is shown in Fig. 6.1*b*. Double heterodyning is when the mixing process is done twice.

Example   An input of 55- to 300-MHz signals from a cable television system is adjusted by step attenuator for on-scale display meter reading. Suppose the tunable local oscillator is set for 455.250 MHz; then the mixer will have a 455.250-MHz signal plus all the incoming signals. However, the high-frequency IF amplifier filter is fixed-tuned at 400 MHz. Now the only incoming frequency that will produce a 400-MHz signal is 455.250 − 400 = 55.250 MHz, which is television channel 2. This 400-MHz signal, which is now at channel 2 signal level, is connected to the second mixer.

The fixed local oscillator runs at 445.750 MHz, producing a difference frequency of 45.75 MHz, which is then selected and amplified by the low-frequency IF. This signal is then detected and displayed. This technique is referred to as *double conversion* (double heterodyning) because the frequency to be measured has been converted to 400 MHz (high-frequency IF) and then down-converted to 45.75 MHz. The process is also called up-down conversion. More sophisticated instruments use triple conversion, in which the third IF operates at 10.7 MHz, which happens to be the standard FM receiver IF frequency. The up conversion that is performed at the input will have what is known as an *image frequency* (2 × IF frequency) that is outside the band of frequencies to be measured, as well as the first local oscillator, which runs (455.25 MHz, for example) at frequencies outside the measurement band as well. Notice also that the second and third IF frequencies are also not in the instrument's measurement bandwidth (55 to 300 MHz).

Just as important as the frequency selection techniques is the accuracy of the detection circuits and the input attenuator. The attenuator should be rugged and easy to use because it is one of the most used of the operator controls. The attenuator basically consists of a switching network with an array of precision resistances in increments of 10 so as to change the input level in a precise logarithmic fashion. The level measurement is actually made in the signal level detection circuitry. Since the signal level to be measured is amplitude-modulated with

NTSC standard television signals, the signal level meter has to measure the peak carrier level. This peak carrier level appears at the horizontal tips of the video input signal and hence occurs during a short interval, i.e., the horizontal synchronizing signal pulse. A peak detector of proper design is used in most modern signal level meters. Essentially the peak detector is a diode feeding a capacitor, as shown in Fig. 6.2.

The diode rectifies the RF signal, and the high frequencies are connected to ground by the capacitor C (short-circuited to ground). The capacitor C is charged to the peak value after a sufficient number of horizontal sync pulses have occurred. This capacitor should be able to hold the charge and yet supply this level to the meter and display circuits. It should be realized that the meter and display circuits should not discharge this capacitor. Usually an operational amplifier with extremely high input impedance (open circuit) is used to read out the peak level signal.

Another area of consideration is the bandwidth of the meter's frequency response. This is important because if the meter is going to be able to read the sound carrier level of a television channel, the upper adjacent channel video carrier signal is 1.5 MHz away. Since the bandwidth of a filter is specified as the difference between the frequency of 3 dB below maximum value points, the filter still should have steep enough sides that upper adjacent video carrier is attenuated enough that it does not interfere with the audio carrier measurement. This phenomenon is shown in Fig. 6.3a and b.

6.2.1 Summary

The signal level meter is the most important piece of equipment used in today's cable television plant. Most manufacturers use heavy-duty ruggedized taut band meter movements for reliability and accuracy. High-quality components and design are used by most to attain excellent temperature stability and low noise characteristics.

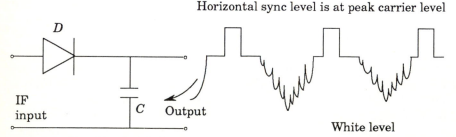

Figure 6.2 Signal level peak detector.

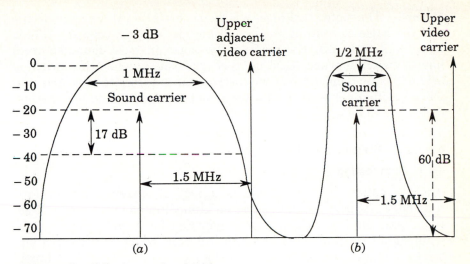

Figure 6.3 Signal level meter bandwidth.

6.3 System Noise Testing

Many manufacturers offer various options that in most cases are extremely useful. One such useful option is the ability to measure system noise and hence carrier-to-noise ratio. The technique is illustrated in Fig. 6.4.

First, use the signal level meter to measure video carrier level A, then tune to the area of missing video carrier B, remove attenuation and read the noise level on the meter. Subtract this reading from the first to get the C/N for this particular point in the system.

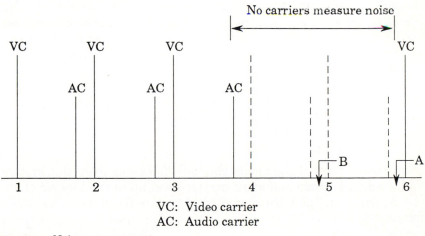

Figure 6.4 Noise measurement.

Example   The meter indicates that video carrier level A is $+26$ dBmV and the noise as read in the second part is -22 dB; then the C/N is $26 - (-22) = 48$ dB. For reading the noise in most manufacturers have a switch marked "C/N" or "S/N" or "Noise"; when this switch is activated, the bandwidth of the instrument is extended to 4 MHz, the FCC-specified value. If this switch is not available, then a correction factor has to be applied.

6.4   Low-Frequency Disturbances

6.4.1   Hum testing

Another option often available is a test of system *hum*, or low-frequency disturbance. In this test an unmodulated video carrier is selected and the input attenuator is adjusted for a mid to upper half-scale signal reading. Then the hum selector switch is activated and the percentage hum is indicated on the meter scale marked "% hum." This value simply means that the video-detected signal contains 60- or 120-Hz components. An unmodulated carrier is used because it is supposed to have no signal components on it. Therefore, whatever is modulating this carrier should not have signal components. The detector output is passed through a low-pass filter and then to the display section.

Modern signal level meters often have a voltmeter as an option that is extremely useful for system maintenance. System power, in the form of 60 Hz, 60 V, can be read on the voltmeter option; thus one instrument can make several measurements.

6.5   Spectrum Analyzers

The modern spectrum analyzer is an extremely useful and valuable piece of equipment; it is also quite expensive. Some instruments cost about $8000 to start and on upward to $50,000 for very wideband, accurate computerized instruments. The ones often used in cable television work are usually in the $8000 to $15,000 class. Essentially most of the ingredients of a spectrum analyzer are found in many signal level meters. One supplier of instruments offers a spectrum analyzer option that uses an oscilloscope as the indicator.

6.5.1   Spectrum analyzer primer

If one were to start at the lowest channel on a cable television system and record each video and audio carrier level up to the highest channel frequency and plot the signal level versus frequency, the whole spectrum would be displayed. Now since the tuning in this case is manual, as are the attenuator settings, an automatic continuous procedure could be used with a wideband display to produce a spectrum display. Essentially this is the method used to construct a spectrum analyzer. A block diagram of a simple spectrum analyzer is shown in Fig. 6.5. Notice the inclusion of many sections appearing in the signal level meter block diagram.

The input signals from the cable system are adjusted for an on-scale setting by the input attenuator-amplifier combination. The input amplifier essentially increases the sensitivity, thus extending the range of input signal levels. The voltage-controlled oscillator generates a sawtooth voltage waveform, whereby the slope and direct current level select the center frequency and the range on each side of the center frequency. The resulting IF also sweeps through the frequency range and drives the detector. The detector amplifier provides the level measurement of the selected frequency carriers at a level that will control the vertical (y-) axis of the display. At the same rate the sawtooth voltage generator causes the horizontal sweep to move from left to the right side of the CRT display while the vertical signal from the detector is driving the vertical deflection of the display. The result is a plot of amplitude (signal level) versus frequency on the vertical-horizontal display. As mentioned, a *spectrum* is a plot of signal level versus frequency.

Modern spectrum analyzers, like modern signal level meters, employ double- or triple-conversion techniques, video processing, digital memories, etc. One important point to remember is that, considering all the features, costs, etc., one has to feel comfortable with the instrument and learn how to use it effectively and often. Operating this type of instrument is complicated and easy to forget; using it often keeps one sharp and familiar with the instrument controls and capabilities. Sample spectrum displays are shown in Fig. 6.6a and b.

Spectrum analyzers with other pieces of equipment can measure or

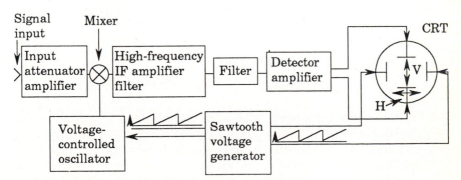

Figure 6.5   Block diagram of elementary spectrum analyzer.

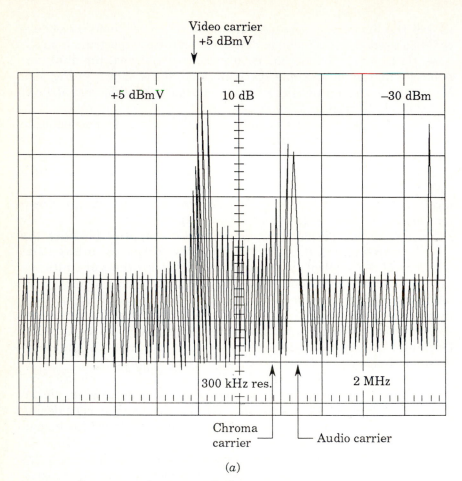

Figure 6.6 Spectrum analyzer screen displays.

test such cable parameters as composite triple beat, intermodulation beat, carrier-to-noise ratio, and hum modulation, as well as signal level.

6.5.2 Spectrum analyzer measurement

The equipment connection diagram for these tests is shown in Fig. 6.7.

6.5.2.1 Composite triple beat. The basic technique is to provide to the cable television distribution system all video carriers at normal level with no modulation. By consulting the system beats table in Chapter 3 (Table 3.4), e.g., for a 60-channel system with all channels active the maximum buildup of beats is at channels 32 and 33, with 1184 pre-

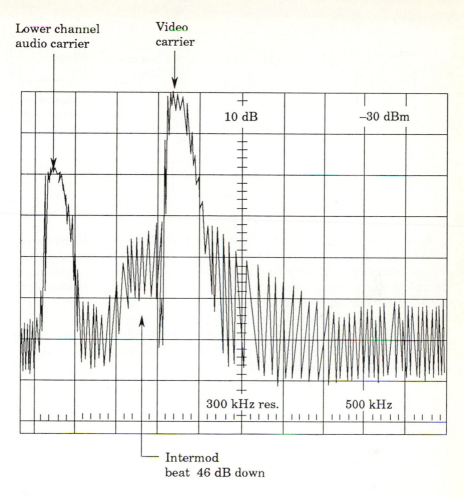

Lower channel
audio carrier

Video
carrier

10 dB

−30 dBm

300 kHz res.

500 kHz

Intermod
beat 46 dB down

(b)

Figure 6.6 (*Continued*).

dicted beats. Therefore, channel 32 or 33 should be tested for composite triple beats, which should be worst case. The bandpass filter at the test site should be tuned to channel 32 or 33. This filter will only pass the test channel so the other signals will not overload the spectrum analyzer. The spectrum analyzer should be tuned for a center frequency for channel 32, for instance, and the span control on the analyzer should be set so the display only covers channel 32. The attenuator should be adjusted so the channel 32 test carrier is at full screen height but no higher. The amplitude scale should be set for 10 dB/division. Now, by means of two-way radio or telephone the channel 32 signal should be removed to clear the screen. Now the sensitivity

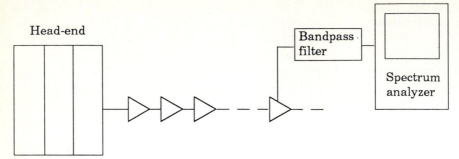

Figure 6.7   Spectrum analyzer system.

can be increased by the input attenuator control 10 or 20 dB while observing the screen. If a beat in the channel 32 space is seen, it should be measured by observing how many decibels down there are from the top of the screen after taking into account any change in the input attenuator. For example, if a beat is observed on the screen that is 4½ divisions down from where the top of the screen was at carrier level, it should be determined that 4½ divisions at 10 dB/division corresponds to a 45-dB difference. Now, if it was required to increase the input sensitivity by decreasing the input attenuator by 10 dB then the actual difference between the carrier level and the beat level is 55 dB. This is the correct measurement. Beat levels of this amplitude will not cause any picture impairment. The threshold of visibility of beats is approximately a 51- to 52-dB difference.

6.5.2.2 Crossmodulation. Crossmodulation can be measured by a spectrum analyzer. However, the span has to have enough range to decrease to a low enough value or, if available, zero scan mode should be used so the video can be measured. The test setup is essentially the same, except that all carriers except the test carrier should be modulated. The same modulation source has to be used to modulate all the carriers. For most cable television systems this is a difficult task. Many technicians and engineers feel that since crossmodulation is a third-order distortion, if this distortion takes place then composite triple beat and intermodulation will take place as well, and it is easier to test for composite triple beat and intermodulation. Essentially the measurement is also made near channel 32 or 33, which is unmodulated. So, now any video modulation appearing on channel 32 or 33 that should have no modulation is due to modulation contained on the beats appearing in the bandwidth of channel 32 or 33. The instrument instruction manual should be consulted for procedures and correction factors when making any of the tests using this type of instrument.

6.5.2.3 Carrier-to-noise ratio. The spectrum analyzer can be connected at almost any system test point, provided there is enough system signal level. Usually +20 dBmV is enough, but a higher level would give more accurate results. Essentially a test carrier that has no other carrier on either side of it is needed. The carrier is selected by the analyzer center frequency control and the span adjusted to cover this one channel and the adjacent vacant channel. Once the carrier level has been measured, the video filter contained in most spectrum analyzers should be selected and placed in the 300-Hz position. Now the noise is essentially averaged by the filter, thus raising the baseline in the vacant carrier area. Since the carrier level was measured at the top screen graticule, the noise measurement is made by simply counting the graticule lines from the top to where the raised baseline intersects, which will be the carrier-to-noise ratio (C/N) in decibels uncorrected. If the analyzer bandwidth was set, for example, at 300 kHz, then a correction factor can be calculated by the following formula:

$$\text{Correction factor (dB)} = 10 \log \left(\frac{4 \text{ MHz}}{0.300 \text{ MHz}} \right)$$

$$= 10 \log 13.33 = 11.25 \text{ dB}$$

For example, if the raised noise floor is 5½ graticules down at 10 dB/graticule, then the uncorrected C/N = 55 dB. Applying the correction factor:

$$\text{C/N corrected} = 55 - 11.25 = 43.75 \text{ dB}$$

6.5.2.4 System hum. The measurement of system hum, like the noise test, does not require any change in the head end and can be performed anywhere in the system with enough signal level. Essentially the spectrum analyzer is tuned to a carrier and placed in a low-scan or zero-scan mode. The video modulation is now seen on the screen and the tuning reset slightly to peak the signal. The instrument scale should be set to linear mode and the amplitude adjusted to full screen height. The percentage ripple may now be calculated by counting from the peak-to-peak ripple to the signal peak-to-peak level. A screen display diagram, as shown in Fig. 6.8, illustrates this measurement.

Therefore, each square represents 12.5 percent hum so one square represents 12.5 percent and measurement is outside the 5 percent maximum allowed value of FCC specifications.

6.5.2.5 Intermodulation. Since the intermodulation test is for beat frequencies that fall into the category of $2f_1 \pm f_2$ then, recalling the example given for channel 3 video and audio carriers, a beat can be produced at 56.75 MHz. This beat falls 1.5 MHz above the lower adjacent video carrier channel 2 of 55.250 MHz. Since this beat is caused by fre-

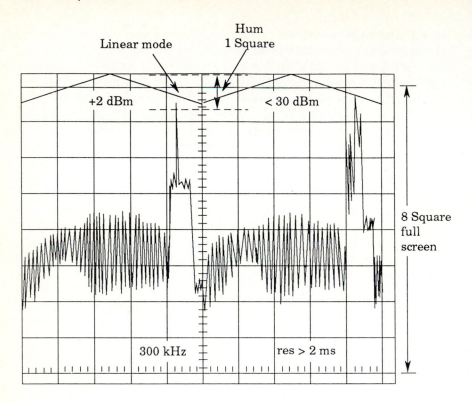

Figure 6.8   Spectrum analyzer measurement of hum.

8 Squares / 100% = 1 / X

$$X = \frac{100}{8} = 12.5\% / \text{square}$$

Hum = 12.5% > 5% allowed

quencies 61.250, 65.75, and 122.50, three frequencies are involved; thus there is a third-order distortion. Also the nonlinearity portion of the amplifiers transfer function plot produces crossmodulation and composite triple beats and also causes intermodulation distortion.

Use of the spectrum analyzer to locate and measure these beats, which are located close to video carriers, is a simple task in that the spectrum analyzer is simply connected to the system at sufficient signal level. Using a low span so the screen is one channel wide and then slowly varying the center frequency control to move the analyzer across the spectrum from channel to channel will allow the intermodulation beats to be seen. If a 30-kHz video filter is available, using it will diminish some of the noise and allow the beats to show up. Fig. 6.6*b* indicates how an intermodulation beat should look.

6.5.2.6 Hints and tips. When using any instrument, caution should be taken so as to not overdrive the instrument. If it is suspected that the input signal is too high, then increasing the input attenuator by, for example, 3 dB should make the signal decrease 3 dB. If the signal decreases by more than 3 dB, it is overdriven. An overdriven instrument will cause beats and noise that come from the instrument not the system to be observed on the screen.

The correct input impedance should also be observed. Since cable television systems are operated at a 75-Ω impedance throughout and some industrial and commercial spectrum analyzers have an input impedance of 50 Ω, this problem can be very common. To use the instruments accurately, the input impedance match has to be maintained. If an impedance mismatch does occur, an impedance matching transformer can be used; however, the frequency response has to be in the area of measurement. In short, an audio 75- to 50-Ω transformer could not be used if such a device were available. Usually a resistive "L" or "T" pad that is not sensitive to frequency can be used. However, an insertion loss of 5.3 to 6 dB should be taken into account for the tests.

Of course, it also must be assumed that test equipment batteries should be charged and instruments should be protected against shock and physical abuse.

6.6 Instrument Calibration and Testing

Periodically and at least before the FCC Proof of Performance Tests the instruments should be calibrated. A signal generator with a fairly accurate frequency dial and a precision attenuator can be used to level-test signal level meters and spectrum analyzers. Sample frequencies should be selected on the generator throughout the spectrum and level measurements made on the spectrum analyzers and signal level meters. If all instruments are within 1 dB in a 20-dBmV level test then the maximum error will be $\frac{1}{20}$ or 5 percent. If only one of five instruments exhibited all of the error, then the remaining could be assumed correct. The frequency dials can be checked by connecting the instruments to an off-air antenna and checking them against the television station's carrier frequency, which is held to very close tolerance.

6.7 System Analyzers and Measurement Techniques

Some instruments on the market today are called system analyzers. These instruments essentially perform the same measurements as other instruments such as the more sophisticated signal level meters and spectrum analyzers. However, with the addition of today's microprocessor digital electronics the ease and accuracy of making measurements are increased. Since digital tuning is available in modern television receivers such digital tuning is the method of channel selection on microprocessor-controlled analyzers. The digital tuning is stepped through a band of channels to be tested. At each channel stop the detector output is digitized into a digital number corresponding to the signal level at that particular tuning step. This digital value is stored in a solid-state memory that is read out and displayed on a screen. The tuning center frequency and frequency span are also digitally selected on a digital front panel keyboard. If the whole spectrum is selected and recorded in solid-state memory only the lower- and higher-frequency bands can have the channel selected in these bands and displayed in split screen format. Thus the low and high pilots can easily be displayed. Because digital circuitry has memory, keyboard settings can be memorized as well as data. When measuring the appropriate signals using the built-in filters the composite triple beat (CTB), crossmodulation (XMO), and second- or third-order distortion may be calculated and printed on the screen. Self-calibration, split screen displays, and ease of operator settings are all features of modern cable television analyzers. Battery operation, often unavailable on spectrum analyzers, is almost always available.

Some sophisticated signal level meters with features that border on those found on analyzers are now becoming available. The question is where the meter stops and the analyzer begins. A good rule of thumb is to arrange a trial period of use over a few days to become familiar with the equipment before purchase. More manufacturers of test equipment are making this offer.

6.8   System Sweeping Techniques

6.8.1   Introduction

Sweeping a cable system is a technique used to measure the frequency response over the system bandwidth. Recall that *frequency response* is the amplitude versus frequency measurement result, which may be a plot of signal level versus frequency. A rudimentary technique that can be used to measure signal level versus frequency is to inject a signal of known frequency and measure the level with a signal level meter. If this procedure is begun at the low-frequency end of the system bandwidth and enough frequency points are measured up to the high frequency, a plot of these values will give a fair representation of system response. This method is shown in Fig. 6.9.

It should be clear that if there are not enough measurement points taken then the question of what happens between the points where there is no measurement arises. Simple as it may seem this procedure

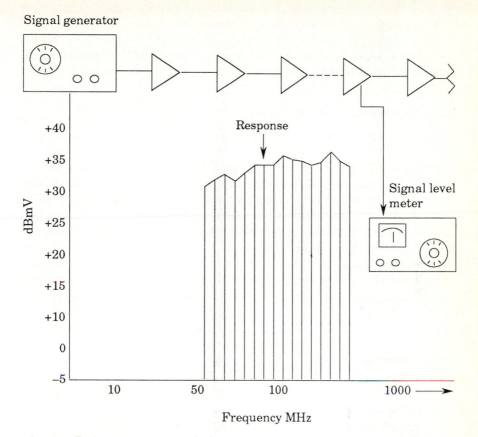

Figure 6.9   Basic system-sweep method.

is very time-consuming and the chance for error is large. It requires two people, one at the head end to select the frequency and level, the other at the test point to make the level measurement. The measurements taken at the test point will be recorded on paper and a plot made later.

Another method, which is a little faster, is to use a spectrum analyzer at the test point. The technique at the head end is for the operator to vary the tuning dial on the signal generator from the beginning low frequency to the high-frequency point at a constant input level. At the test point the operator sets the center frequency to the bandwidth midpoint (50 to 450 MHz midpoint is at 250 MHz) and a resolution of say 50 MHz/horizontal division. The horizontal axis will now have 400-MHz bandwidth spread over eight graticule lines, effectively filling the screen. The level attenuator should be adjusted for a midscreen deflection and this value recorded. At the test point the operator selects a horizontal sweep speed to catch the sweep from the head end. Several tries may be needed, and if the analyzer has either digital or analog memory the good sweep can easily be photographed with a scope camera. By now it should be clear that sweeping simply involves a signal carrier's being swept through the frequency bandwidth at a constant known level.

6.8.2   Device sweeping techniques

System components can have their frequency response measured by the same technique. Passive components, such as signal splitters and directional couplers, and taps, as well as active devices, such as amplifiers and processors, can be sweep-tested as well. Since the equipment to be tested as well as the test instruments can be placed on the same bench, the procedure is not overly cumbersome and only one person is needed to do the work. Still the problem of synchronizing the generator and the signal level meter presents a problem. In this day and age of electronics this is accomplished in a *sweep test system*, which consists of a sweep generator that has an electronic sweep control in that the carrier can be swept from some selected start frequency (low-frequency value) to some stop frequency (high-frequency value) at a constant output level. The speed of the process [number of sweeps of f (start) − f (stop) per second] can also be selected. A synchronizing signal drives the receiving instrument, which makes the level measurement. This can simply consist of a flat frequency response detector attenuator set with a large-screen oscilloscope as the indicator. The detector-attenuator system is often contained in a *comparator*, which provides a baseline on the oscilloscope of known amplitude, thus setting the level scale. This system is outlined in Fig. 6.10. The frequency calibration on the display oscilloscope is usually accomplished by small trace interruptions on the response trace called *birdy markers*, which can be selected on the sweep generator, for example, every 10-MHz or every 50 or 100 MHz, thus giving an indication of frequency. High-quality sweep systems of this type can produce very accurate results. If large-screen oscilloscopes are used, photographing the screen is very difficult, but the alignment technique is performed easily. System amplifiers are aligned in this manner before they are installed in the system, making overall system balancing much easier. Any amplifiers that fail the test are never installed in the system; thus troubleshooting a new system can be prevented or held to a minimum.

6.8.3   Simultaneous-systems sweeping

The sweeping system device just discussed, although accurate, is difficult and cumbersome to use for system sweeping. The problem is

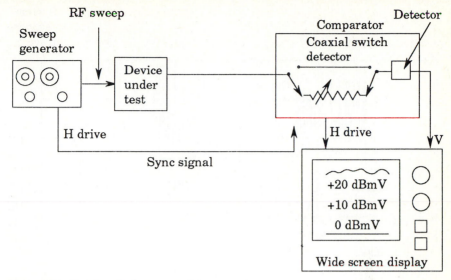

Figure 6.10 Synchronized sweep system.

that the H trigger needed by the detector uses an extremely long cable. A system that places the trigger system on an RF carrier and sends it through the cable has been developed. Now the level measuring equipment can be placed away in the system at some location and receive the synchronizing signal signifying the start and stop frequencies. This technique is known as *simultaneous sweeping* or *simul-sweep*. Simultaneous sweeping systems fall into two categories, low-level sweeping and high-level sweeping. The difference is the level of the sweep signal that should not cause interference with the subscriber's picture. Both types of systems have advantages and disadvantages.

6.8.3.1 Low-level sweeping. The level of the sweep generated signal is usually 20 to 30 dB below the system signal level, thus causing a small amount of picture interference and no disturbances to the system pilot carriers. Slow sweep speed can also be used to make the sweep more visible. The main disadvantage is that the sweep signal falls below the carriers and above the noise floor on the oscilloscope showing the detected output. Some technicians use a grease pencil on the screen to trace out the sweep signal. Fig. 6.11 illustrates the display.

6.8.3.2 High-level sweeping. The level of the sweep signal is usually 15 to 20 dB higher than the system signal level. To prevent over-

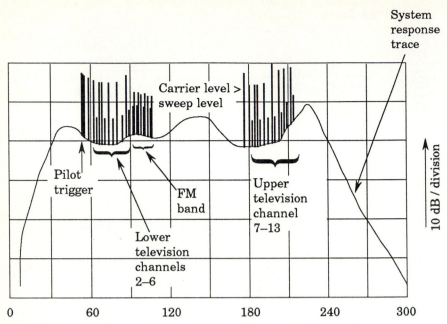

Figure 6.11 Simultaneous low-level sweep display for a 12-channel system.

driving the amplifiers, this signal has to be extremely rapid (short duration) and usually not very frequent (time). Typically the sweep duration is several milliseconds and a sweep occurs every few hundred milliseconds, thus causing quite a bit of subscriber picture interference and resulting complaints. The benefit is that the display is easily read because the sweep signal is above the carrier level. If the system AGC circuits are fast-acting, then a trapping filter on the output of the sweep generator usually is used to prevent the sweep signal from disturbing the pilot signal. Traps should be used to trap both the low and high pilots. The system connections are diagrammed in Fig. 6.12.

The high-level system presents a clear display of the system frequency response but does cause a lot of interference with subscriber reception, so it is usually performed early in the morning.

6.8.3.3 Sweep recovery method.   Naturally improvements are always being developed. The sweep recovery system is just one example. Basically the sweep duration is very short, on the order of 1, 2, or 3 milliseconds (ms) selectable, and the sweep time is infrequent: 2, 3, 4, 5 seconds (s) also selectable. The optimum seems to be a 1-ms sweep occurring every 5 s. One such system is digital in operation; in it the sweep generator (transmitter) is stepped through the spectrum at a fixed number of steps over the selected start-stop frequencies. A data

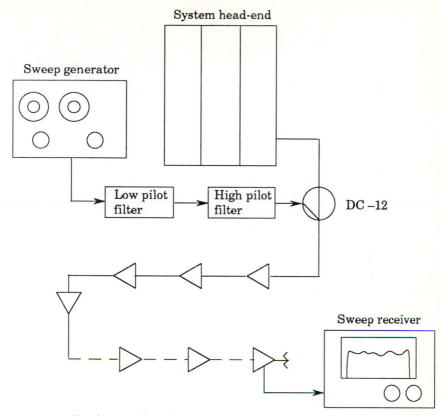

Figure 6.12 Simultaneous high-level sweep.

carrier, usually just below the lower band edge, for example, 50 MHz, contains the start and stop frequencies and the stepping rate, time, and level information which it carries to the receiver at the test point. The battery-operated receiver takes this information and sets its circuits to respond. The first signal tells the sweep receiver to step its tuning from f_1 to f_2 and at each step to record the measured signal level value in digital form at a location in its solid-state memory. This signal consists of the system signals so that this particular section of the solid-state memory in the receiver contains this information. The second signal tells the receiver to do the same thing; however, this time the transmitter sends the sweep. The receiver stores these stepped signal level values in another solid-state memory bank. At the end of this signal the receiver subtracts the first memory section from the second memory section and displays the result on the screen of the receiver. In this resulting display the signal carriers are subtracted from the second sweep-containing carrier plus sweep, thus leaving the sweep to be displayed. Hence it is called a *sweep recovery*, in that the sweep signal is recovered from the system signal carriers. Systems such as these are rugged and easy to use and provide accurate results with a minimum of subscriber interference. The work can be performed by only one person. Usually when instruments of this nature are available they are used more often and the system is better maintained. When instrument purchases are being considered an on-site system demonstration is very helpful.

One of the most useful and important things an instrument manufacturer can and must provide with the instrument is a comprehensive and well-written, concise instruction manual. Before purchasing any instrument ask to see the instruction manual. A sample sweep photo of the sweep recovery system is shown in Fig. 6.13.

6.8.3.4 Sweeperless sweep. Remember that a *sweep system* was defined as the measurement of a constant level carrier varied in frequency (sweeping) through the system across the bandwidth. Now if there is no carrier being swept through the system, it really is not a sweep system. A sweep system is a method used to measure the frequency response and a sweeperless sweep system will do just that. The question is, How? Presently there are two similar systems commercially available. No doubt many more will be marketed in the future.

The way such a system works is that the instrument is placed at the head end, where levels of the whole spectrum of carriers are measured

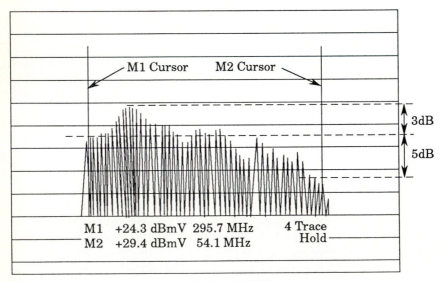

Figure 6.13   Sweep recovery system display: 300-MHz system.

and each value is digitally stored in a solid-state memory. When the instrument is placed out in the system at a test point of sufficient level (usually +20 dBmV minimum), the system levels are measured again. The values previously measured at the head end are subtracted from the values taken at the test point, leaving the difference signals to be displayed.

It should be understood that in the areas where no signal is available no measurement will take place, so system frequency response will only be measured and displayed where signals are present. In other words, where there are no carriers present (unused channel space) no measurement will take place. Therefore, system response for unused channel space will not be known. If there are means to inject simultaneously the missing channels via a multicarrier signal generator or spare modulators, then the sweeperless sweep system will provide good results over the band.

6.8.3.5 System linearity and peak-to-valley ratio.

The frequency response as provided by a sweep system or sweeperless sweep display screen shows the signal variations as a function of frequency. Such data display system slope as well as *peak-to-valley ratio*, which is a measure of system linearity. Recall that the specification of peak-to-valley ratio is calculated by the formula

$$\left(\frac{N}{10} + 1\right) \text{dB}$$

where N = number of like amplifiers in cascade

A sample sweep response indicating some of the measurement parameters is shown in Fig. 6.14.

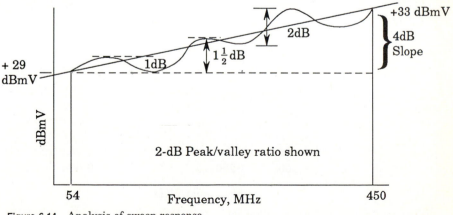

Figure 6.14 Analysis of sweep response.

6.8.3.6 Noise power as a test method.
As can be seen on a wideband spectrum display for a cable television distribution system, the noise level is quite constant over the system bandwidth (50 to 450 MHz). If this noise is passed through a narrowband filter, then the noise is said to be *band-limited* and the spectrum display will show the shape of the passband of the filter. This phenomenon was put into practice in a measurement system that consists of a high-level flat (constant-level) wideband noise source. A plot of noise level versus frequency is depicted in Fig. 6.15.

Fig. 6.15 shows that noise level is constant from 5 to 500 MHz at +20 dBmV and down 3 dB at 5 and 600 MHz. Therefore, the noise bandwidth is 595 MHz. Noise-generating instruments usually have a couple of crystal-controlled frequencies at a precise level to compare the noise level to the standard level at two or more frequencies in the bandwidth.

Flat- (constant-) level noise of +10 to +20 dBmV can be used to test such parameters as signal level meter flatness, frequency response of filters, and cable loss. When it is coupled with a bridge circuit, measurements of return loss can be made easily across the band of frequencies. Such techniques are shown in Fig. 6.16.

6.9 Frequency Measurements of Cable Carriers

The FCC proof of performance tests calls for the measurement of video and audio carrier frequency for all "off-air" television stations where

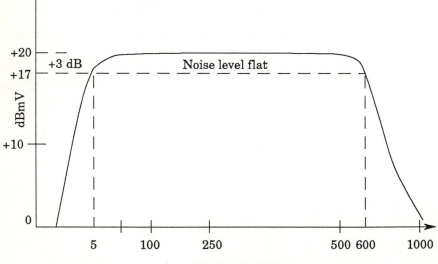

Figure 6.15 Flat noise through a filter shows frequency response of filter.

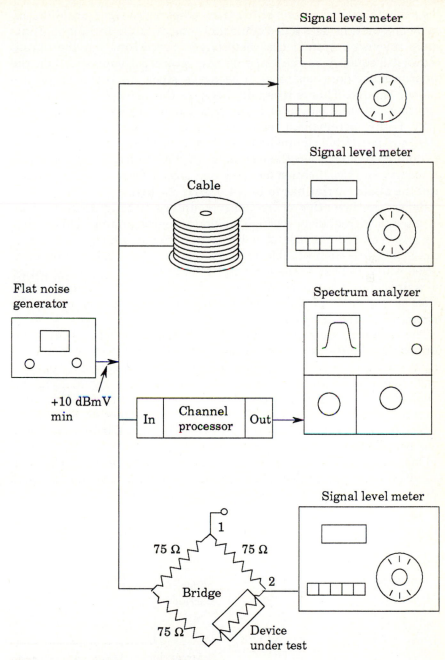

Figure 6.16   Device testing using flat noise.

the cable television system either transposes these signals to different cable channels or uses a demodulation-remodulation technique. If the cable television system uses heterodyne processors and the output channel frequency is the same as the received frequency, then the measurement does not have to be made since there can be no frequency change. This is true only because the heterodyne processor's input-output local oscillator is the same, hence there is no frequency shift. If the incremented regulated carrier (IRC) or harmonic regulated carrier (HRC) channelizing system is used with phase locking to a comb generator, then the channels have to be frequency-tested for accuracy. The specification for the video carrier frequency is ± 25 kHz and the audio carrier has to be +4.5 MHz ±1 kHz.

Before the invention of frequency counters making accurate measurements of frequency was tedious and time-consuming. This earlier process required the use of a frequency meter, which consisted of mixer and precision-tunable local oscillator. The carrier to be measured was zero beat to the local oscillator, which was then measured on a precision frequency dial. An auxiliary oscillator called an *interpolation oscillator* provided vernier measurement plus or minus the measured carrier frequency.

The modern frequency counter provides accurate results for unmodulated pure carriers of sufficient signal level. However, obtaining the unmodulated condition means removing the channel from service. One test instrument manufacturer produces a signal stripping amplifier that tunes to a channel and removes the modulation, thus presenting a clean signal carrier to the frequency counter. Such a system is shown in Fig. 6.17. Notice also that the 4.5-MHz intercarrier is also provided for the audio carrier measurement.

One such manufacturer provides a frequency counter with a built-in signal stripping amplifier. This type of instrument provides the most easy-to-use meter with the necessary accurate results.

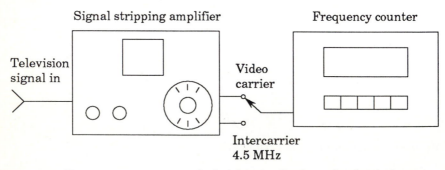

Figure 6.17   Frequency measurement of television signal using a signal stripping amplifier.

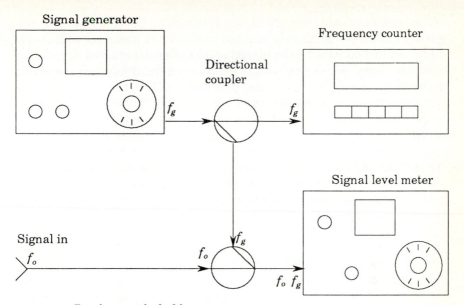

Figure 6.18 Zero beat method of frequency measurement.

If modulation cannot be removed, a method of beating (mixing) the carrier to be measured with a variable signal generator is used. When zero beat is detected on a signal level meter, then the variable oscillator is at the same frequency as the one to be measured and the variable oscillator (generator) can be measured with a suitable frequency counter. This method is shown in Fig. 6.18.

When using this method care should be taken so the signal input is nearly equal in level to the generator signal at the signal level meter input. Also, the counter should have a high enough generator level to make the measurement accurately. Usually the signal level meter should have the carrier levels at about +10 dBmV for each signal. Most frequency counters need +35 to +50 dBmV minimum to count the frequency accurately. Essentially the frequency counter counts either the positive peaks or the negative peaks (selectable) of the input signal sine wave over a 1-s period. The counter circuits have to generate this highly accurate time base to perform the frequency tests (cycles/second) accurately. The frequency counter also has to display enough digits to provide the necessary accuracy. For example, ±25-kHz accuracy in reading for a video carrier of megahertz in the hundreds needs at least six digits of display, as in the following example:

$$
\begin{array}{ccc}
\text{100 MHz} & \text{1 MHz} & \text{10 kHz} \\
\downarrow & \downarrow & \downarrow \\
\end{array}
$$

$$213 \cdot 250$$

$$
\begin{array}{ccc}
\uparrow & \uparrow & \uparrow \\
\text{10 MHz} & \text{100 kHz} & \text{1 kHz}
\end{array}
$$

When the zero beat method is used, careful adjustment of the frequency dial of the generator and the signal level meter as well as the audio volume control on the signal level meter is required to obtain a clear zero beat. When zero beat is approached the difference frequency between f_g and f_0 becomes less, producing a low tone heard in the speaker. When the tone reaches zero, then $f_g = f_0$ and f_g should then be measured by the frequency counter. Often a pair of good-quality headphones can be helpful if the signal level meter has headphone jacks. A spectrum analyzer can be used as a zero beat indicator instead of the signal level meter. The display of the analyzer can give a good indication of approaching zero beat, but the precise zero beat point is still not easy to identify. The analyzer instruction manual should be consulted for such a use.

6.9.1 Instrument calibration procedures

As previously mentioned, accuracy of cable television signal level measuring equipment can be checked by comparing various signal level meters to a common value. Any difference in readings can be caused by an inaccurate instrument. The largest difference in reading can be interpreted as the maximum error. If this value is quite small, then it may be assumed the instruments are accurate in a practical sense. Some manufacturers offer a signal generator with a calibrated output level. Often this instrument is not very expensive and is good for testing signal level meters. This generator provides a calibrated output level, also selectable, to act as the standard level for all the television frequencies. Since the frequency tuning is from a dial, the frequency accuracy is not very great. This instrument can only be used as a standard level test and not a frequency test.

6.9.2 Frequency testing

Frequency testing is usually performed by a frequency counter, which actually counts the number of cycles in a second. The accuracy of the frequency counter is determined by the accuracy of its time base. After all, if the period of 1 s provided by the counter is not accurate, then the number of counts in a second will not be accurate. Since commercial television broadcast stations have to maintain a highly accurate frequency tolerance, the presence of the AM modulation makes it impossible for a frequency counter to lock on to the carrier frequency. FM stereo broadcast stations have to maintain a carrier frequency accuracy of ± 2000 Hz. This relates to an accuracy, for a 100-MHz carrier, of $(2 \times 10^3) \div 1 \times 10^8 = 2/10^5 = 20$ parts per million. Again, since the carrier is frequency-modulated the counter does not detect a stable carrier. Therefore, unless modulation is not present, the counter cannot be tested.

A method used by test and calibration people is the direct comparison method, which uses the carrier frequency of the calibrated time broadcast station of the U.S. government. Such stations are WWV and WWVB with calibrated carrier frequencies with National Bureau of Standards (NBS) accuracy. These carriers are used to check the usual 1-MHz or 5-MHz time base in the frequency counter. The method is shown in Fig. 6.19.

The calibrated carrier frequency that is tuned by the receiver is mixed with a sample of the time base frequency. If zero beat is detected by the high-quality headphones the counter time base is accurate. If not, an adjustment of the time base is usually available on the counter. This control is adjusted to zero beat. Since zero beat is essentially "0" frequency or dc, the phones can most likely only detect down to 20 or 30 cycles. If an audio voltmeter is available the zero beat below 20 cycles can usually be seen on the meter. If the 10-MHz or 15-MHz WWV signals are used, it may be necessary to insert a small signal diode in series with the time base insertion coil at point x. This diode's nonlinearity will provide the required harmonic content to the counter's time base to beat against the higher WWV signal.

Some frequency counter manufacturers have fast calibration service; one that I know of will trade the time base frequency determining module on a fast turnaround basis with accuracy at NBS traceable standards. Another manufacturer offers a frequency counter with a

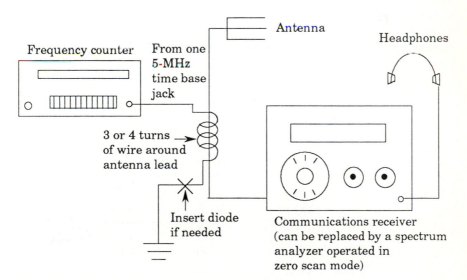

Figure 6.19 Calibration of frequency counter using WWV.

built-in calibration receiver. Since the FCC has precision frequency offset requirements that move conflicting television cable channels from the air traffic communication and navigational frequencies by a precise amount, highly accurate frequency counters have to be available to the cable technical people.

6.10 Amplifier Balance and Alignment

Before amplifier balance and alignment start the manufacturer's instructions and alignment procedures should be studied. Because many of the procedures are more or less specific to a given manufacturer's equipment, the basic philosophies will be discussed.

As was discussed on the chapter on cable, the loss in decibels per 100 ft is a function of frequency. The higher the signal frequency transmitted through the cable the greater the loss, as diagrammed in Fig. 6.20.

Since most cable television amplifiers are flat wideband amplifiers, the high signal input at the low-frequency point can overdrive the amplifier, thus causing signal distortion. A network used to level the signal has the loss characteristics that essentially are the reverse of those of the cable. That is, such a network will have significant loss at the low signal frequencies and negligible loss at the high signal frequencies. This is demonstrated in Fig. 6.21a.

If the amplifier has any built-in equalization then the input equalizer will have to have less compensation, hence a smaller equalizer value. The manufacturer's instructions should be consulted to prevent overequalization.

Most trunk amplifiers have their output signal level sloped upward at the high frequencies. For example, at 450 MHz the output level is

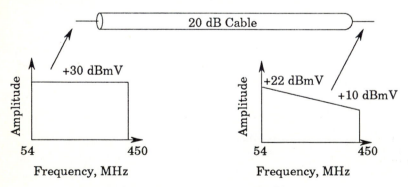

Figure 6.20 Signal level variation through coaxial cable.

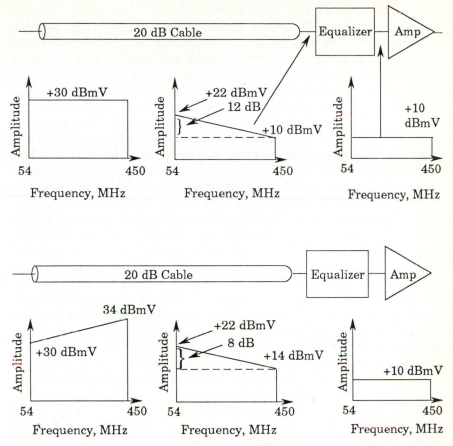

Figure 6.21 (*a*) Effects of the cable equalizer. (*b*) Effects of input slope.

often 4 dB higher than the lowest-frequency 54 MHz, as shown in Fig. 6.21*b*.

The equalizer has to compensate for only 8 dB, instead of 12 dB as before. It must be remembered that the smaller feeder-type cable has a steeper loss characteristic at the high frequency, and hence usually an 8- to 9-dB upward slope is used on the feeder system. Now the equalizers used in the feeder amplifiers, i.e., bridging and line extender amplifiers, will not have to compensate for large cable tilt.

Most manufacturers of amplifiers make a large variety of equalizers ranging from 0 dB (flat) to 30 dB in 1.5-dB steps for upper frequency limits of 216-, 270-, 300-, 330-, 400-, 450-, and 550-MHz systems. For every 1.5-dB step only about 0.9 dB of actual compensation at the low-frequency end (54 MHz) will result.

To align a section of system trunk, the usual procedure is for the technician to examine the system design maps, noting the pad and equalizer values calculated by the design engineer. At the amplifier location the selected pads and equalizers are installed and the signal is measured at the input and output test points at the lowest and highest signal frequency carriers, i.e., 54 and 450 MHz. This procedure is started from the head end and proceeds outward along the amplifier cascade until the end of the cascade run is reached. The technician then returns to any trunk splits and proceeds to the end of that cascade, continuing on until the system is completely aligned.

Now at this point it may seem that alignment has been completed. However, this procedure is considered rough balancing. The next procedure is to leakage-test the system to make sure that the connectors, housings, and devices are tight and no standing waves at various frequencies cause signal level variation between the low- and high-frequency limits. Several leakage detection systems that are commercially available enable a technician to test the cable plant for leakage. All schemes use the basic technique of placing a known and easily recognizable signal on the cable systems and with a sensitive receiver and antenna system travel along the cable plant hunting for any sign of signal emitting from the cable plant. This procedure should be performed before signal sweeping the plant or connecting subscribers to the system. Subscriber drops and/or illegal subscriber connections of home equipment to the interior wiring can cause leaks of significant magnitude that make proper plant leakage testing difficult.

Most commercial leakage testing equipment signal sources are narrowband, with the carrier frequency away from the usual strong off-air signals. The signal detectors are again narrowband and have high sensitivity to detect even the smallest leaks, less than the 20-μV/m standard.

6.10.1 Automatic gain and slope control principles

Automatic gain and slope control of trunk amplifiers is necessary because of the change in cable characteristics as a function of temperature. As discussed in the chapter on coaxial cable (Chapter 3), the attenuation of coaxial cable increases 1 percent for every 10°F rise in temperature. It has been proved that in actuality 0.11 percent/°F change is indeed more accurate, but a 1 percent/10°F change is a good rule of thumb. To examine the problem more accurately, consider the following example. For a 60-channel, 450-MHz HRC cable television distribution trunk system spaced at 20 dB, the median design temperature is 68°F or 20°C. Also, for 20-dB average spacing then, a 22-dB operational amplifier gain is normally used. Therefore, 20 dB of cable plus 2 dB of accumulated flat loss could be allowed or padded for if not allowed. Now, one could calculate the cable loss at the low frequency end channel 2 at 54-MHz video carrier frequency and the upper frequency limit of 450 MHz. However, because the variation with temperature is needed, we shall proceed with a shortcut. If the 20 dB of spacing is at channel 62, then the loss at the low end will vary by

$$A_{lo} = A_{hi} \sqrt{\frac{f_{lo}}{f_{hi}}}$$

Substituting

$$A_{lo} = 20 \text{ dB} \sqrt{\frac{54}{450}} = 6.93 \text{ dB}$$

If the temperature gets very high and increases $+50°F$,

$$68°F + 50°F = 118°F$$

For

$$\frac{1\%}{10°F} \times +50°F = 5\% \text{ change in attenuation}$$

For channel 2 at 54 MHz:

$$6.93 \text{ dB} + (0.05 \times 6.93) = 7.28 \text{ dB}$$

For channel 62 at 450 MHz:

$$20 \text{ dB} + (0.05 \times 20) = 21 \text{ dB}$$

Suppose that the temperature drops from 68°F to $+18°F$, i.e., $-50°F$. Now the normal average 20-dB spacing changes to

$$\text{channel 2: } 6.93 \text{ dB} - (0.05 \times 6.93 \text{ dB}) = 6.58 \text{ dB}$$

$$\text{channel 62: } 20 \text{ dB} - (0.05 \times 20 \text{ dB}) = 19 \text{ dB}$$

In summary, a $-50°F$ to a $+50°F$ temperature change constitutes a total change of 100°F. The change at channel W is 2 dB for a single-trunk span of cable and at channel 2 of 0.7 dB. Now if this can happen for a single amplifier cable span of the unity gain building block, the question becomes, What happens to a 20-amplifier cascade? The answer is

$$\text{channel 2: } 20 \times 2 \text{ dB} = 40 \text{ dB}$$

$$\text{channel W: } 20 \times 0.7 \text{ dB} = 14 \text{ dB}$$

By now it should be plain that help is needed. The accumulated effect from the cascade of amplifiers can make the picture quality completely unacceptable. When temperature varies 100°F, the total signal level variation at channel W is 40 dB and 14 dB for channel 2. This can cause disastrous results in picture quality. Unfortunately, the actual change in cable attenuation as a function of temperature is not a linear function of frequency because cable attenuation is not a linear function of frequency. In short, the cable slope or tilt changes with temperature. To illustrate, consider that for a 20-dB, 450-MHz cable span at 118°F loss at channel 2 (54 MHz) is 7.28 dB and at 450 MHz is 21 dB, so 21 − 7.28 = 13.72. Repeating the procedure at +18°F gives 19 − 6.58 = 12.42, a 1.3-dB difference in slope. To say it another way, a 20-dB, 450-MHz span of cable at 68°F has 6.93-dB loss at channel 2 and if the temperature increases 50°F to 118°F, the cable loss increases 5 percent at all frequencies. Now 5 percent for 20-dB spacing gives a 1-dB change and at channel 2 a change of 0.35 dB, giving us a 1.35-dB change in slope over 68°F. Consider the diagram in Fig. 6.22.

In actuality these lines are not straight because the cable loss as a function of frequency is not linear. If the amplifier cable span is compensated for, the nonlinear loss characteristics of the cable, this is called *true tilt* instead of slope control. Some manufacturers approximate the true tilt in the automatic correction process.

6.10.2 Automatic gain and slope correction

To control level and slope accurately to within acceptable limits two video carriers are chosen as pilot carriers. One pilot carrier is always chosen in the lower band edge, usually channel 2, 3, or 4, and is called

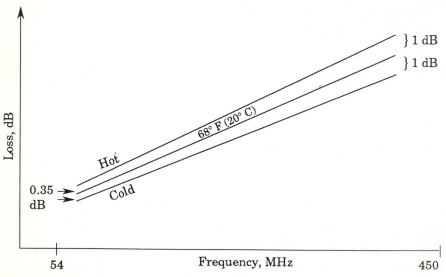

Figure 6.22 Slope change as temperature varies.

the *low pilot*. The high-pilot carrier is usually in the middle to upper limit of the band edge, usually channel 11 for 12- and 21-channel systems, channel N (27) for 35-channel systems (300 MHz), and channel 35 or 62 for 450-MHz systems. The pilot carriers act as the level reference points over the system bandwidth. The automatic gain and slope modules in the trunk amplifiers select these carriers and measure the levels. The level voltage from the pilot carrier detectors controls the gain of the trunk amplifiers for the low and high ends of the system bandwidth, thus controlling system gain and slope. This circuitry is also temperature-compensated so any errors in control will be extremely small. Controlling the gain at the low and high ends of the signal band compensates for level changes due to cable temperature changes, amplifier level changes, and noise that is not constant over the band. Caution should be observed when setting head-end system signal levels so the pilot carriers are precisely level-maintained. These pilot carriers should be set to a level that agrees with the proper head-end tilt. However, if the head end is adjusted flat, then all carriers will be at the same level. One way to set the head end is to place a technician at the first amplifier with a signal level meter connected to the input test point. The head-end technician can then set all the system carriers one at a time as instructed by two-way radio by the distribution system technician. Another way is to calculate the cable system path loss from head end to the first amplifier for each carrier frequency. Then the level desired for each carrier can be calculated and either plotted on a graph or simply listed. Then the head end can be set up according to the calculated levels and verified at the first amplifier with a signal level meter or a spectrum analyzer.

If the amplifier manufacturer requires the pilot carrier system to be field-tuned, then a bench test setup should be used according to the amplifier instruction manual. Precise pilot carrier tuning should be maintained by the automatic control modules in the trunk amplifier stations.

If the pilot carrier tuning is not maintained precisely, then the control module may instead be using one of the adjacent channels as a pilot carrier. One simple way to test for this problem is to take signal level readings at the various trunk amplifiers on the ends of the systems for band edge signals while turning off and on the signal carriers adjacent to the pilot carriers. For instance, for the low band edge where the pilot carrier is channel 4, if channel 6 carrier level is monitored at the system ends, then channel 5 is turned off, then on; if channel 6 level remains unchanged, then it means the pilot on channel 6 is not being interfered with by channel 5. Next, channel 3 will be turned off, then on, and if channel 6 remains unchanged, then channel 3 does not interfere with channel 4. Now if channel 4 is turned off, then channel 6 should increase in level. When channel 4 is turned back on, then channel 6 should return to its original level. The time it takes to return to its proper level is referred to as the *response time*.

This procedure should be repeated for the high pilot point. If at any system end a significant variation in system levels results from turning on and off the pilot carrier adjacent signals or the pilot carriers themselves, then a problem with a control module in one or more of the amplifier stations on that system branch has occurred. The proper way to locate the faulty module is to work back along the system branch having the adjacent signals to the pilots turned off and on, observing the amplifier output levels. When an amplifier station where no variation occurs is reached, then the problem is at the previous tested amplifier.

In many systems, the use of automatic gain and slope control modules at every trunk amplifier station is preferred. Then only basic thermal compensation is used for the line extender amplifiers with only two allowed in cascade. At one time, it was recommended that automatic control should be placed in every other amplifier station with the intermediate ones just having thermal compensation. These systems used automatic control in the second of a three-line extender cascade. Such decisions often are made by the manufacturer on the basis of temperature chamber tests of amplifier cascades.

6.11   End Level Tests and Subscriber Signal Level

Once a cable system has been aligned, balanced, leakage-tested, and sweep-tested, tests of the signal level at the ends of the system should be made. The last tap at the feeder system ends is the *system end*. Here the signal level on all channels should be made and logged at these end points. This test gives some assurance that the other taps in the system have proper signal level; it should be completed before subscribers are connected to the system. The installation crews should have few surprises when connecting subscribers to the system. End level testing as an ongoing procedure is discussed in detail in Chapter 7.

Subscribers should have a minimum signal level on the weakest channel of 0 dBmV at the back of the converter or set. Usually converters have a signal gain of 3 to 5 dB so the television receiver has a minimum of 3 to 5 dBmV on the weakest channel. Subscriber drop cable, for example, RG-6 foam type, has a loss at channel 2 of 1.5 dB/100 ft and at channel 62 (450 MHz) of approximately 4.5 dB/100 ft. For a drop length of 250 ft the example in Fig. 6.23 shows the results at a subscriber's set.

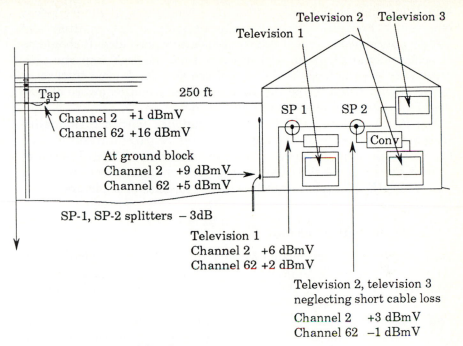

Television 2   Television 3

Television 1

Tap   250 ft

Channel 2   +1 dBmV
Channel 62 +16 dBmV

At ground block
Channel 2   +9 dBmV
Channel 62 +5 dBmV

SP-1, SP-2 splitters − 3dB

SP 1   SP 2

Conv

Television 1
Channel 2   +6 dBmV
Channel 62 +2 dBmV

Television 2, television 3
neglecting short cable loss

Channel 2   +3 dBmV
Channel 62 −1 dBmV

Figure 6.23   Signal level drop losses.

It should be evident that with a tap signal level as shown, the drop length is extremely important and in this example of a three-outlet installation, the end set signal level is on the edge. Longer drops should be serviced by using larger drop cable, such as RG-11. The example can be worked out in one's head if certain rules are followed. Recall that the RG-6 foam drop cable has a channel 2 loss of 1.5 dB/ 100 ft and channel 62 of 4.5 dB/100 ft, which gives a downward tilt of 3 dB/100 ft of signal level. For 250 ft the tilt accumulates to 7.5 dB (3 + 3 + 1.5 = 7.5 dB). Now the tap level has an upward signal level tilt of 3 dB, so the signal level at the end of 250 ft will be sloped down only 4.5 dB (7.5 dB − 3 dB = 4.5 dB). Therefore, the signal levels of channel 2 and channel 62 are only 4.5 dB apart, and if the signal level is calculated for the end television set at either channel 2 or channel 62, the remaining signal level is easily calculated.

Example   Loss at channel 62 is (4.5 + 4.5 + 2.25 = 11.25 ≈ 11 dB). Signal level is + 16 dBmV − 11 dBmV = 5 dBmV. Level at channel 2 is 4 dBmV + 4.5 dBmV = 8.5 dBmV ≈ 9 dBmV. Signal splitters often have flat loss of about 3.5 dB. However, since the usual cable system has thousands of signal splitters in use, a good idea is to sweep-test samples to analyze flatness and loss as a function of frequency before purchasing. From time to time random testing of drop splitters and directional couplers for new shipments is also a good idea. It should be remembered that these subscriber levels are immensely important because they are the basis of the rest of the system. A good chief technician should listen to the installers' problems and be aware of how they find the signal levels throughout the system.

The HDTV receiver will have to be supplied with a high-quality strong cable signal to assure a good C/N figure. HDTV receivers will have a width-to-height (aspect) ratio on the order of 16:9. Therefore, picture snow will be more like falling popcorn. Subscribers who own wide-screen HDTV sets certainly will not tolerate poor pictures.

6.12   System Signal Leakage

Signal belonging to the cable system can leak to the surrounding space from loose or improperly installed connectors, devices with improperly installed covers and gaskets, as well as cracked or broken cable. This signal propagates through space just like a group of television transmitting stations. The intensity of the leak depends on the looseness of the connectors and/or device covers or the size of the hole in the cable or the magnitude of the signal at the break point. Since such leaks propagate through space, the leakage measurement is in the same units as radio transmission, microvolts per meter. Since cable people operate in the decibel-millivolt environment and measure signal level with a signal level meter, a short discussion of signal transmission through space will follow to show the relationship between the measure quantities.

6.12.1   Field strength theory of wave propagation

Consider a transmitter of radio frequency energy P_{tr} driving a transmitting antenna. This energy emanates from a point source, as shown in Fig. 6.24.

$$\text{At receiver site, received power } P_r = \frac{P_{tr}}{4\pi r^2} \text{ W/m}^2$$

$$\text{Also } P_r = \frac{E_r \text{ (rms V/m)}}{120\pi} \text{ W/m}^2$$

where $E_r = P_r 120\pi$ V/m
    120π = impedance of free space (Ω)

$$\text{Area of receiving antenna } A_r = \frac{G\lambda^2}{4\pi} \text{ (m}^2)$$

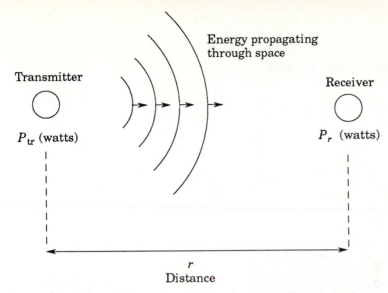

Figure 6.24 Radiated energy through space from transmitter to receiver.

Since at the receiving site we have energy of P_r W/m^2, with an antenna A_r m^2 the power in the antenna can be calculated as follows.

$$\text{Combining and substituting } \frac{V_{\text{rec}}^2}{Z_o} = \frac{E_r^2 G \lambda^2}{480 \pi^2}$$

$$\text{Recall } \lambda = \frac{300 \text{ m/s}}{f_{\text{MHz}}}$$

Substituting in the equation and solving for E_r

$$\frac{E_r^2 G \lambda^2}{480 \pi^2} = \frac{V_r^2}{Z_o} \qquad E_r^2 G \lambda^2 = \frac{480 \pi^2 V_r^2}{Z_o}$$

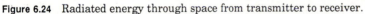

$$E_r^2 = \frac{480 \pi^2 V_r^2}{G \lambda^2 Z_o} = \frac{480 \pi^2 V_r^2 f_{\text{MHz}}^2}{G \, 300^2 Z_o}$$

$$E_r = \sqrt{\frac{480 \pi^2 V_r^2 f_{\text{MHz}}^2}{300^2 G Z_o}} = \frac{\pi V_r f_{\text{MHz}}}{300} \sqrt{\frac{480}{G Z_o}} \times \frac{1/4}{1/4}$$

$$= \frac{V_r \pi f_{\text{MHz}}}{75} \sqrt{\frac{30}{G Z_o}}$$

Let $Z_o = 75 \ \Omega$ and $G = 1.64$, which is antenna gain of a dipole.

$$E_r \ \mu\text{V/m} = V_{r \ \mu\text{V}} \times 0.0419 \times \sqrt{\frac{30}{1.64 \times 75}} \times f_{\text{MHz}}$$

$$= 0.0207 V_{r \ \mu\text{V}} f_{\text{MHz}}$$

The number 0.0207 is rounded to 0.021, which appears in the usual formula.

6.12.2   Cable system radiation
measurement

The Federal Communications Commission (FCC) has specific rules about cable system signal leakage levels and the records CATV companies are required to keep on file. In brief, signal leakage control is addressed in the following sections of Part 76, Subpart K.

76.601 (a) Cable operators have to make sure their systems comply with the specifications and be prepared to show that they do so.

76.601 (b) A list of current cable television channels has to be kept at its local office.

76.601 (c) System performance tests have to be conducted yearly at intervals not to exceed 14 months. This data should be on file at the local office, and these records must be kept on file for 5 years. These tests must be performed at three widely separated points on the system, one of which has to be at the longest cable distance from the head end. A description of instruments and the measurement procedure and the qualifications of the technician performing the test has to be in the records. All records have to be on file and ready for FCC inspection.

76.601 (d) If the above tests are successful still it does not in any way relieve the system operator from the technical standards at all subscriber terminals.

76.605 (a) (12) Radiation is the only measurement not required to be measured at the subscriber's terminal and is measured in accordance with procedures in Section 76.609(h).

Frequencies	Radiation limit, μV/m	Distance, ft
Up to and including 54 MHz	15	100
Over 54, up to and including 216 MHz	20	10
Over 216 MHz	15	100

76.609 (a) Measurements are made to demonstrate conformity with the performance requirements in sections 76.701 and 76.605 and shall be made with the system operating normally.

76.609 (h) Measurements to determine field strength of radiation from a cable system shall be made using standard engineering practices. Measurements made on frequencies above 25 MHz shall include the following:

1.   A field strength meter of adequate accuracy using a horizontal dipole antenna shall be employed.

2.   Field strength shall be expressed in rms value of synchronizing peak for each television channel for which radiation can be measured.

3.   The dipole antenna shall be placed 10 feet above ground and positioned directly below the cable system. The dipole antenna should be positioned no closer to the system than 10 feet.

A dipole antenna is equal to one-half wavelength in distance for a given frequency. However, the actual measurement is usually something different than the calculated measurement of wavelength. Manufacturers of commercially available dipole antennas usually provide a chart for the dipole length in inches for each of the television picture carrier frequencies and any correction factors desired. Consider the diagram of a half-wave (dipole) antenna in Fig. 6.25. Therefore, if the meter reads -36 dBmV, then the system does not pass the test.

Some manufacturers provide the voltage level that the dipole presents to the meter in decibels above 1 μV (dBµ).

Example   At channel 6 the dipole should have at the FCC limit 21 dBu (microvolt). To convert one must recall that this level is referenced to

$$1 \; \mu V = \frac{1}{1000} \, mV \quad \text{or} \quad 1000 \; \mu V = 1 \, mV$$

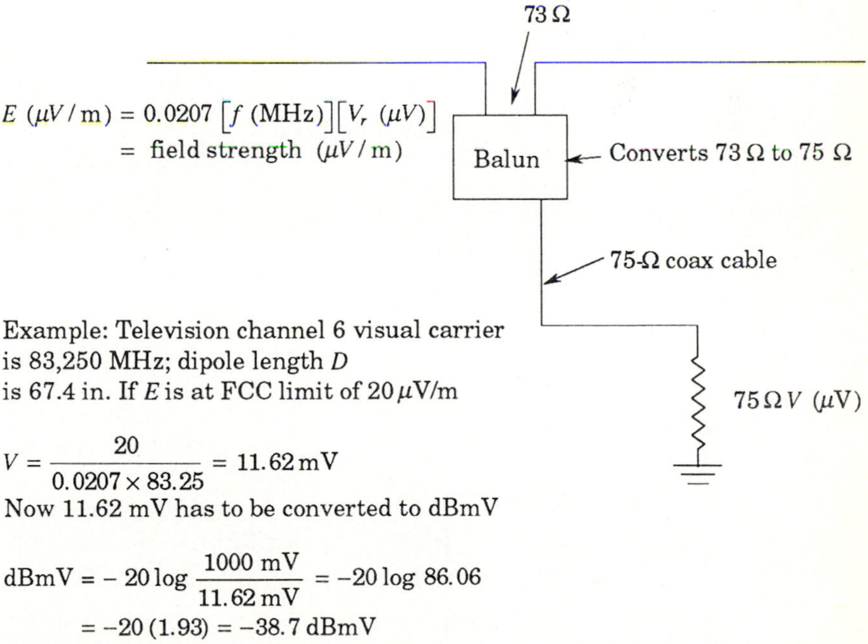

$73 \, \Omega$

$E \; (\mu V / m) = 0.0207 \left[f \; (MHz) \right] \left[V_r \; (\mu V) \right]$
$\qquad = \text{field strength} \; (\mu V / m)$

Balun $\quad\leftarrow$ Converts $73 \, \Omega$ to $75 \, \Omega$

$75\text{-}\Omega$ coax cable

Example: Television channel 6 visual carrier is 83,250 MHz; dipole length D is 67.4 in. If E is at FCC limit of $20 \, \mu V/m$

$75 \, \Omega \, V \; (\mu V)$

$V = \dfrac{20}{0.0207 \times 83.25} = 11.62 \, mV$

Now 11.62 mV has to be converted to dBmV

$dBmV = -20 \log \dfrac{1000 \; mV}{11.62 \; mV} = -20 \log 86.06$
$\qquad\quad = -20 \, (1.93) = -38.7 \, dBmV$

Figure 6.25   Dipole antenna at channel 6.

$$\text{dBu} = 20 \log \frac{V}{1\ \mu\text{V}}$$

and

$$21 = 20 \log \frac{V}{1\ \mu\text{V}}$$

$$\frac{21}{20} = \log \frac{V}{1\ \mu\text{V}} \qquad 1.05 = \log \frac{V}{1\ \mu\text{V}} \qquad V = 11.22\ \mu\text{V}$$

This is similar to 11.62 μV, as before. Now 11.22 μV converted to decibel-millivolts is

$$\text{dBmV} = -20 \log \frac{1000\ \mu\text{V}}{11.22\ \mu\text{V}} = -20 \log 89.13$$

$$= 20 \times 1.95 = -39\ \text{dBmV}$$

which is close to what we had before.

When using commercially made dipoles, the manufacturer's correction charts should be consulted. The correct placement of a test dipole is given in the sketch in Fig. 6.26.

The cumulative leak index method is used to relate the cumulative signal leak effects to detectable interference with aircraft communications and navigational systems in the air space. Cable television system leakage causing interference with critical aircraft frequencies

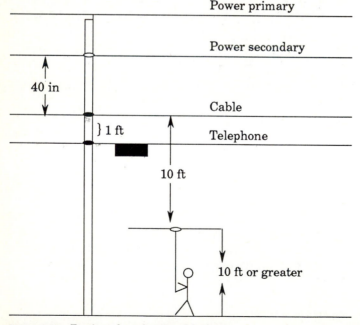

Figure 6.26 Testing plant for signal leakage.

could cause a serious accident. The Federal Aviation Administration (FAA) is insisting that the FCC enforce the prohibition of cable leakage and the FCC is complying. July 1990 is the deadline by which all cable television systems must comply with the FCC requirements. Failure of a cable system to comply could cause loss of television channels and/or fines.

Of course, one method would be to mount an antenna on an airplane and with a calibrated receiver and a signal level meter fly over a cable system and record the position of detected leakage. Mounting anything such as an antenna on an airplane will require a certificate of air-worthiness by an FAA license person. Some companies are offering such fly-over services to cable operators using fixed wing airplanes or helicopters. However, it must be remembered that if a leak is detected in the air, it is caused on the ground and a follow-up ground search will have to be made to locate and repair the cause of the leak.

6.12.3 System leakage test procedure and cumulative leak index

Testing for system leakage will be conducted on the complete system three times a year. Seventy-five percent of the complete system will be measured on a quarterly basis. The cumulative leak index (CLI) method will be used to provide a figure of merit for system leakage, as given by the following formula:

$$\text{Cumulative leak index} = 10 \log \left[\frac{\text{total plant miles}}{\text{test plant mileage}} \times \text{sum of each leak}^2 \right]$$

The instrument used to locate leaks may be the Comsonics Corporation model Sniffer* or other equivalent instrument operating at a frequency of, for example, 116.625 MHz. First the transmitter will be activated at the head end at normal video carrier level. The receiver will be vehicle-mounted in the truck cab and activated by either its internal batteries or the truck 12-V_{dc} power using the cigarette lighter adapter cord. The magnetic mounted whip antenna on the truck cab roof will be connected to the receiver using the calibrated coaxial cable. When a leak is located, it will be measured by using the Wavetek[†] RD_1 dipole antenna or equivalent and 16-dB amplifier connected to a Sam III[†] signal level meter or equivalent. The dipole antenna will be extended and tuned to 116.625 MHz. The length of each rod will be found by the following formula:

*Trademark Comsonics Company.
†Trademark Wavetek Company.

$$1 \text{ Wavelength (ft)} = \frac{984}{f_{\text{MHz}}} \qquad \text{Dipole} = \frac{1 \text{ wavelength}}{2}$$

Therefore, each rod will be one-quarter of a wavelength.

$$\text{Length of each rod (ft)} = \frac{984}{4 \times f_{\text{MHz}}} = \frac{246}{116.625} = 2.11 \text{ ft}$$

Therefore, 2.11 ft × 12 in/ft = 25.32 in ≈ 25⅓ in.

Once the dipole has been adjusted it will be connected through the 16-dB gain amplifier using the short jumper cable and then connected to the signal level meter using a suitable piece of drop cable.

The RD_1 amplifier and signal level meter will be turned on and the meter tuned to a frequency of 116.625 MHz. The distinguishing chirp of the Sniffer signal should be heard when the meter is properly peaked on the signal and the audio volume level set at a comfortable level. The signal level will now be measured and recorded on the report form in decibel-millivolts.

To translate this value to microvolts/meter (μV/m) the following formula is used:

$$E \ (\mu V/m) = 0.021 \times f \ (\text{MHz}) \times V(\mu V)$$

$$\text{for } E = 20 \ \mu V/m = 0.021 \times 116.625 \times V(\mu V)$$

$$V(\mu V) = \frac{20}{0.021 \times 116.625} = 8.17 \ \mu V$$

Therefore, 8.17 μV corresponds to −42 dBmV. Taking into account 16 dB of gain this signal will be read by the signal level meter as −26 dBmV:

$$-42 \text{ dBmV}$$

$$\underline{16 \text{ dB}}$$

$$-26 \text{ dBmV}$$

Using the formula and a chart of conversion of decibel-millivolts to microvolts a table such as Table 6.1 can be made.

6.13   Video Measurements and Instruments

The video signal waveform for the NTSC system is a complicated waveform consisting of many components. Such components are the synchronizing pulses and the luminance and chrominance complex waveforms. This video waveform is the modulating signal that modu-

TABLE 6.1   Decibel-Millivolt Equivalents to Microvolts and Microvolt Meter

dBmV (SLM + 16 dB)	μV	μV/m
−30	5.01	12.3
−28	6.31	15.5
−26	8.17	20.0
−24	10.00	24.5
−22	12.59	30.8
−20	15.85	38.8
−15	28.18	69
−10	50.12	122.8
−5	89.16	218.4
0	158.5	388.3

lates an RF carrier that is broadcast over the air through either a cable system or a satellite or microwave transmission link. Preservation of this waveform is extremely important so the receiving television sets can provide good pictures.

The NTSC video waveform can be measured on an oscilloscope or what is known as a *waveform monitor*, which is a special form of the more general-purpose oscilloscope. The basic oscilloscope will be discussed first so that the waveform monitor will be more easily understood. The vector scope is also a special case of the oscilloscope. Both the vector scope and the waveform monitor are used extensively in video waveform measurements.

6.13.1   Basic oscilloscope

The basic element of the oscilloscope is the cathode-ray tube (CRT) display. This is actually an electron tube but is not of the amplifier type. A diagram of the basic CRT is shown in Fig. 6.27.

In Fig. 6.27 when current flows through the filament of the CRT the generated heat causes a lot of electron activity around the cathode area. The high negative charge on the cathode repels the electrons from the cathode toward the more positively charged grid. The more positively charged focus anode speeds up as well as forces the electrons into a beam that is accelerated by the highly positively charged accelerating anode. If there is no charge on the deflection plates, the electron beam will strike the fluorescent screen, causing a small bright spot of light. Many fluorescent screens are green because of the chemical content of the screen material. The Aquadag coating is a conductive coating painted on the tapered interior of the tube surface and is connected to the accelerating anodes for a positive charge. This

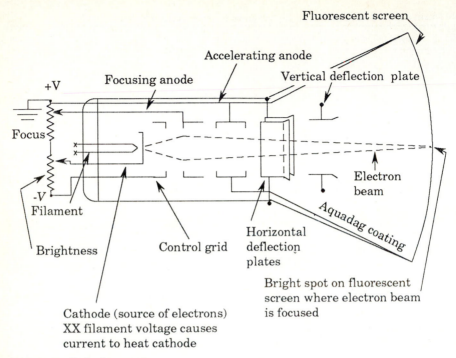

Figure 6.27   Cathode-ray tube.

coating, being positively charged, attracts any negative electrons that bounce off the screen, thus improving the focusing and deflection action. The charge on the grid controls the brightness of the spot, and the charge on the focusing anode controls the size of the spot. To cause the spot to move off the center of the screen, the charge on the deflection electrodes has to change. There are two pairs of *deflection plates*, as they are called. One pair, the *vertical deflection plates*, causes the bright spot to move in an up-down (vertical) direction. The other pair is the horizontal *deflection plate*, which causes sideways (horizontal) movement of the bright spot. For the standard oscilloscope instrument a sawtooth voltage waveform is applied to the horizontal deflection plates, causing the beam to move from the left to right. An end view of the screen of the CRT showing the position and charge on the deflection plates is given in Fig. 6.28.

At $T = 0$ the left plate suddenly gets a positive charge, attracting the electron beam, while the right-hand plate gets a large negative charge, repelling the beam toward the left. Thus the spot is nearly instantaneously moved to the left-hand side of the screen. Then the left plate becomes less positive and the right plate becomes more positive, moving the beam toward the right. At time $T/2$ both plates have a 0-V

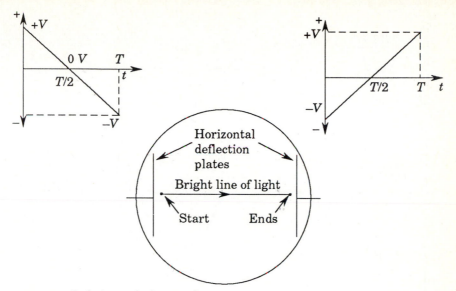

Figure 6.28   End view cathode-ray tube.

charge. Hence the beam is at the center of the screen. The short time
it takes to repeat this procedure when the beam has to be suddenly
moved from the rightmost position to the leftmost position is the *retrace period*. During this time a pulse of negative charge is applied to
the grid to shut off the beam so the retrace period will not be viewed.
The charges appearing on the horizontal deflection plates are referred
to as the *sweep voltages*; they cause the spot to be viewed as sweeping
smoothly from left to right, i.e., sweeping across the screen. The
waveform for the horizontal deflection plates is often referred to as the
push-pull sawtooth sweep voltage. The display now appears as a green
line drawn across the center of the screen. This line can be shifted up
and down by placing fixed voltages on the vertical deflection plate.
The oscilloscope instrument has a vertical and horizontal centering
control, which regulates the fixed dc bias voltages on the deflection
plates.

Consider now, what would the screen display if we applied a
sinusoidal voltage to the vertical plates with our regular sweep voltage on the horizontal plates? The answer is simple: we would see a
picture of the sine wave, that is, if the frequency of the sweep has the
same period as the sine wave. This is illustrated in Fig. 6.29.

The input sine wave to be measured is split into a push-pull voltage
that is applied to the vertical deflection plates, causing the electron
beam to trace out the sine wave form. By now, with the horizontal signal a linear sweep signal, the screen will trace out a picture of the ver-

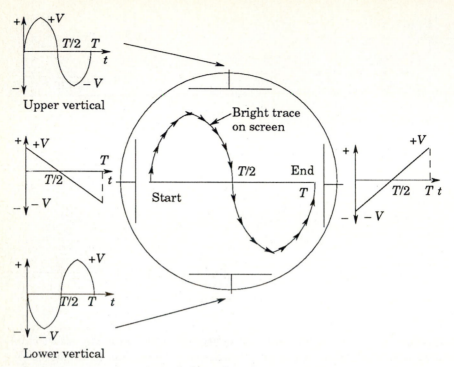

Upper vertical

Lower vertical

Figure 6.29   Sine wave on vertical plates gives sine wave on screen.

tical signal. By speeding up the horizontal sweep signal the vertical signal will be stretched out. If the horizontal sweep is slowed down then more cycles of the vertical signal will be displayed. Most modern oscilloscopes use solid-state devices to provide an accurate calibrated sweep signal as well as condition the input (vertical) signal to the proper level needed to produce the desired deflection. This type of CRT display is called the *electrostatic* type since electrostatic fields cause the beam movement. Typical voltage amplitudes for the deflection circuits are in the range of several thousand volts. Ten thousand or more volts often is used for the tube-accelerating anode.

The larger the CRT becomes the higher the voltage needed for proper operation. For larger screen sizes magnetic deflection is used instead of electrostatic deflection. Tubes using magnetic deflection techniques are limited to lower sweep speeds and hence lower-frequency uses. High-speed oscilloscopes for pulse and digital work use CRTs of the electrostatic type and have flat rectangular screens. Over the tube screen is usually placed a plastic sheet graticule with crosshatch markings. Often these markings are 1-cm squares. Therefore, vertical deflection is often referred to in volts per centimeter and horizontal deflection in seconds, milliseconds, or microseconds per centimeter. The vertical input is usually connected to the circuit to be measured by a test probe. This probe essentially isolates the oscilloscope instrument from the circuit to be measured, thus preventing errors caused by upsetting the circuit. The probe feeds the signal to be measured to the high-impedance input of the vertical amplifier. The vertical amplifier gain is controlled by a precision attenuator, usually calibrated in volts and millivolts. The usual vertical amplifier input impedance is on the order of 1 MΩ and the probe often has a direct mode or 10:1 mode changed by a selector switch on the probe. In the 10:1 mode the input impedance to the vertical amplifier is increased by 10 MΩ. Many scopes have a dual scale on the vertical amplifier attenuator dial, giving the direct reading in volts/centimeter and the 10:1 reading in volts/centimeter.

To enhance the accuracy of the instrument further, a precision square wave signal of known amplitude, usually 1 V peak to peak, is generated by a built-in calibration circuit. This circuit provides the calibrated signal to a front panel test jack. The probe tip can be touched to this jack to feed the calibration square wave to the scope. Now the vertical input amplifier attenuator can be set to the 1-V/cm range and a square wave of 1 V peak to peak should appear on the screen. When the attenuator is changed to 0.5 V/cm, this signal should appear over 2 cm of deflection. This is illustrated in Fig. 6.30.

So far the oscilloscope we have discussed is still fairly simple and more "bells and whistles" are needed. One important improvement is signal synchronization for the sweep circuits. If we use the scope as it is, the input signals will slide across the screen and more detailed waveform analysis will be impossible. The problem at hand is to control the start of the horizontal sweep at the same time as the start of the input waveform. This technique is performed by an internal triggering circuit that starts the sweep on a positive or negative slope of the input signal voltage. The positive or negative choice is usually switch-selectable on the scope's front panel. Often a trigger level control is provided with the center position as zero and to the right as positive and to the left as negative, as shown in Fig. 6.31.

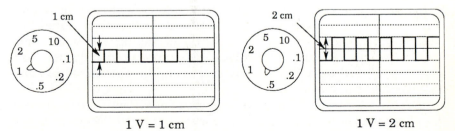

1 V = 1 cm   1 V = 2 cm

Figure 6.30 Oscilloscope vertical axis.

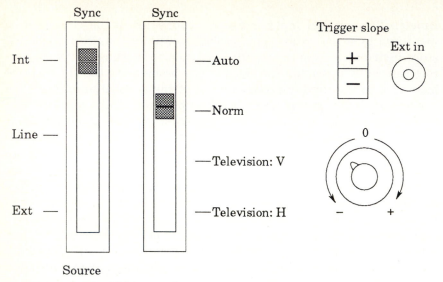

Figure 6.31　Synchronizing controls.

Stable displays of waveform information rely on the quality of the trigger circuits used to start the sweep. If the trigger circuits contain inherent noise, then the point of triggering will vary, causing the displayed waveform to jitter. Accurate measurements of the waveform are difficult if jitter is present, presenting a headache to the operator. The trigger circuits have to be able to trigger at a frequency about twice that of the displayed waveform. Some scopes have special features that make video waveform measurements much easier. One such feature is a special control that sets, e.g., the sweep rate (horizontal time per division) to the vertical frame rate and adjusts the trigger to start at the end of the vertical blanking signal. Two fields of video information are then displayed. This same feature can be set for the horizontal synchronizing rate to give a line of video information.

Another useful feature is the delayed sweep. The input signal is fed to the trigger circuits to start the sweep and the input signal from the vertical amplifier is fed to a delaying amplifier that delays the input signal until the sweep is started. This allows more of the input waveform to be displayed and hence is useful for transient signals.

Selecting an oscilloscope is as difficult as buying a new car and sometimes more difficult. A good idea is to make a list of desired features and special uses. Then many scopes can be eliminated to simplify the selection process.

Since the CRT is one of the principal elements of any oscilloscope, it should be discussed further. So far we have discussed the single CRT. Some of the more sophisticated CRTs have more than one beam, which are like two CRTs in one glass envelope sharing a common screen. Instruments using a dual-beam CRT have two input vertical amplifiers and sometimes two time bases. But more often they have a common time base and triggering controls that synchronize the two horizontal traces, making comparison of two waveforms easy. At the expense of speed, a single sweep can be chopped, i.e., sampled, by a square wave, producing a dual trace from a single electron gun tube. Consider the diagram that illustrates the arrangement, Fig. 6.32.

Essentially dual-input vertical amplifiers are switched to their respective trace positions. For example, input amplifier A is connected to the time point where the beam is making trace A. The so-called sampling theorem still applies. The input waveform and sweep time require that the chopper speed be at least 5 times the input waveform; 10 times is better to produce decent waveforms on the dual traces. Therefore, dual-beam and dual-trace instruments are different. Of course, the dual-beam tube is more complicated and hence more expensive but has advantages.

Another important feature is the chemical content of the phosphor on the CRT screen, which contributes to the persistence of the trace. If the input signal to be analyzed is transient, possibly a longer persistance of the trace will be desirable. For ordinary general-purpose work at P31 phosphor provides a good practical display. On-screen alphanumerics (letters and numbers) corresponding to scale parameters and instrument control settings, although useful, do not enhance the basic accuracy of the instrument but definitely add to the cost. Some oscilloscope instruments offer digital memories that essentially digitize the incoming waveform and store the digital data in solid-state memory. The CRT then can display the contents of the memory. Such instruments can subtract, add, multiply, and divide input waveform data and memorized data and display the results. The

Blanked out

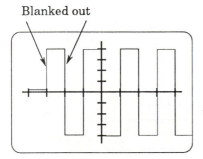

Slow chop   Fast chop

Figure 6.32 Difference in chop rate.

input data can update the memorized data by changing control settings on the front panel. Screen prompting alphanumerics can aid in the operational setup of such instruments.

By now it should be evident that the oscilloscope instrument is extremely valuable in signal waveform analysis. The display on the CRT screen is actually an amplitude versus time plot of the input waveform. Amplitude versus time information can aid in measuring peak values, root mean square (rms) values, and phase differences between two signals on dual-beam or dual-trace instruments. Time and hence period measurement can allow frequency to be easily calculated. Basic amplitude accuracy depends on the precision of the input amplifier gain and attenuator, but more on the screen graticule scale and operator skill. Consider an oscilloscope screen 8 cm high and 10 cm wide with five divisions for each centimeter, as shown in Fig. 6.33.

For a vertical scale of 1 V/cm, each of the small marks is 0.2 cm, corresponding to 0.2 V. Therefore, 0.2 V is 20 percent of 1 V. So if the input signal is 6 V peak to peak it will be spread over ±3 cm; the 0.2 cm corresponds to 10 percent error. If one can estimate to 0.1 cm or if more input can be displayed, more accuracy can be obtained. If the input signal uses the whole screen height of 8 cm exactly, then each 1 centimeter will correspond to 12½ percent of full scale and each 0.2 cm to 1¼ percent. This accuracy is in the same ballpark as a high-quality moving coil voltmeter.

The horizontal scale of one cycle can be displayed over 10 cm; if each cm is divided into 0.2-cm spaces, then there will be fifty 0.2-cm lines in the 10-cm distance. Thus each 0.2-cm mark corresponds to 0.5, or ½ percent, horizontal accuracy. To be able to analyze signals to the basic graticule accuracy, the trace focus should be as fine as possible with reasonable brightness. Traces that are too bright tend to become thick, making waveform analysis more difficult. If more accuracy is demanded, other instruments and techniques may be needed. ·

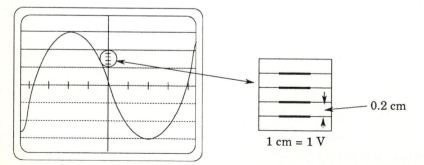

Figure 6.33 Oscilloscope graticule.

6.13.2   Video waveform monitor

The waveform monitor used in television baseband signal measurements is in essence a specialized form of the oscilloscope. All circuits and their corresponding controls are geared to making video waveform measurements. This instrument for use in the United States conforms to the NTSC standard video signal for color television transmission. However, the instrument is available for foreign country use and labeled PAL or SECAM (European television signal standards). When using an instrument designated, for example, NTSC and PAL, make sure the controls are properly set for NTSC signals. When in doubt always refer to the instrument's operations manual. The synchronizing modes are switch-selectable, e.g., vertical at one frame (two fields) or one full field and horizontal at line rate. More elegant waveform monitors have a line selection control that synchronizes the horizontal axis to horizontal lines; for example, 9 through 22 of these are the lines appearing at the very top of the television picture out of sight and usually contain such signals as color bars and special calibrated test signals referred to as *vertical interval test signals* (VITS). Certain lines in the vertical interval contain video text information and closed caption signals for people with hearing problems. The vertical interval is approximately 20 horizontal lines in each field, which amounts to nearly 1.27 ms calculated as follows:

$$1/15{,}735 \text{ Hz} = 63.6 \ \mu s \qquad 63 \ \mu s \times 20 = 1270 \ \mu s = 1.27 \text{ ms}$$

The lines 14 to 21 appear at the very top of the television set screen, and the presence of test signals can be verified by decreasing the television height control. A single line of color bars can be identified easily on the television screen. A waveform monitor with line selection can give close scrutiny to the test signals.

The horizontal axis of the waveform monitor screen is usually expressed as time in milliseconds and microseconds. The vertical scale is in IRE units, which were established by the Institute of Radio Engineers (IRE), a forerunner of the well-known professional society the Institute of Electrical and Electronic Engineers (IEEE). The vertical graticule corresponds to 140 IRE units, which corresponds to 1.0 V peak to peak. Therefore, 100 IRE units is equivalent to 0.707 V peak to peak. The standard video baseband signal specified for camera output level, modulator or transmitter input level, is 1 V peak to peak, or 140 IRE units. Satellite receivers at CATV earth stations provide 1.0-V peak-to-peak output; the output level from receiver descramblers is also 1.0 V peak to peak, as required by the cable modulators. This is the standard level that should appear across 75-Ω output and input impedances. Any gain control adjustments should only be used to overcome any of the minimal losses from cabling and connectors.

The screen graticule appearing on the waveform monitor is marked for the standard test signals used to establish video signal quality. Such signals are usually transmitted by television station broadcasters in the vertical interval during normal programming hours and as a full-field signal often just before or after the programming day. Suppliers of programming to cable television systems via satellite often have test signals on the VITS during normal programming hours and full-screen at off hours. Cable television systems with the more sophisticated line-selectable waveform monitors can measure the test signal during the normal programming hours. Communication between the cable TV system technical people and the program supplier engineers can be helpful in establishing the on-off time of the test signals for full-field tests. Actually a good oscilloscope can also be used as a waveform monitor during field test signal times. Of course, the use of the special graticule on the waveform monitor makes the measurements much easier. This graticule is shown in Fig. 6.34.

The standard test signals appearing on every line (full-field) or on selected lines on the VITS are used to make measurements of video signal distortion. Such distortions, if severe enough, can cause picture

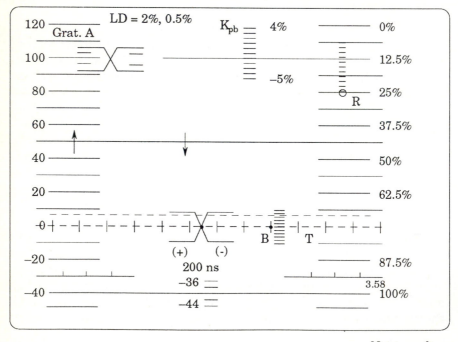

Not to scale

Figure 6.34   Waveform monitor graticule.

impairment. Cable engineers and technicians should take care in keeping the pictures distortion-free, and proper observation of the test signals with a good calibrated waveform monitor is necessary to head off trouble before it starts. Usually the test signals are inserted at the programming source, so when the test signals are measured at a cable television head-end facility, the overall transmission link can be measured.

The waveforms for the three test signals are shown in the Fig. 6.35*a, b, c.* Any departure from the wave shapes of these signals is a form of distortion. These signals were designed to indicate certain forms of distortion and hence the possible source of distortion clearly. Usually a log of such measurements is kept on a per-channel basis for reference in case of any problem. Such log forms can be kept in a ring binder and accompanied by a list of procedures used to make the measurements. A sample log form appears in Form 6.1. It is assumed that a waveform monitor and a vector scope are used to make the tests (Fig. 6.36). A weighting filter may also be needed to make the random noise tests. A short discussion of each section of the video waveform test log with reference to the waveform and the equipment used will follow. It is assumed that the source of the test signals will be full-field (every horizontal line).

6.13.2.1   Test 1: video gain (NTC 7, Sec. 3.2).   The measurement is made to ascertain that the signal is as it should be, at 140 IRE units, and that the BAR is indeed at 100 IRE units. The BAR signal appears on all three signals but is longest on the composite signal (18 μs). The tolerance in the BAR height is ± 3 IRE units. The black level should be exactly on the 0-IRE baseline and the sync tip should be at -40 IRE. In other words, the signal should look like the figures describing the test signals. Figures containing errors for these tests are shown in Fig. 6.37.

6.13.2.2   Test 2: line time distortion (NTC 7, Sec. 3.4).   The measurement is also made on the BAR signal, beginning at the leading edge to the trailing edge. Bar tilt is the indication of a problem with this particular test. BAR tilt indicates a midfrequency response problem.

6.13.2.3   Test 3: short time distortion (NTC 7, Sec. 3.5).   The composite signal is used again. The $2T$ pulse and the bar edge are observed. This test indicates any high-frequency problems. The measurement is made on the difference in amplitude between the height of the $2T$ pulse and the height of the bar center. If the centering control or a sweep fold back control is available on the waveform monitor, the measurement can be made a little simpler. If a centering control is available, the bar height can be set to the center of the screen at 100

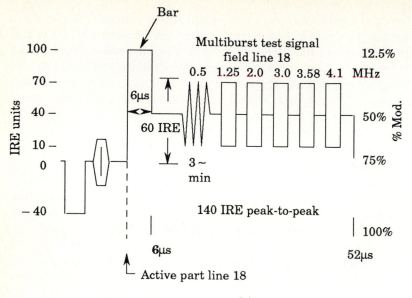

(a)

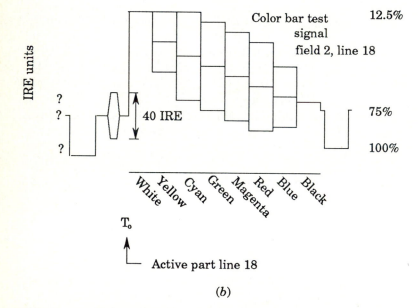

(b)

Figure 6.35   Television video test signals.

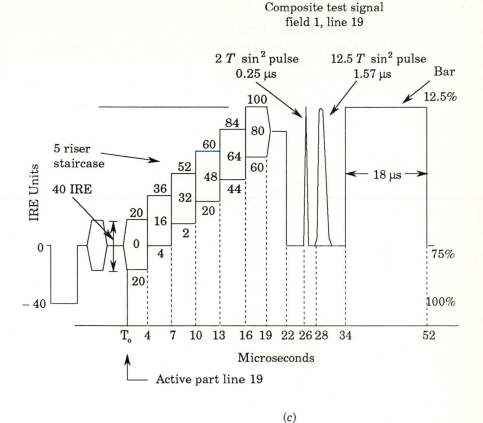

Composite test signal
field 1, line 19

$2\,T\,\sin^2$ pulse 0.25 µs

$12.5\,T\,\sin^2$ pulse 1.57 µs

Bar 12.5%

5 riser staircase

40 IRE

IRE Units

18 µs

(c)

Figure 6.35 *(Continued).*

IRE. Then the pulse can be moved over and the difference in IRE units recorded. The Kell factor (K_{pb}) scale can be used; the difference in amplitude will be given as percentage of K factor of picture impairment. (See waveform monitor screen in Fig. 6.34.)

6.13.2.4 Test 4: field time distortion (NTC 7, Sec. 3.3). Failure of test 4 indicates a problem with low-frequency response. Poor field time distortion causes a picture to vary in shading from top to bottom. Poor power supply filtering (ac ripple present), coupling capacitors that have lost some of their capacitance, can cause the system to fail. This test has to be made on an out-of-service test by removing the program from the channel tested, leaving just a carrier, indicating a gray screen. The waveform monitor is set so that either one or two fields are visible. Any tilt of the whole-field signal indicates a problem.

Video Waveform Test Log

Date_____ Location_____ Company_____ Tech Initials_____

Test no.	Test Parameter	Signal	Limit		Reading	Error	Remarks
1	Insertion gain NTEC sec 3.2	Composite bar height	+3 IRE (100 IRE)				
2	Line time distortion NTC7 sec 3.4	Composite bar edge	±2 IRE				
3	Short time distortion NTC7 sec. 3.5	Composite 2T pulse Bar edge	±6 IRE 10 IRE				
4	Field time distortion NTC7 sec. 3.3	Bar tilt of alternative field	±3 IRE				
5	Chroma-lum. gain NTC7 sec. 3.6	12.5 T chrom pulse	±6 IRE				
6	Chroma-lum delay NTC7 sec. 3.7	12.5 T chrom pulse	±40 ns				
7	Gain-frequency NTC7 sec. 3.8	Multiburst	46 to 54 IRE	0.5 1 2 3 3.58 4.1			
8	Differential gain NTC7 sec. 3.11	Staircase	11 IRE				
9	Differential phase NTC7 sec. 3.14	Staircase	3°				

Equipment	Model no.	Serial no.	Date last cal.

Form 6.1   Video waveform test log.

6.13.2.5 Test 5: chrominance-to-luminance gain (NTC 7, Sec. 3.6).   The $12.5T$ modulated pulse (modulated with 3.58-MHz color subcarrier) of the composite test signal is used. To make this test the pulse height is set for 100 IRE units and the baseline for 0 IRE. Any change in the area along the pulse baseline indicates chrominance-to-luminance gain and delay errors. A nomograph is given in Figure 6.37 to facilitate the measurement. If the errors are too large, differences in color saturation (reds redder, etc.) will cause poor pictures.

6.13.2.6 Test 6: chrominance-to-luminance delay (NTC 7, Sec. 3.7).   Again the $12.5T$ modulated pulse is used to make the test. This is a test of any

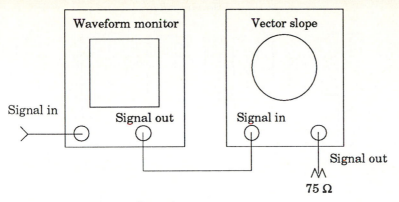

Figure 6.36 Video test configuration.

changes in timing of the chrominance elements to the video waveform. Color smearing or bleeding indicates a problem in this area, often referred to as color ghosting. If television monitor chrominance control is turned off and the ghost goes away, a chrominance-to-luminance delay problem is indicated. This measurement is made along with the previous one; the nomograph gives the delay value in nanoseconds (1 ns $= 10^{-9}$ s).

6.13.2.7 Test 7: gain frequency (NTC 7, Sec. 3.8).   The test is made on the multiburst signal. Basically this signal is a white BAR level followed by a series of six bursts at 60 IRE units high and at frequencies of 0.5, 1.25, 2.0, 3.0, 3.58, and 4.1 MHz. Any changes in amplitude of the different frequency bursts indicate a problem. The maximum deviation in amplitude from the 60-IRE level is the value recorded on the log form.

6.13.2.8 Test 8: differential gain (NTC 7, Sec. 3.11).   The test is conducted on the five-riser staircase signal of the composite test signal. The horizontal controls should be set to spread the staircase signal to about half screen width. The input amplifier filter (3.58 MHz) is switched on. Now, since the staircase signal is (phased to burst signal) 3.58 MHz at increasing brightness levels, the filter passes only the 3.58-MHz signal. Since this signal actually is at various brightness levels any ripple of the area of modulation indicates a problem. For better accuracy the gain of the waveform monitor is increased so the top of the modulation area is at 100 IRE units. Now any departure along the top can be noted. The maximum departure should be 11 IRE.

6.13.2.9 Test 9: differential phase (NTC 7, Sec. 3.14).   The staircase portion of the composite test signal is used to measure differential phase.

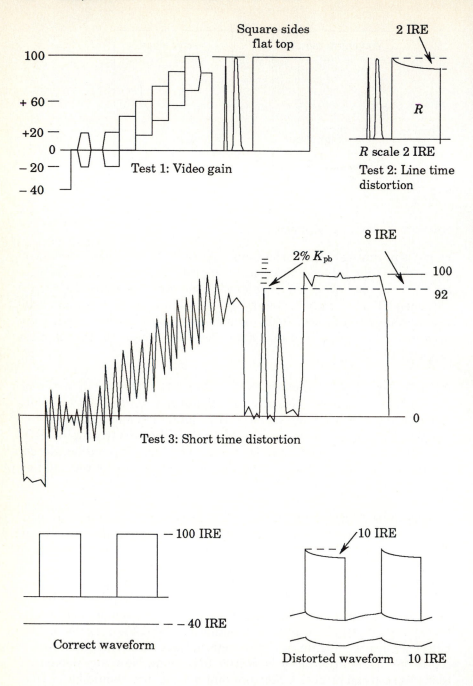

Test 1: Video gain

Test 2: Line time distortion

Test 3: Short time distortion

Test 4: Field time distortion

Figure 6.37

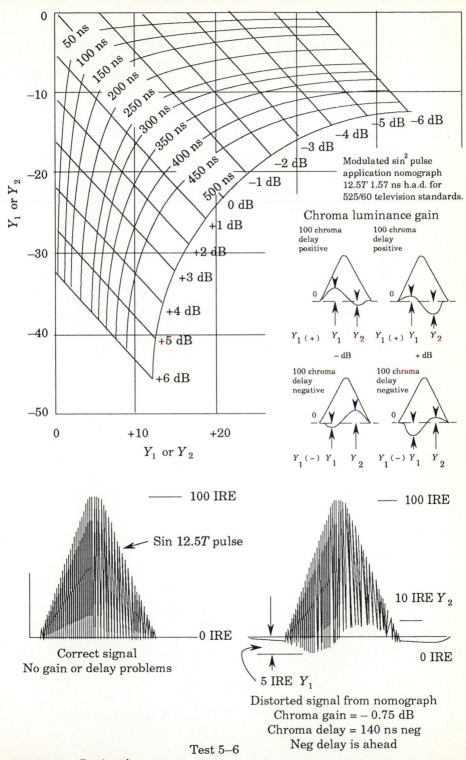

Modulated sin^2 pulse application nomograph 12.5T 1.57 ns h.a.d. for 525/60 television standards.

Chroma luminance gain

100 chroma delay positive

100 chroma delay positive

$Y_{1(+)}$ Y_1 Y_2 $Y_{1(+)}$ Y_1 Y_2

– dB   + dB

100 chroma delay negative

100 chroma delay negative

$Y_{1(-)}$ Y_1 Y_2 $Y_{1(-)}$ Y_1 Y_2

—— 100 IRE

Sin 12.5T pulse

0 IRE

Correct signal
No gain or delay problems

—— 100 IRE

10 IRE Y_2

0 IRE

5 IRE Y_1

Distorted signal from nomograph
Chroma gain = – 0.75 dB
Chroma delay = 140 ns neg
Neg delay is ahead

Test 5–6

Figure 6.37 *(Continued).*

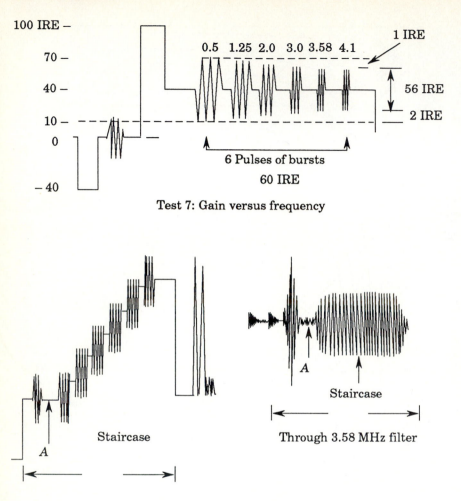

Test 7: Gain versus frequency

Staircase

Through 3.58 MHz filter

Test 8: Differential gain

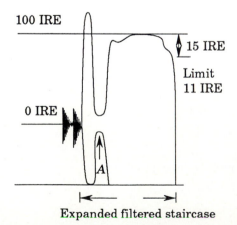

Expanded filtered staircase

Figure 6.37   *(Continued)*.

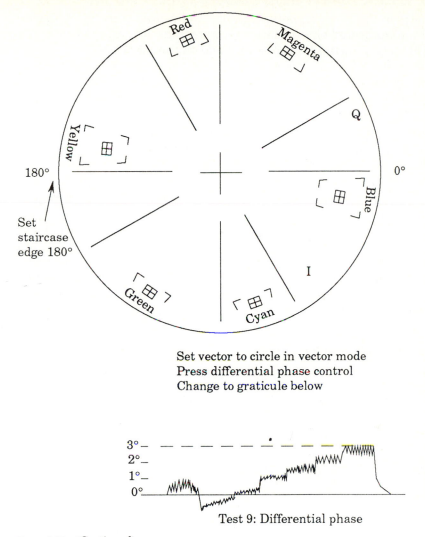

Set vector to circle in vector mode
Press differential phase control
Change to graticule below

Test 9: Differential phase

Figure 6.37   (*Continued*).

Any change in phase between the 3.58-MHz signals as a function of
stair height is the differential phase measurement. Usually the maximum phase change occurs between the lowest and highest stair steps.
The procedure is to loop through the signal from the waveform monitor to the vector scope and terminate the signal at 75 Ω, as shown in
Fig. 6.36. The controls on the vector scope should be set to position the
staircase chrominance signal on the 0° and 180° lines to the edge of
the circle. The differential phase control button is activated, and the
special graticule gives the value to the nearest degree. If the vector

scope is of the state of the art variety, the double trace mode display and the calibrated phase control give readings accurate to tenths of a degree. The 3° usual limit can be made on the graticule.

6.13.2.10 Noise testing.   To make accurate signal-to-noise measurements a noise test (calibrated source) instrument is needed. The technique is to compare the system noise level (grass area on scope) with that of the calibrated noise source. The instrument operations manual should be consulted for the step-by-step procedure. The minimum signal-to-noise level (limit) should be 52 dB. This is the baseband signal amplitude-to-noise amplitude power in decibels. The concepts of signal-to-noise ratio and carrier-to-noise ratio are often incorrectly interchanged.

Figure 6.37 contains instrument screens showing common errors for the tests described. Quality of video signals is very important, and some of the problems are beyond the control of the cable operators. However, having the instruments and knowing how to make the measurements are definitely advantages for recognizing problems, aiding in troubleshooting, and chasing down the causes of the problems.

Selected Bibliography

Bartlett, George W., ed. *Engineering Handbook*, ed. 6. National Association of Broadcasters, Washington, D.C., 1975.

Benson, K. Blair, editor in chief. *Television Engineering Handbook*, McGraw-Hill Book Co., New York, 1986.

Buchsbaum, Walter H., Sc.D. *Buchsbaum's Complete Handbook of Practical Electronic Reference Data*, ed. 2. Prentice-Hall, Englewood Cliffs, N.J., 1980.

Cunningham, John E. *Cable Television*, ed. 2. Howard W. Sams & Co., Indianapolis, 1985.

Freeman, Roger L. *Telecommunication Transmission Handbook*, ed. 2. John Wiley & Sons, New York, 1981.

Grant, William. *Cable Television*, ed. 2. GWG Associates, Fairfax, Va., 1988.

Jordan, Edward C., editor in chief. *Reference Data for Engineers: Radio, Electronics, Computers and Communications*, ed. 7. Howard W. Sams & Co., Indianapolis, 1985.

Taub, Herbert, and Schilling, Donald L. *Principles of Communication Systems*, ed. 2. McGraw-Hill Book Co., New York, 1986.

Television Operational Measurements, Video, and Radio Frequency for NTSC Systems, ed. 3. Tektronix, Beaverton, OR, 1984.

Terman, Frederick Emmons, Ph.D., Petit, Joseph Mayo, Ph.D. *Electronic Measurements*, ed. 2. McGraw-Hill Book Co., New York, 1952.

Standards, Records, and Reference Data

7.1 System Performance Standard

The Federal Communications Commission (FCC) was the first government agency to impose rules and regulations on cable operating systems. Through the Communications Act of 1934 it was initially formed to regulate radio transmission use in the public interest. At first the FCC was primarily involved with frequency assignments for ships' communications at sea; later it began to regulate AM (standard broadcast) radio, military radio communications, and amateur radio activities. As the FM radio and television broadcast industries developed, they too came under the FCC rules and regulations. Today the commission is concerned with selecting recommended standards for HDTV systems. It should come as no surprise that any recommended standard will be compatible with present-day NTSC television systems. Obviously it would not be in the public interest to make millions of television sets obsolete. It has also been suggested that the FCC recommend standards for scrambling and descrambling methods so that laws could be set in motion to make it a federal offense to defeat the system. Areas of interest in which scrambling and descrambling are needed are cable television signals; satellite broadcast, including multiple distribution systems (MDS); and VCR tape duplication. The Cable Communications Policy Act of 1984 has extended the 1934 act, authorizing the FCC to control the unauthorized reception of scrambled signals as well as the trafficking of illegal decoders. This extension of the act makes it a crime to receive unauthorized signals (signal piracy) or buy or sell illegal decoders. Use of the mail, which is common, constitutes the additional offense of mail fraud.

The section of the FCC rules and regulations pertaining to the cable television industry is Part 76. All cable operators are required by the commission to have a copy of the Rules and Regulations, Part 76, in their possession. Cable operators may obtain a copy by simply writing to the FCC in Washington, D.C., or through the National Cable Television Association (NCTA), also in Washington, D.C. The Society of Cable Television Engineers also makes it available for a small fee to cover costs.

Subpart K is the section containing the rules and regulations governing the technical standards for cable systems. This section is further divided into Performance Tests, Technical Standards, Measurements, and Operation in the frequency bands of 108 to 136 and 255 to 400 MHz. Also covered are operation near certain aeronautical and marine emergency radio frequencies, interference from cable television systems, and responsibility for receiver-generator interference. An addendum to this subpart contains the FCC Pole Attachment Regulations of April 16, 1980.

The section on performance tests, in brief, states that the cable system must comply with the rules if so ordered by a representative of the FCC. The cable system also has to keep a file containing a list of television channels and broadcast stations carried on those channels. A cable system has to conduct performance tests at three specified points (one the longest cascade) at intervals not exceeding 14 months at the three monitoring points located in the system.

The section on technical standards is summarized in Table 7.1. Clearly some of the FCC standards may seem outmoded compared to industry standards. Therefore, a few comments might clarify matters.

The television frequency boundary parameters are still good today. Initially they were specified because the bandwidth of most television tuners resembled a hill (not very selective). Therefore, if the frequency accuracy and video carrier and audio carrier levels were within the allowable limits, no adjacent channel interference would result. This feature protects the broadcaster from providing viewers with poor pictures as a result of the cable system. Cable operators certainly realized that the standards were a benefit to all concerned.

The allowable value of 36-dB carrier-to-noise ratio may seem questionable because this is at the original Television Allocations Study Organization (TASO) limit for picture impairment. The rule specifically states that this is the overall antenna terminal to subscriber receiver value of C/N. If a preamplifier at the antenna is used, the input must be terminated before making the measurement. Some cable operators have placed the preamplifier on the tower, and the rules state that it is permissible to remove the channel from service while making this measurement. Most cable operators remove the antenna lead from the signal processor on the ground and make the test, taking into

TABLE 7.1 Television Signal Parameters versus Specifying Standard

Parameter	FCC standard 76.605	Industry standard
Video carrier frequency	1. The video carriers will be 1.25 MHz above the lower boundary of the channel ± 25 kHz.	N/A
	2. Using a converter, the converter output carrier will be 1.25 MHz above the lower boundary of the channel ± 25 kHz.	N/A
	3. The frequency parameter for television broadcast frequency boundaries (Part 73.603)	N/A
Sound carrier frequency	The sound carrier will be 4.5 MHz above the video carrier ± 1 kHz	
Video carrier level at subscriber location	1. At the set: 0 dBmV at 75 Ω 2. At the top: 12 dBmV 3. Level of any adjacent channel: 3 dB 4. Maximum difference of level at any channel: 12 dB 5. The signal cannot be at a level great enough to cause receiver overload.	N/A
Sound carrier level	1. At a level 13 to 17 dB lower than the video carrier level. 2. At a level 7 to 17 dB lower than the video carrier level with no adjacent channels present.	N/A
Channel frequency	In the band limits of 0.75 to 50 MHz the channel response will not vary ± 2 dB.	N/A
Carrier-to-noise ratio (C/N)	1. For television broadcast stations in the grade B contour C/N = 36 dB. 2. For direct video feed also 36 dB.	C/N = 44 dB below carrier level (dBc)
Carrier to cross-modulation and/or intermodulation	46 dB below carrier level (dBc)	C/XMO = 49 dBc
Composite triple beat (CTB)	N/A	52 dBc
Second-order beat	N/A	60 dBc
Carrier to low-frequency disturbances (hum)	5%	40 dBc, which is 1%
Carrier to co-channel	36 dBc	N/A
Subscriber signal isolation	1. 18 dB between any subscribers. 2. Low enough so a subscriber mismatch will not cause another subscriber poor pictures	30 dB
Overall system linearity	N/A	(dB) peak to valley = N 10 + 1 n = no. amplifiers in cascade to test point

account the noise figure of the preamplifier or substituting a spare preamplifier.

The co-channel specification is also at 36 dBc, (decibels below carrier level) and many rural cable systems may have some signals operating at 36 dBc for both carrier-to-noise and carrier-to-co-channel ratios some of the time.

The carrier-to-crossmodulation and carrier-to-intermodulation specification is 46 dBc. However, this is also marginal. Crossmodulation and intermodulation are forms of third-order distortion. Crossmodulation is a difficult parameter to measure and worst case occurs when all carriers' modulation is in synchronism, which occurs only during a short interval. To make this measurement, some means is needed to modulate all the video carriers with the same source of video modulation or substitute an instrument that simulates all the cable-carried video carriers with a common source of video modulation. Such instruments are very costly and few systems can afford them. Often these instruments are used when a cable system is tested for final acceptance after construction is completed. A more convenient and hence better measure of third-order distortion is the composite triple beat. The test and measurement procedures in Chapter 6 should be followed to confirm the system operating parameters.

Carrier to second-order beat is now becoming important. Early systems used single-ended amplifiers that had larger second-order distortion components. The push-pull solid-state amplifier had low second-order distortion; consequently, third-order distortion became the limiting factor for the cascade number. Now feed-forward amplifiers use distortion cancelling techniques that principally lower third-order distortion. Therefore, second-order distortion is a limiting factor again. All of these specifications are acceptable for NTSC video and audio modulation of radio frequency carriers contained in a 6-MHz band.

Large-screen NTSC receivers with their 4:3 aspect ratio really need a C/N of 47 dB instead of 44 dB. The proposed HDTV systems with the 16:9 aspect ratio will certainly need C/N ratios of 50 dBc or better composite triple beat (CTB) at 54 dBc, and second-order distortion at 63 dBc.

The low-frequency (hum) parameter of 5 percent of the signal level is by today's standards quite large. Since hum or power line voltage, infiltrating the system signal is caused mainly by corroded connections and defective amplifier power supply filter capacitors, proper maintenance usually limits this problem.

It must be remembered that the FCC rules and regulations are law, and violation can result in a fine. The column in Table 7.1 labeled Industry Standards is based on standards set forth by the National Cable Television Association (NCTA). This organization has consistantly

TABLE 7.2 Video Signal Parameters versus Specifying Standard

Parameter measured	NCT7	NCTA	HBO	EIA RS250B
Video insertion gain	±3 IRE	____	±3 IRE	±2 IRE
Line time distortion	4 IRE	____	4 IRE	1 IRE
Field time distortion	4 IRE peak to peak	Linear	4 IRE peak to peak	3 IRE peak to peak
Short time distortion	10 IRE peak to peak	15 IRE peak to peak	7 IRE peak to peak	4 IRE peak to peak
Chrominance-to-luminance gain inequality	±6 IRE	____	±3 IRE	±4 IRE
Chrominance-to-luminance delay inequality	±75 ns	±150 ns	±40 ns	±26 ns
Chrominance-to-luminance intermodulation	3 IRE	____	3 IRE	2 IRE
Frequency response	+0.5 dB, −0.9 dB	____	±0.5 dB	±0.7 dB
Luminance nonlinear distortion	10 IRE	____	____	6 IRE
Chrominance nonlinear gain	±10%	____	±6%	±2%
Chrominance nonlinear phase	5°	____	3°	2°
Differential gain	15%	10%	8%	4%
Differential phase	5°	5°	3°	1.5°
Weighted video signal-to-noise ratio	53 dB	55 dB	52 dB	56 dB

set, revised, and added standards and measurement procedures established by their Engineering Subcommittee on Standards and Practices. A manual of Recommended Practices and Measurements on Cable Television Systems was published by the NCTA in 1983 and is available from them. A more convenient way of obtaining a copy is through the Society of Cable Television Engineers. A nominal charge to cover costs is required and certainly fair.

7.2   Video Performance Standards

The industry itself has set standards on the NTSC video baseband signal parameters. The standards are set to ensure that the video signal is of high quality so no picture impairment will result. Originally the NTC7 Report was the result of a study conducted by the Network Transmission Committee. This committee was a joint venture by the major television networks and the Bell System.

The Bell System was acting as carrier of video signals for the networks through their microwave links. The NCTA essentially tightened the specifications from the original NTC7 report and Home Box Office (HBO), which was one of the first suppliers of premium programming to the cable television industry, tightened the specifications further. The latest standards and the more stringent are those of the Electronic Industry Association (EIA), RS-250B. The measured parameter and the specification for the principal four industry organizations are given in Table 7.2.

7.2.1   Cumulative effects of video signal parameters

Each piece of equipment must pass a test of its parameters, and since most systems are made up of various connected pieces of equipment the combination, or cumulative effects, of, e.g., a distortion parameter should be investigated. It has been found that the cumulative effects of the various parameters vary. Table 7.3 lists the parameters and the manner in which the effects combine. The method of combining is illustrated in Fig. 7.1.

The previous examples are essentially for video equipment in cascade; for example, d_1 could be a microwave receiver output, d_2 a video decoder, d_3 a video amplifier, d_4 a video routing switch, and d_5 a monitor amplifier.

TABLE 7.3   Distortion Combining Method for Video Signal Parameters

3/2 Power method	Linear method	Square root of sum of square method
Differential gain	Field time distortion	Chrominance-luminance
Differential phase		Delay inequality
Chrominance-luminance		Chrominance-luminance
Intermodulation		Gain inequality
Luminance nonlinear distortion		Line time distortion frequency response
Chrominance nonlinear gain		Video gain
Chrominance nonlinear phase		Weighted video S/N

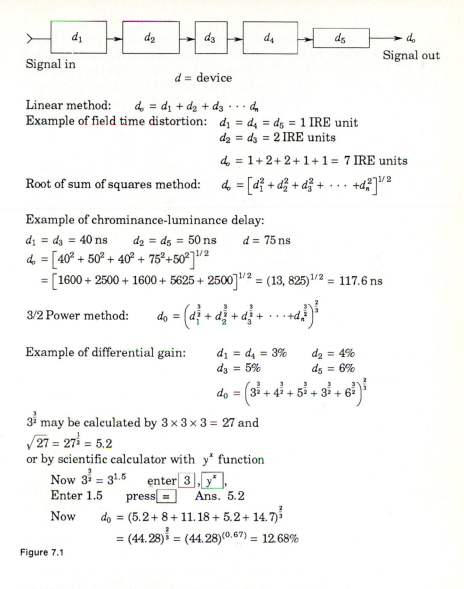

Signal in

$$d = \text{device}$$

Linear method: $d_o = d_1 + d_2 + d_3 \cdots d_n$

Example of field time distortion: $d_1 = d_4 = d_5 = 1 \text{ IRE unit}$
$$d_2 = d_3 = 2 \text{ IRE units}$$

$$d_o = 1 + 2 + 2 + 1 + 1 = 7 \text{ IRE units}$$

Root of sum of squares method: $d_o = \left[d_1^2 + d_2^2 + d_3^2 + \cdots + d_n^2 \right]^{1/2}$

Example of chrominance-luminance delay:

$d_1 = d_3 = 40 \text{ ns}$ $d_2 = d_5 = 50 \text{ ns}$ $d = 75 \text{ ns}$

$$d_o = \left[40^2 + 50^2 + 40^2 + 75^2 + 50^2 \right]^{1/2}$$
$$= \left[1600 + 2500 + 1600 + 5625 + 2500 \right]^{1/2} = (13,825)^{1/2} = 117.6 \text{ ns}$$

3/2 Power method: $d_0 = \left(d_1^{\frac{3}{2}} + d_2^{\frac{3}{2}} + d_3^{\frac{3}{2}} + \cdots + d_n^{\frac{3}{2}} \right)^{\frac{2}{3}}$

Example of differential gain: $d_1 = d_4 = 3\%$ $d_2 = 4\%$
$$d_3 = 5\% \qquad\qquad d_5 = 6\%$$

$$d_0 = \left(3^{\frac{3}{2}} + 4^{\frac{3}{2}} + 5^{\frac{3}{2}} + 3^{\frac{3}{2}} + 6^{\frac{3}{2}} \right)^{\frac{2}{3}}$$

$3^{\frac{3}{2}}$ may be calculated by $3 \times 3 \times 3 = 27$ and
$\sqrt{27} = 27^{\frac{1}{2}} = 5.2$
or by scientific calculator with y^x function

   Now $3^{\frac{3}{2}} = 3^{1.5}$ enter $\boxed{3}$, $\boxed{y^x}$,
   Enter 1.5   press $\boxed{=}$ Ans. 5.2

   Now $d_0 = (5.2 + 8 + 11.18 + 5.2 + 14.7)^{\frac{2}{3}}$

   $$= (44.28)^{\frac{2}{3}} = (44.28)^{(0.67)} = 12.68\%$$

Figure 7.1

7.2.2   System maintenance and records

7.2.2.1   Initial records and procedures.   Like the medical profession, the CATV industry maintains complete and updated records. The longer the performance history is maintained, the more the technical staff can see what has happened, could happen, and most likely will happen so problems can be headed off before a calamity strikes. Again, like the medical profession, the cable industry practices preventive medicine. Unfortunately many cable television systems are operated only from the end of the construction to whenever a failure causes an outage. Often the squeaky wheel attitude reigns supreme. Many of us feel that the proper way to maintain a system is first to make sure that it has been constructed of the finest equipment, built with the most careful and qualified labor, and tested well with the results properly written and filed. This is indeed the best way a system can start out providing service and the most likely way to ensure that it will continue for a long time. Sometimes there may be disagreement about methods and supplies. Most engineers and well-trained technical people know either through education or practical experience the proper steps for putting a system in service. The following steps commence after a system has been spliced and is ready for turn-on.

1. Turn on the head end with pilot carrier frequencies at the lowest and highest limit frequencies and the leakage detector transmitter set at proper video carrier levels.

2. Start from the head end on each trunk leg, turning on the power supply and monitoring the voltage and current. Any problems should be corrected before proceeding.

3. Start at the first amplifier to verify the design-specified input carrier levels for the test carriers.

4. Set the proper pad and equalizer values and set the output levels for the test signals on the system. Record these values on a proper form (an example is shown in Form 7.1).

5. Repeat for all the trunk amplifiers.

6. Perform the same adjustments and record the same information for the line extenders.

7. Now, once the preceding steps are completed, leakage-test the whole system. This will assure that the system is tight and that the cable and device impedance match is correct so as to minimize any standing waves that may cause ripple in the system frequency response. This way the final balance and sweeping will not merely compensate for a poorly spliced system. All improperly made connections should be corrected before proceeding to the next step.

8. Final sweep and balance can commence at this time. The input and output signals can be verified and adjusted by using an accurate signal level meter and recorded on the form. The results of the system sweep can be photographed and pasted on the form. Form 1 is an example of a trunk amplifier log sheet.

9. All of the initial forms can be punched and placed in a ring binder with copies in a fireproof container. Many of the more progressive

Trunk amplifier log

Amplifier location

Forward _____ Design
cascade no. _____ Actual

Reverse _____ Design
cascade no. _____ Actual

Town _____

Street _____ Pole #_____

Pad dB	EQU dB

Date _____

Temp °F _____

Tech _____

Trunk amp type / manuf. no. _____

Bridger amp type / manuf. no. _____

Level dBmV video carrier

Ch 2H   Ch 4H   Ch 36H   Ch 62H   Ch 69H

Forward input
Forward output

Reverse input video carrier dBmV   T-7_____ T-11 _____

Reverse output video carrier dBmV   T-7 _____ T-11 _____

Sweep photographs

Centerville CATV
74 Main Street
Centerville, TS 04464
Tel. 543-2100

Form 7.1

cable companies are keeping their records in computer-type storage in the form of a word processing spreadsheet system and recorded on floppy disc.

The positions for the ground level test points are marked on the tree diagram for reference in case any out-of-tolerance readings are experienced, so that the faulty amplifier can often be pinpointed. A sample tree diagram is shown on Form 7.2. An end level sample form is shown in Form 7.3.

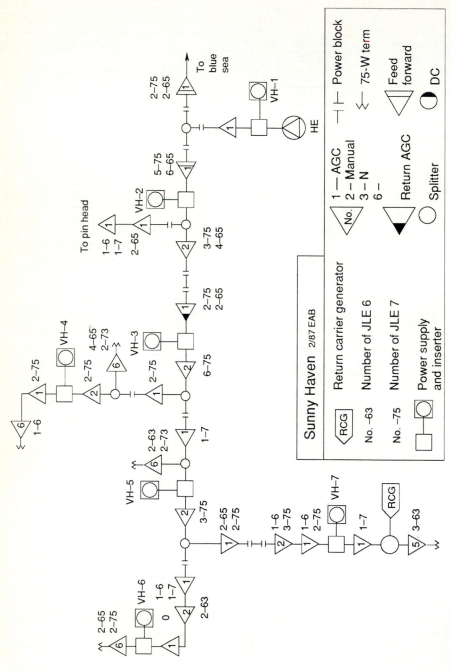

Form 7.2 Tree diagram.

TEST POINT LOCATION			LEVEL DBMV								
TOWN	STREET	POLE #	CH 2	CH 4	CH 14	CH 22	CH 12	CH 27	CH 36	CH 40	CH 62

Down-Wind Cable TV
Big Butte, MS 07711
876-678-5432

Form 7.3 End level test.

Another good reference technique employs a system summary form. This form should contain information on the designed input-to-output amplifier levels and slope. A system cable loss table records the kinds of cable used in the system. Tap loss as well as splitter and directional coupler losses should be included. This form can be placed in the vehicle map book so technicians testing and troubleshooting will have it available for reference. An example of this form is shown as Form 7.4.

7.2.2.1.1 Power supply testing. Power supplies used to power cable television distribution systems are either the nonstandby or the standby type. The *nonstandby type* is usually a ferroresonant transformer with two secondary windings. One secondary winding is the 2:1 winding, which steps down the 120-V commercial alternating power to 60 V_{ac}. The second winding is a high-voltage winding that is connected across a high-voltage capacitor. The value of this capacitor is such that it resonates with the transformer inductance. The current flowing in this resonant parallel circuit saturates the iron core of the transformer. If the input voltage drops for any reason and the drop is not catastrophic, the core will remain saturated and the secondary 60-V_{ac} will not drop. Therefore, the ferroresonant transformer will maintain a degree of output voltage regulation. Essentially this type of power supply needs very little maintenance. However, it is often advantageous to have an ammeter connected to the output, which is a usual option offered by the manufacturer. If this is not available and a fuse block in series with the cable output is available, an ammeter of appropriate range can be placed across the fuse. Removing the fuse will cause the current to flow through the ammeter so system current supplied by the power supply can be read and recorded. Replacing the fuse and removing the ammeter in that order will cause no interruption of system power.

Another method is to use a clip-on ammeter, which may be placed around the jumper wire feeding current to the cable connector. In both these instances system output current is measured, by placing the clip-on ammeter around the input black wire (hot) and taking the reading. This reading is usually less than the output current because the input voltage is higher than the secondary voltage. The input and output voltage can be measured by using a standard voltmeter. Since the output wave shape is not a sine wave there will be an error in reading the output voltage and current. A true rms voltmeter and ammeter will give the correct results. However, since these readings are used to monitor any changes, the small difference in reading is not important. From a maintenance point of view, each power supply should be visited at least once a year; twice a year, of course, is better. During these visits, the input and output voltages and currents should be

SPT TAP 2 PT.

No.	Nom. Val.	Loss 50 MHz	Loss 450 MHz
4T	4.3		
7	7.2	3.0	4.1
10	10.3	1.7	2.2
12	12.2	1.0	1.5
14	13.8	0.5	0.8
17	16.9	0.5	0.7
20	20	0.4	0.5
23	23	0.3	0.5
26	26	0.2	0.4
29	29.2	0.4	0.6
32	32	0.4	0.6
35	35	0.4	0.6

SPT TAP 4 PT.

No.	Nom. Val.	Loss 50 MHz	Loss 450 MHz
7T	7.2		
10	10.2	3.2	4.3
14	14.3	1.4	1.7
15.5	15.3	1.4	1.4
17	17.3	0.5	1.0
20	20.1	0.4	0.8
23	23	0.4	0.7
26	26	0.3	0.5
29	29.6	0.3	0.5
32	32.5	0.3	0.5
35	35	0.3	0.5

Cable ATT dB/100' @ 68°

Type	54 MHz	450 MHz
0.860 QR	0.32	0.95
0.500 QR	0.52	1.55
RG – 11	0.96	2.75
RG – 6	1.60	4.40

FFT/G TAP 8 PT.

No.	Nom. Val.	Loss. 50 MHz	Loss 450 MHz
10	10.8		
14	14.2	3.5	4.1
17	18.6	1.3	1.6
20	20	0.7	1.0
23	22.5	0.7	0.8
26	26.1	0.7	0.7
29	29.2	0.7	0.7
32	32.2	0.7	0.7
35	34.7	0.7	0.7

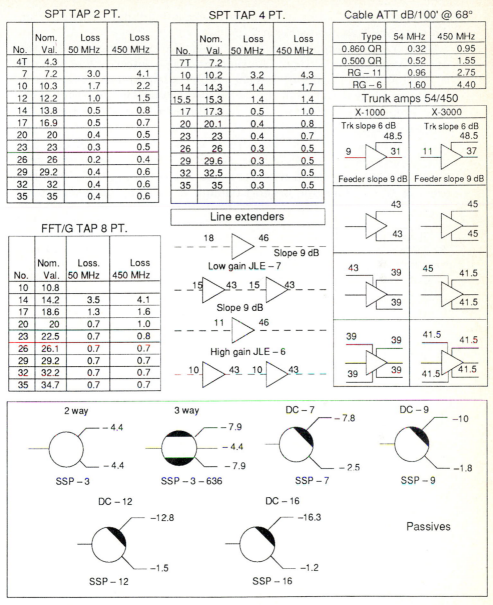

Form 7.4 Information summary.

measured and recorded. A can of spray paint and a wire brush should be used to remove rust and paint any spots needing a touch-up. A thorough visual inspection should complete the visit. The time, date, technician's initials, input and output voltages and currents, and any abnormalities should be noted on a power supply maintenance form. An example is shown as Form 7.5.

Standby power supplies are of two types: the switching type and the uninterruptible (UPS) type. The *switching* type detects a loss of input commercial power and switches the batteries to power an inverter circuit to make ac power from the dc battery power. The *uninterruptible* type keeps the inverter floating and takes over if input power fails. The uninterruptible supply is, of course, more expensive and requires more maintenance simply because it is more complicated and has more components. The switching supply is usually considered adequate for cable television needs because the switching time is on the order of 10 to 20 ms and is short enough so the amplifiers do not re-

ID NO.	STREET NAME	POLE PED. NO.	VOLTS OUT	AMPS OUT	AMPS IN	REMARKS ON CONDITION	DATE	TECH. INITIALS	PASS YES / NO

Form 7.5   Power supply maintenance log.

spond to the power drop. Added to the measurements are the battery voltage and current, which should be measured and recorded during maintenance visits. These measurements can be done with a multimeter using the dc mode if power supply metering is not available. If standby power supplies are used, it is a good idea to spend a little more and get the metering option and possibly some diagnostic features. An example of a standby power supply maintenance log form is shown as Form 7.6.

STANDBY POWER SUPPLY LOG_____

NO_____ TOWN_____ST. _____ POLE # ____

DATE/TECH TEMP °F		SEQUENCE	AC VOLTS	AC AMPS	NORM. PWR. LT.	NORM. SUB. LT.	INV. RDY. LT.	CHARGE LT.	LOW BAT. LT.	BAT. VOLTS DC	REMARKS
	NORM										
	STBY.										
	3 MIN.										
	6 MIN.										
	NORM										
	NORM										
	STBY.										
	3 MIN.										
	6 MIN.										
	NORM										
	NORM										
	STBY.										
	3 MIN.										
	6 MIN.										
	NORM										
	NORM										
	STBY.										
	3 MIN.										
	6 MIN.										
	NORM										

Form 7.6

It should be noted that one important test is to place the power supply in the standby mode either by using the test switch if available or by removing the input commercial power. The battery voltage and current are monitored over a time sample to establish an observable battery discharge rate. This information can be used to determine the condition of the batteries. The batteries used by most standby power supplies are usually maintenance-free; therefore, they cannot be tested by a hydrometer. As with the nonstandby power supply, a visual inspection for corrosion, cracked or damage terminals, and/or loose wires should be made and problems corrected and duly noted on the maintenance form.

7.2.2.1.2 Radiation and system signal leakage.
Radiation and signal leakage tests are always required (FCC) and records of the test results should be kept as long as coaxial cable systems exist. From a signal quality point of view, poor connectors or poorly made connections cause cable impedance mismatches that cause standing waves on the cable. These standing waves cancel or reinforce the signal level at various points along the cable, ultimately causing the signal level to vary across the spectrum, as exhibited by poor peak-to-valley, i.e., flatness or linearity. Recall that short sections of coaxial cable terminated in a short-circuit or open-circuit condition at certain frequencies, where the length of these short sections of cable is equal to a quarter or half or any multiple of a quarter-wavelength, can completely trap out the frequency. This is termed by cable technical people as *signal suck-out*, where one frequency or narrow band of frequencies is completely removed from the system. Loose or improperly made connectors can cause noise hum and signal ingress, which can create interference with the desired signals on the cable. The cable signals that escape from the system can interfere with other licensed services; in the case of aircraft navigational and communications frequencies, a catastrophic situation can result.

The testing procedures for system leakage are covered in Chapter 6. Proof that the tests were made and test results should be recorded on a form and filed. These tests have to be carried out over at least 75 percent of the cable plant four times per year. The measurements have to be made with the leaks measured in microvolts per meter, the pole number or area defined, and the information recorded on a form. After the field work the calculations for the cumulative leak index (CLI) are made and recorded. After repair of the leaks a retest should be made and noted on the log form. This information can be filed in a ring binder or filing system for inspection by a representative of the FCC if necessary. Currently there is on the market computer software (programs) that can be run on PCs to keep such records and make the

Bayberry CATV Inc.
Pine Is. FO
999-876-5544

DATE _____   TEMP. _____   TECH. _____

START MILES _____

STOP MILES _____

MILES TOTAL _____

MILES PLANT _____

$$\text{CLE} = 10 \text{ LOG } \frac{\text{MI PLANT}}{\text{MI TOTAL}} \sum \text{LEAKS}^2$$

Form 7.7   System leakage log.

CLI computations with the added benefit of more sophisticated leakage analysis performed by the computer. A sample of a leakage test report form is shown as Form 7.7.

7.2.2.1.3 The maintenance manual.   For the distribution system a maintenance manual can be simply fashioned by making copies of the measurement procedures and placing these copies in plastic covers in a ring binder. After each procedure blank forms for the particular procedure can be placed in the ring binder. Throughout the year the information for the specific tests can be recorded on the forms for study by the system engineer. If a system is tested in this manner, the system engineer should know the condition of the system and feel confident that any more stringent tests, e.g., annual proof of performance, will be passed with flying colors. The tests for system leakage can be made and the results filed in the maintenance log ring binder. Some systems start a new maintenance log for each year, and more progressive systems place this log information on a computer disc for data storage and analysis by system software.

7.2.2.1.4 Head-end log.   Like the distribution system, the head end should be tested and the results filed by any appropriate filing system. A head-end log book is the usual means used by many cable television systems. This is probably an outgrowth of radio and television broadcast station requirements. The signal level leaving the system head end to drive the cascade of amplifiers is very important and should be extremely stable. Of principal importance is the level of the pilot carriers used as a reference to control the AGC and ASC circuits in the trunk amplifiers. One pilot carrier, the *low pilot* (low-frequency), is usually cable channel 2, 3, or 4. The high pilot carrier for a 220-MHz upper-frequency-limit cable system is often channel 12 and for 300-MHz systems may be channel 12 or channel N (241.24 MHz IRC or 240 MHz HRC) and for 400-MHz systems again channel N or M. For 450-MHz systems the high pilot frequency is usually somewhere around 400 MHz. The reason for having a choice of pilot frequencies is that any strong off-air station could cause interference with the pilot carrier, which in turn could cause system instability. This could happen for the low-frequency pilot. At nearly 400 MHz it is improbable that interference by any off-air signals could cause a problem. When an off-air television channel is heterodyne-processed by a signal processor, this processor should have a standby or signal substitution system installed. If the station goes off the air for any reason, e.g., normal operating hours, transmitter failure, or catastrophic signal fading, loss of the pilot carrier can cause the whole distribution system serious signal level shifts and noise buildup. The substitution signal at the same level instantaneously switched to the system, will take care of any loss of pilot signal. It should be obvious that from time to time the standby or substitution signal should be activated and its level tested to make sure it is at the same level as the normal pilot signal.

7.2.2.1.5 Signal level tests.   The signal level of all carriers leaving the head end should be tested at an appropriate test point placed close to the cable system input. The diagram in Fig. 7.2 gives a good example. This test point, being close to the cable, will give an accurate reading (minus 20 dB) of the signal driving the first trunk amplifier. A log of the signal level should be kept in a head-end log book for examination by the system engineer or technical director. An example of such a log form is shown as Form 7.8. This form has a column for the comb gen-

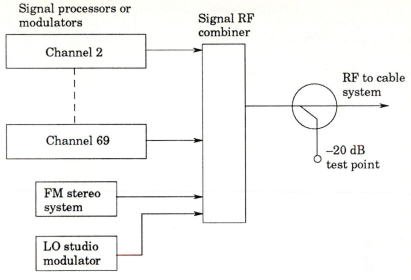

Figure 7.2

erator level, which is often used in 400-MHz and higher incremental regulated carrier (IRC) or harmonic regulated carriers (HRC) systems. This comb provides a carrier frequency at appropriately spaced intervals across the frequency spectrum that is used to phase lock the cable television signal processors and modulators. Usually the television modulators have a panel meter or test point on the panel to provide a measurement of the comb level. The signal processors often have a lamp to indicate the presence of the phase lock signal and the condition of phase lock. Some systems indicate only that the condition is OK by placing a ✔ in the comb G and/or ϕ lock columns. (A comb generator provides equally spaced carriers of fixed constant level where the spectrum display resembles a hair comb.) If just a reference signal (REF) light is used, then the presence of the comb signal can be checked in this column. For the case of a heterodyne signal processor fitted with a signal input test point, a test of the off-air signal level can be made and the value placed in the In Level column. For satellite-generated signals the 1-V peak to peak can be measured and set with a checkmark placed in this column. Usually these tests are made weekly, biweekly, or monthly. When a head end is new, weekly tests should be made. When signal level stability is attained, this test can be made monthly. Seasonal changes of off-air signal level can be tracked by making a plot of signal level (per channel) versus months. Any possible signal level adjustment, e.g., padding or amplification, can be anticipated by an examination of this information. The infor-

County Cable Television

DATE _____   TIME _____   TEMP. _____   TECH _____

CONV. CH.	IN LEV./SG	φ LOCK	REF.	COMB. G	OUT LEV.	REMARKS

Form 7.8

mation provided by a good head end log can be very valuable in maintaining a high-quality signal driving the cable system.

7.2.2.1.6 Head-end frequency tests.   In the past the FCC proof of performance tests required that the frequency of any class A television broadcast station carried on a cable system be maintained within ± 25 kHz. If head-end heterodyne signal processors were used and the input frequency was the same as the output frequency (on channel), then no error would result and the frequency test was not required. In other words, the cable operator did not have control over the television frequency. If the television channel was transposed to another channel by a signal processor, the frequency had to be maintained and tested to prove that it was within ±25 kHz. Essentially this is still the case. However, at the present frequency limit (550 MHz) a phase-lock system is usually used and for the present-day FCC offsets of +12.5 kHz the comb frequency source has to be maintained to a high degree of accuracy. The frequency accuracy for the signal source is 12 parts per million, which at 450 MHz is

$$\frac{12 \text{ parts}}{10^6} \times 450 \times 10^6 = 5400$$

which is 5400 Hz or 5.4 kHz. Therefore, one has to measure at 450 MHz, 450,005,400 Hz. The frequency counter should have a minimum of seven digits and nine is more usual. The clock or time base of the counter also has to be very stable and accurate. For the comb accuracy of HRC systems the 6-MHz clock has to be accurate to 6.0003 MHz ± 1 Hz, which means 6,000,300 ± 1 Hz. Accuracies such as this are for many systems beyond the capabilities of the technical staff and/or system instrumentation. Most head-end equipment manufacturers recommend that the clock generator be rechecked every 6 months; that means that a cable system would need two precision frequency sources, one in calibration, the other in operation. The rubidium-based clock is the type needed as the precision source and is more stable and accurate than most frequency counters. The more wealthy cable operators might purchase a precision frequency counter, and large system operators might set up a calibration laboratory. The cost of precision equipment will most likely decrease as more cable operators have to comply. The precision comb generator sources have decreased to about one-half of what they cost when they first appeared on the market. The technical staff of any cable television system should through self-education keep abreast of the laws, rules, regulations, methods, and practices required to maintain and improve the level of service to the subscribers.

The technical file and system log books of the results of proper and accurate tests should satisfy the FCC requirements. This information should also direct attention to any system weaknesses before any problems that could cause a system outage or catastrophic situation surface.

Also noted on the log form is the temperature for the head-end electronic equipment room. This in the past has been an important parameter because some of the earlier equipment was not very temperature-stable. Modern equipment can usually stand temperature swings of at least ±10°F around a normal 68°F (20°C) mean temperature. However, particularly at high temperatures, usually over

85°F, solid-state equipment can malfunction. Usually the head end is monitored either by daily visits or by some remote alarm system for temperature, smoke, fire, and intrusion. For head ends where the off-air stations are received at a remote mountain site and the satellite receiving antenna systems are located at a more convenient site, a more sophisticated alarm or telemetered surveillance or performance system is recommended. This could save a lot of difficult daily visits to such remote sites.

A good halon fire extinguishing system also is recommended. If personnel are near the area when a fire breaks out and this type of automatic fire extinguishing goes off, it is much safer and personnel are not affected by as much as some other systems. Consultation with the local fire chief is a good idea before a system is chosen for installation.

The cable industry should, and very soon, start preparing for HDTV. Doing so will require cable television operators to improve their systems through increased maintenance and or rebuilding. Through its CLI monitoring enforcement plan the FCC will pressure cable operators to perform more cable plant maintenance. It has been proved by some cable operators that dealing with the CLI by repairing the signal leaks improves system return loss and that better system frequency response (linearity, peak-to-valley ratio) and general subscriber service result. Thus service calls usually decrease significantly. The large aspect ratio of proposed HDTV systems will show more clearly any ghosting due to return loss deterioration and snow due to degraded carrier-to-noise ratio. Amplifier distortion effects will also be more noticeable to HDTV viewers. Demonstrators of HDTV systems have to be very cautious to keep the video signal used for demonstration purposes extremely high in quality. Cable operators will have to do the same.

Selected Bibliography

Bartlett, George W., ed. *Engineering Handbook*, ed. 6. National Association of Broadcasters, Washington, D.C., 1975.

Benson, K. Blair, editor in chief. *Television Engineering Handbook*. McGraw-Hill Book Co., New York, 1986.

Buchsbaum, Walter H., Sc.D. *Buchsbaum's Complete Handbook of Practical Electronic Reference Data*, ed. 2. Prentice-Hall, Englewood Cliffs, N.J., 1980.

Cunningham, John E. *Cable Television*, ed. 2. Howard W. Sams & Co., Indianapolis, 1985.

Freeman, Roger L. *Telecommunication Handbook*, ed. 2. John Wiley & Sons, New York, 1981.

Grant, William. *Cable Television*, ed. 2. GWG Associates, Fairfax, Va., 1988.

Jordan, Edward C., editor in chief. *Reference Data for Engineers: Radio, Electronics, Computers, and Communications*, ed. 7. Howard W. Sams & Co., Indianapolis, 1985.

NCTA Legal Staff. *Rules and Regulations, Part 76—Cable Television Service*. Federal Communications Commission, Washington, D.C., 1984.

Orr, William I. *Radio Handbook*, ed. 23. Howard W. Sams & Co., Indianapolis, 1988.

Index

ABOUT THE AUTHOR

Eugene R. Bartlett is Senior Staff Engineer with
Tele-Media Corporation in Pleasant Gap, Pennsylvania.